Comparative Genomics

METHODS IN MOLECULAR BIOLOGY™

John M. Walker, SERIES EDITOR

METHODS IN MOLECULAR BIOLOGY™

Comparative Genomics

Volume 2

Edited by

Nicholas H. Bergman

Bioinformatics Program and Department of Microbiology and Immunology, University of Michigan Medical School, Ann Arbor, MI

HUMANA PRESS ✷ TOTOWA, NEW JERSEY

Softcover reprint of the hardcover 1st edition 2007
999 Riverview Drive, Suite 208
Totowa, New Jersey 07512

www.humanapress.com

This publication is printed on acid-free paper. ♾
ANSI Z39.48-1984 (American Standards Institute) Permanence of Paper for Printed Library Materials

Cover illustration: From Figure 1, Volume 1, Chapter 10, "PSI-BLAST Tutorial," by Medha Bhagwat and L. Aravind. Ribbon diagrams comparing the three-dimensional structures of the human PCNA protein and the E. coli DNA polymerase III beta subunit. The coordinates for these structures are taken from a public database.

Cover Design: Karen Schulz

Production Editor: Christina M. Thomas

For additional copies, pricing for bulk purchases, and/or information about other Humana titles, contact Humana at the above address or at any of the following numbers: Tel.: 973-256-1699; Fax: 973-256-8341; E-mail: humana@humanapr.com; or visit our Website: www.humanapress.com

10 9 8 7 6 5 4 3 2 1

Preface

Over the last ten years the amount of biological sequence data available to researchers has increased by several orders of magnitude, and complete genome sequences (nearly nonexistent ten years ago) have become commonplace. The techniques involved in analyzing these sequences have evolved almost as rapidly, and several (e.g, BLAST) have become so commonly used in molecular biology that their names have become verbs. Even so, a number of extremely powerful tools and techniques developed for comparative genomic analysis remain unfamiliar to molecular biologists, and thus are underutilized.

The primary aim of these volumes is to provide a set of tutorials that will be useful to molecular biologists beginning to use comparative genomic analysis tools in a number of different areas. Volume I contains the first four of seven sections: In the first section, the reader is introduced to genomes via a number of visualization tools that allow one to browse through a particular genome of interest. The second and third sections deal with comparative analysis at the level of individual sequences, and present methods useful in sequence alignment, the discovery of conserved sequence motifs, and the analysis of codon usage. The fourth section deals with the identification and structural characterization of non-coding RNA genes—this class of genes is particularly difficult to predict, and discovery of these elements is almost completely reliant on comparative genomics. (Note that the much larger question of identifying protein-coding genes is not addressed here, because there a separate volume in the MiMB series devoted to this issue).

In the second volume, the fifth section describes a number of tools for comparative analysis of domain and gene families. These tools are particularly useful for predicting protein function as well as potential protein-protein interactions. In the sixth section, methods for comparing groups of genes and gene order are discussed, as are several tools for analyzing genome evolution. Finally, the seventh section deals with experimental comparative genomics. This section includes methods for comparing gene copy number across an entire genome, comparative genomic hybridization, SNP analysis, as well as genome-wide mapping and typing systems for bacterial genomes.

Each chapter includes not only detailed instructions for using a particular tool or method, but also an introduction to the theory behind the technique. Importantly, there are also a number of Notes at the end of each chapter that guide the beginning user through commonly encountered difficulties, and provide key tips for using the method most efficiently. Readers are encouraged to note that although some of tools presented in a given section are quite similar in aim, they are often designed quite differently, and will have different strengths and weaknesses. This is particularly true in considering the computational tools, where the same overall goal (e.g., discovery of conserved motifs) can be pursued using a number of very different statistical approaches. Users should therefore explore several different options in attempting comparative analyses—a combined approach is often best.

These volumes are the collective effort of many people. I would like to extend a special thanks to all of the contributors, and to the staff at Humana Press, who helped at every stage of the publication process. I would also like to especially thank Erica Anderson and Ellen Swenson at the University of Michigan Medical School and Tim Read at the US Naval Medical Research Center for valuable advice, and help in putting these books together.

Nicholas H. Bergman, PhD

Table of Contents

Contributors

FARNAZ ABSALAN • *Stanford Genome Technology Center*
KIYOKO AOKI-KANEHISA • *Department of Bioinformatics Soka University, Faculty of Engineering*
ROLF APWEILER • *European Bioinformatics Institute*
MEDHA BHAGWAT • *National Center for Biotechnology Information, National Library of Medicine, National Institutes of Health*
VOLKER BRENDEL • *Department of Genetics, Development, and Cell Biology, Iowa State University*
BRIAN BRUNELLE • *Virus and Prion Diseases of Livestock Research Unit, National Animal Disease Center, USDA Agricultural Research Service*
YI CHU • *University of Medicine and Dentistry of New Jersey, Robert Wood Johnson Medical School, Department of Molecular Genetics, Microbiology & Immunology*
MARK L. CROWE • *Genetic Solutions Pty Ltd, Australia*
XIANGFENG CUI • *University of Medicine and Dentistry of New Jersey, Robert Wood Johnson Medical School, Department of Molecular Genetics, Microbiology & Immunology*
AARON DARLING • *Department of Computer Science, University of Wisconsin-Madison*
PETER DE RIJK • *Department of Molecular Genetics, University of Antwerp*
JURGEN DEL-FAVERO • *Department of Molecular Genetics, University of Antwerp*
TODD DELUCA• *Department of Systems Biology, Harvard Medical School*
MEETA DESAI • *Applied and Functional Genomics, Health Protection Agency, United Kingdom*
ROBERT FINN • *Wellcome Trust Sanger Institute*
DMITRIJ FRISHMAN • *Technical University of Munich, Department of Genome Oriented Bioinformatics, Institute for Bioinformatics/MIPS, GSF—Research Center for Environment and Health*

RICHENG GAO • *University of Medicine and Dentistry of New Jersey, Robert Wood Johnson Medical School, Department of Molecular Genetics, Microbiology & Immunology*
DANIELLE M. GREENAWALT • *University of Medicine and Dentistry of New Jersey, Robert Wood Johnson Medical School, Department of Molecular Genetics, Microbiology & Immunology*
JOEL GRESHOCK • *GlaxoSmithKline, Abramson Family Cancer Research Institute, University Pennsylvania School of Medicine*
GUOHONG HU • *University of Medicine and Dentistry of New Jersey, Robert Wood Johnson Medical School, Department of Molecular Genetics, Microbiology & Immunology*
YEN LIN HUANG • *Department of Computer Science, National Tsing Hua University*
MINORU KANEHISA • *Kyoto University Bioinformatics Center, Human Genome Center, Institute of Medical Science, University of Tokyo*
STEFAN KURTZ • *Department of Genetics, Development, and Cell Biology, Iowa State University*
HONGHUA LI • *University of Medicine and Dentistry of New Jersey, Robert Wood Johnson Medical School, Department of Molecular Genetics, Microbiology & Immunology*
JAMES Y. LI • *University of Medicine and Dentistry of New Jersey, Robert Wood Johnson Medical School, Department of Molecular Genetics, Microbiology & Immunology*
YING CHIH LIN • *Department of Computer Science, National Tsing Hua University*
BJØRN-ARNE LINDSTEDT • *Norwegian Institute of Public Health*
CHIN LUNG LU • *Department of Biological Science and Technology, National Chiao Tung University*
CHRISTIAN C. LUEBBERT • *Genomics and Proteomics Group, Institute for Biological Sciences, Canadian National Research Council*
MINJIE LUO • *University of Medicine and Dentistry of New Jersey, Robert Wood Johnson Medical School, Department of Molecular Genetics, Microbiology & Immunology*
XAVIER MESSEGUER • *Department of Software, Technical University of Catalonia-Barcelona, Barcelona Supercomputing Center (BSC)*
JAINA MISTRY • *Wellcome Trust Sanger Institute*
NICOLA MULDER • *European Bioinformatics Institute*
KNUD NAIRZ • *Institute of Neuropathology, University Hospital of Zurich*

YASUHITO NANYA • *University of Tokyo, Department of Regeneration Medicine*
JOHN H.E. NASH • *Genomics and Proteomics Group, Institute for Biological Sciences, Canadian National Research Council*
TRACY NICHOLSON • *Respiratory Diseases of Livestock Research Unit, National Animal Disease Center, USDA Agricultural Research Service*
TIANHUA NIU • *Division of Preventative Medicine, Department of Medicine, Brigham and Women's Hospital, Harvard Medical School*
MATTHIAS OESTERHELD • *Institute for Bioinformatics/MIPS, GSF—Research Center for Environment and Health*
SEISHI OGAWA • *University of Tokyo, Department of Regeneration Medicine*
PHILIPP PAGEL • *Technical University of Munich, Department of Genome Oriented Bioinformatics, Institute for Bioinformatics/MIPS, GSF—Research Center for Environment and Health*
XIOAKANG PAN • *Department of Genetics, Development, and Cell Biology, Iowa State University*
SOPHIE PASEK • *Laboratoire Statistique et Génome, CNRS*
GIULIO PAVESI • *Dipartimento di Scienze Biomolecolari e Biotecnologie, University of Milan*
NICOLE T. PERNA • *Department of Animal Health and Biomedical Sciences Genome Center, University of Wisconsin-Madison*
GRAZIANO PESOLE • *Dipartimento di Biochimica e Biologia Molecolare, University of Bari and Istituto Tecnologie Biomediche del C.N.R. (sede di Bari)*
MOSTAFA RONAGHI • *Stanford Genome Technology Center*
THOMAS SCHMIDT • *Technische Fakultät, Universitat Bielefeld, International NRW Graduate School in Bioinformatics and Genome Research, Germany*
MANUEL SCHNEIDER • *Kantonsschule Zug*
JENS STOYE • *Technische Fakultät, Universitat Bielefeld, Germany*
NORMANN STRACK • *Technical University of Munich, Department of Genome Oriented Bioinformatics*
VOLKER STÜMPFELN • *Institute for Bioinformatics/MIPS, GSF—Research Center for Environment and Health*
KARSTEN SUHRE • *Information Génomique et Structurale, CNRS*
EDUARDO N. TABOADA • *Genomics and Proteomics Group, Institute for Biological Sciences, Canadian National Research Council*
CHUAN YI TANG • *Department of Computer Science, National Tsing Hua University*

IRINA V. TERESHCHENKO • *University of Medicine and Dentistry of New Jersey, Robert Wood Johnson Medical School, Department of Molecular Genetics, Microbiology & Immunology*
TODD TREANGEN • *Department of Software, Technical University of Catalonia-Barcelona*
PAUL VAN den IJSSEL • *VU University Medical Center*
DENNIS P. WALL • *Department of Systems Biology, Harvard Medical School*
HUI-YUN WANG • *University of Medicine and Dentistry of New Jersey, Robert Wood Johnson Medical School, Department of Molecular Genetics, Microbiology & Immunology*
DAVID T.W. WONG • *UCLA School of Dentistry*
GO YAMAMOTO • *University of Tokyo, Department of Regeneration Medicine*
SIAMAK P. YAZDANKHAH • *Norwegian Institute of Public Health*
BAUKE YLSTRA • *VU University Medical Center*
XIAOFENG ZHOU • *UCLA School of Dentistry*
PEDER ZIPPERLEN • *Tecan Schweiz AG*

Table of Contents

I

Comparative Analysis of Domain and Protein Families

1

Computational Prediction of Domain Interactions

Philipp Pagel, Normann Strack, Matthias Oesterheld, Volker Stümpflen, and Dmitrij Frishman

Summary

Conserved domains carry many of the functional features found in the proteins of an organism. This includes not only catalytic activity, substrate binding, and structural features but also molecular adapters, which mediate the physical interactions between proteins or proteins with other molecules. In addition, two conserved domains can be linked not by physical contact but by a common function like forming a binding pocket.

Although a wealth of experimental data has been collected and carefully curated for protein–protein interactions, as of today little useful data is available from major databases with respect to relations on the domain level. This lack of data makes computational prediction of domain–domain interactions a very important endeavor.

In this chapter, we discuss the available experimental data (iPfam) and describe some important approaches to the problem of identifying interacting and/or functionally linked domain pairs from different kinds of input data. Specifically, we will discuss phylogenetic profiling on the level of conserved protein domains on one hand and inference of domain-interactions from observed or predicted protein–protein interactions datasets on the other. We explore the predictive power of these predictions and point out the importance of deploying as many different methods as possible for the best results.

Key Words: Conserved domains; function prediction; protein interactions; data integration.

1. Introduction

Biological networks have long been an important topic in computational biology and they constitute the focal point of the emerging field of systems biology. Just like metabolic pathways, regulatory or transcriptional networks,

From: *Methods in Molecular Biology, vol. 396: Comparative Genomics, Volume 2*
Edited by: N. H. Bergman © Humana Press Inc., Totowa, NJ

physical interactions (PPI), and functional associations between proteins can be represented in the form of large graphs.

In many cases, protein function and, specifically, interactions are carried out by conserved domains, which are reused in different proteins throughout an organism. This modular architecture has been a focus of interest for a long time ***(1)***. The problem of formal description, biological annotation, and classification of these domains has been addressed by several groups leading to valuable resources such as SMART ***(2)***, BLOCKS ***(3)***, Pfam ***(4)***, and the recent integration endeavor InterPro ***(5)***.

Because of their important role as biological modules carrying function and/or mediating physical binding to other proteins the prediction of domain–domain interactions (DDI) and interaction domains is an exciting challenge for both experimental labs and computational biologists.

The availability of data on protein networks was greatly improved in recent years. High-throughput PPI mapping, expression analysis and ChIP-Chip experiments as well as large annotation efforts by PPI databases such as DIP, BIND, HPRD, and MIPS ***(6–9)*** have contributed significantly to the goal of creating a comprehensive picture of the interactome and regulome in model organisms. Many methods such as phylogenetic profiling, coexpression analysis, or genomic neighborhood studies have been developed and have recently been integrated in the STRING resource ***(10)***. Nevertheless, computational prediction of physical binding and functional relations between proteins remains a hot topic because our picture of biological networks is highly incomplete.

In contrast to PPI, data on DDI is currently very sparse rendering *in silico* predictions even more important. Individual publications often narrow down interaction sites to some extent but little of this information has made its way into the databases and no comprehensive large-scale experiments addressing DDI have been published so far. The only large-scale data on the topic is based on domain contacts derived from experimentally determined three-dimensional structures of proteins (PDB ***[11]***): iPfam ***(12)*** and 3did ***(13)*** have generated such data using sligthly different methods. A variety of approaches to the problem of identifying yet unknown domain interactions (and interaction domains) have been developed by different groups ***(14–18)***.

In this chapter, we will describe some of these approaches, assess their performance, and demonstrate the benefits of integrating different methods and data sources.

2. Data and Methods

2.1. Experimental Data

Narrowing down the region of a protein that contains an interaction site or other functionally important feature is usually done by stepwise removal or alteration of different sections of the protein chain followed by experimental assessment of binding or other functional properties of the mutant protein. For individual proteins or interactions this is done frequently in published articles but no large-scale data has been published so far for any of the usual model organisms. We expect this situation to change in the near future because many groups have recognized this lack of data. Nevertheless, comprehensive coverage will not be reached anytime soon. Even the available small-scale data has not been collected systematically in the manually curated interaction databases although efforts to change this situation are under way.

Recently, different groups have started to use protein structure data from the PDB database *(11)* for the identification of domain–domain contacts: iPfam is a novel resource well integrated into the Pfam domain database *(12)* at the Sanger Center (http://www.sanger.ac.uk/Software/Pfam/iPfam/), whereas 3did represents independent work done at EMBL (http://3did.embl.de) *(13)*. Based on the primary sequence of all proteins in the PDB dataset conserved domains are located using standard tools (hmmer). Contacts between different sections of the same protein chain (intraprotein contacts) and among different chains (interprotein contacts, PPI) are identified using the three dimensional atom coordinates contained in the database. Both datasets can be accessed through a web interface and are available for download.

Except for rare crystallization artifacts this kind of data can be expected to be of the highest quality, as the physical contact is directly observed by structure determination. Differences among the two databases can be attributed to slightly different thresholds used to define contacts. Because of its presumable reliability domain interaction data available from the iPfam and 3did database play the role of the gold-standard against which predictions should be measured.

Nevertheless one has to keep in mind that PDB-derived domain interaction data have some intrinsic limitations. First, structural data are not as easy and cheap to produce as gene and protein sequences. Accordingly, PDB covers a much smaller fraction of the protein space than protein sequence resources such as UniProt. Second, certain proteins are easier to crystallize than others resulting in a significant bias in the data found in PDB. A good example of this bias is represented by the class of transmembrane proteins, which are known to be notoriously hard to crystallize and, thus, underrepresented.

Sometimes, researchers circumvent this and other crystallization problems by using fragments of the proteins that are easier to work with. Of course, this does not solve the bias problem and may introduce additional problems if the excluded sections were significant for forming contacts.

2.2. Domain Phylogenetic Profiling

The idea of phylogenetic profiling has been introduced as a method to predict interactions and functional relations among proteins ***(19)*** and has since been widely used for genome annotation. The underlying idea of the method is that proteins (or other entities), which require each other for their function are only useful to the organism if both are present. If one is lost during evolution the other looses its function. Although this is an oversimplification of the actual situation the approach appears to work well.

Building on this general idea, we have recently introduced the concept of domain phylogenetic profiling (DPROF) as a means to identify domain–domain relations. We have demonstrated the utility of the method to predict PPI by mapping the predicted domain network onto a proteome of interest and found the results to add significantly to the results of classic protein profiling ***(18)***.

When performing phylogenetic profiling the first step is the selection of organisms to be analyzed. As the method uses correlated loss or gain of domains it is important to only use genomes, which can be reasonable considered to be completely sequenced. Otherwise the absence of certain domains from a proteome can be easily caused by simple lack of coverage. Another issue is the phylogenetic distribution of selected genomes. Very closely related species (or strains of the same species) should not be used for phylogenetic profiling, because this will give undue weight to loss or gain events occurring in these organisms. For a general purpose application, it is best to use as many organisms as possible from a wide phylogenetic range. Sometimes coverage can be improved by running a profiling analysis on a number of different sets of species and combining the results.

In contrast to protein phylogenetic profiling the domain centered method does not require identification of orthologs by computationally expensive all vs all sequence comparisons. Instead, for each proteome to be included in the analysis all proteins are tested for the occurrence of known protein domains. In our work, we use the Pfam collection of sequence domains represented as hidden Markov models generated from high-quality sequence alignments. The software to detect Pfam domains in genomic proteins (hmmer) is readily available. Precomputed Pfam analyses are often available from the major genome/proteome databases (e.g., **ref. *20***). Of course, other sources of sequence domains can be used if desired.

Given the domain assignments for each of the proteomes we build a matrix $M_{i,j}$ with $i \in$ domains and $j \in$ organisms where $M_{i,j}$ takes a value of "1" if domain i was detected in at least one protein in species j and "0" otherwise. A row M_i of the matrix is called phylogenetic profile of domain i.

Based on the assumption of correlated occurrence of related features a measure of distance or similarity between profiles is required to predict related domains. The naive approach simply uses the number of positions two profiles differ in (Hamming distance d). Although many reasonable predictions can be made using this measure, it also produces large quantities of false-positives. This is mostly owing to the fact that profiles of domains which are found in almost all or only very few organisms have similar profiles although they are hardly related. In both cases, profiles differ in only a few positions and accordingly are treated as similar although the only feature they share is the extremely high or low abundance of the respective domains in the selected organism set. This situation can be remedied by taking into account the information content or entropy H of the profiles ($H = -\sum_{k=0}^{1} p_k \log p_k$; where p_k is the relative frequency of symbol k in the profile) of the profiles (entropy weighted bit difference: $d_w = \frac{d}{\min(H_X, H_Y)}$) or using the *mutual information* between profiles ($I(X;Y) = H(X) + H(Y) - H(X,Y)$ where H is the entropy or joint entropy of a profile or pair of profiles X and Y, respectively; $H(X,Y) = -\sum_{X=0}^{1}\sum_{Y=0} p(X,Y) \cdot log\, p(X,Y)$). **Figure 1** gives an overview of the method.

2.3. Inferring DDI From PPI

Another approach to finding domain interactions uses protein–protein interaction data to infer domain pairs most likely to mediate the observed PPI (PPI2DDI). This idea has been investigated by many different groups using a variety of statistical methods ***(14–17)*** and gradually adding new ideas and concepts to the basic theme.

Today, protein interaction databases cover large numbers of experimentally verified interactions from individually published experiments as well as large-scale surveys of the interactome of various model organisms. The idea of PPI2DDI is that interaction domains, which serve as adapters will be overrepresented in the dataset compared to other domains whose frequencies will lie well within expectation. Essentially, all proposed methods try to identify pairs of domains i, j which are most likely to be involved in the observed PPI x, y.

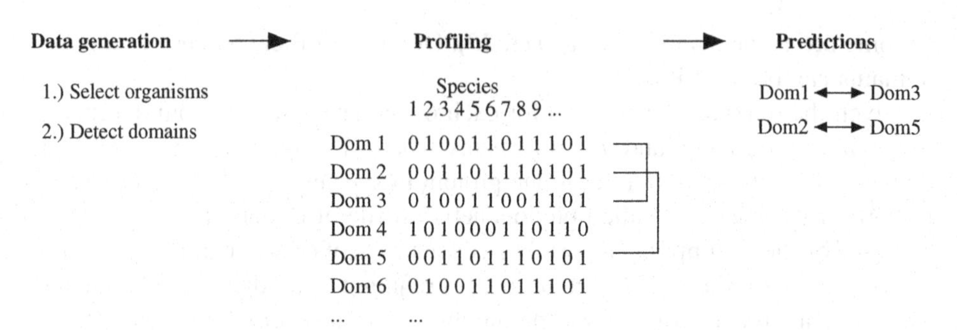

Fig. 1. Phylogenetic domain profiling. After species selection conserved domains are detected in all proteomes. Using this data a domain occurrence matrix is built where rows represent conserved domains and columns species. The rows are called phylogenetic profiles of the respective domains. An element of the matrix is set to "1" if the corresponding domain is present in at least one protein of the respective organism and to "0" otherwise. Domains with similar profiles are called related (interacting and/or functionally linked). In the example, profiles of domains one and three differ only in one position and those of domains two and five are identical, whereas all other combinations exhibit much larger differences.

An early method uses the ratio of observed over expected number of domain-pairs: $S_{ij} = \frac{I_{ij}}{A_{ij}}$, where I_{ij} is the number of PPI where domains i and j occur in the proteins and A_{ij} is the total number of possible protein pairs which contain i and j *(17,21)*. Others have proposed a maximum likelihood estimation approach to improve predictions using the expectation maximization algorithm to maximize the likelihood of the observed interactions ***(15,16)***. In a recent publication, Riley et al. added a modified likelihood ratio statistics to the maximum likelihood estimation method by removing each domain pair individually from the data and recomputing the likelihood by expectation maximization. Through this technique they measure how much influence each domain pair has on the likelihood estimate ***(14)***.

The latter method appears to represent the current state of the art in DDI inference from PPI data.

2.4. Inferring DDI From Predicted PPI

Although large amounts of PPI data are available coverage is still far from complete and focuses on a small set of model organisms. This limits the PPI2DDI approach, which needs observed PPI for inference. Using predicted PPI in addition to the experimental data is a promising way to improve coverage.

The most comprehensive resource for PPI predictions is the STRING database, which integrates a large number of different PPI prediction techniques in a single resource *(10)*. STRING combines predictions and data from gene neighborhood analysis, gene-fusion events, cooccurrence of genes (phylogenetic profiling), RNA coexpression data, experimental evidence (PPI databases), functional association data, and text mining results. The predicted PPI data from STRING can then be used in the same way as the experimental data by the same algorithms for inferring DDI. This can be done by either treating STRING2DDI as a separate prediction method or by simply adding the STRING predictions to the experimental data before analysis.

2.5. Availability

Although there is much literature on the topic, most authors merely provide the results of their own analyses but do not make any software or online resources available. Even more importantly, it would be highly desirable to have a one-stop shopping resource that integrates the results of experimental data and as many prediction methods as possible.

To fill this gap, we provide the Domain Interaction Map (DIMA) resource (http://mips.gsf.de/genre/proj/dima) *(22)*. Although the initial implementation covered only experimental data (iPfam) and the domain phylogenetic profiling method, the approaches previously described have recently been added to the portfolio. Other prediction algorithms and experimental data will be added as they become available.

The website features a simple yet powerful user interface allowing the occasional user to quickly find information on domains of interest. Queries can be performed by Pfam or Interpro IDs, domain descriptions or protein sequences, whereby the latter are then analyzed for conserved domains. Domain neighborhoods are returned as tables, which provide links to external resources (e.g., domain databases), detailed views of individual predictions (if applicable), and visualization of the neighborhood graph. Advanced users may be interested in obtaining the entire domain network for offline analysis with their own tools. As some methods cannot be precomputed for all possible parameter combinations, generation of entire networks is offered as a offline service with the results of a computation delivered by email upon completion.

3. Performance Comparison

3.1. Performance Measures

We use the following performance measures to assess quality and coverage of predictions: true-positive rate (sensitivity, recall) $\frac{tp}{tp+fn}$. Positive predictive value (precision) $\frac{tp}{tp+fp}$. Enrichment $\frac{\text{observed hits}}{\text{expected hits}}$. *tp*, *fp*, *tn*, *fn* stand for true-positives, false-positives, true-negatives, and false-negatives, respectively.

3.2. Hit Enrichment in Top Scoring Pairs

A useful way to investigate the utility of prediction scores is to compute the ratio of observed hits (number of correct predictions) divided by the number of expected hits (number of iPfam pairs in a random sample of the same size). In **Fig. 2**, this ratio (or hit-enrichment) is plotted against the cumulative rank of the different scores. This avoids undue influence of threshold selection and may even guide us in finding good cutoffs. Out of 8,872,578 possible domain combinations (from 4212 Pfam domains) 2733 are found in the iPfam data. That is, we would expect to find one iPfam hit by chance in approximately every 3246th pair. The plot shows that all three prediction algorithms are strongly

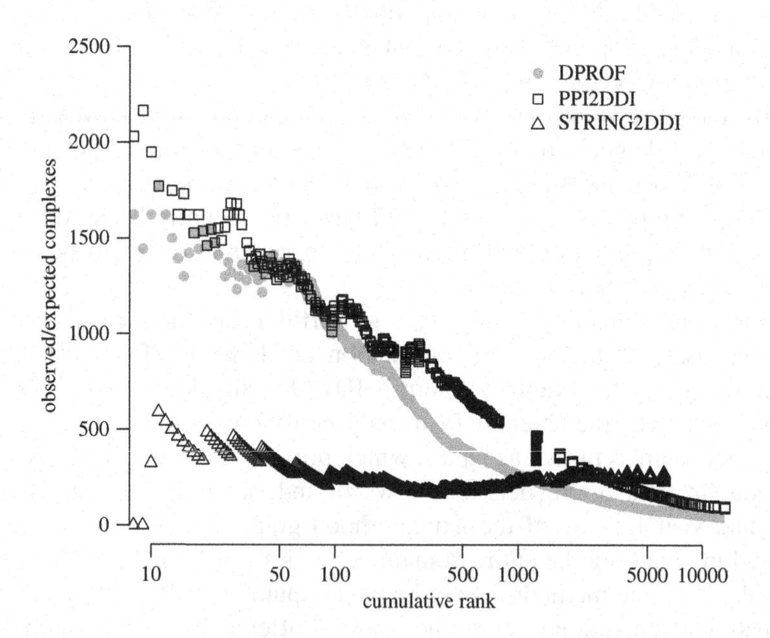

Fig. 2. Hit enrichment of different methods. The *y*-axis shows the hit enrichment, i.e., the ratio of observed over expected number of hits to iPfam reference data for each method. The *x*-axis represents the cumulative rank of the different prediction scores on a logarithmic scale. Domain phylogenetic profiling and PPI2DDI performance is similar, especially for the top 100 scoring pairs. As expected STRING2DDI fairs much lower than the other two because it operates on purely predicted data. Nevertheless, enrichment over random expectation is still large.

enriched in domain contacts. The performance of DPROF and PPI2DDI is similar for approximately the top 100 scoring pairs (with the latter performing slightly better). STRING2DDI fares significantly lower although even this method produces enrichments in the range of 200 to 500.

3.3. Prediction Parameters

In the following analysis, we used empirical score thresholds for the prediction methods which by no means represent the single best choice and we encourage users to experiment with different values for their own applications. For the domain profiling we included domain pairs with a mutual information score of better than 0.4, which yielded a total of 5841 predicted domain pairs which is a sample of less than 1% of all possible pairs. For the PPI2DDI method a threshold of $E \geq 3.0$ (likelihood ratio) has been proposed by the authors of the original publication. Our analysis was limited to the Pfam-A data (the most reliable subset of Pfam) and we increased the score threshold to 5.0. 3264 domain interactions (i.e., 11.5% top scoring pairs) were predicted with this value. For the STRING2DDI predictions a more strict threshold of 10.0 was chosen to account for the more hypothetical nature of the input data which represents all putative PPI with a combined STRING score of $\geq$0.9 (computed without considering experimental data). Under these conditions STRING2DDI produced 1169 predicted domain pairs.

3.4. Coverage and Overlap With iPfam

As no single method will be able to cover the entire domain space, we will first look at the total number of predictions made by each method and their mutual overlap as well as the agreement with the experimental reference data. The results are shown in **Fig. 3** and confirm the notion that the different approaches produce complementary results thus improving total coverage. The data also show that the majority of predictions is not found in iPfam, which is not surprising given the low coverage of this database. At this point there is no way to determine what fraction of these can be attributed to missing data in the reference set and how many represent false-positives.

Four hundred-twenty-two out of 2733 iPfam interactions were predicted by at least one of the three algorithms which translates to a sensitivity of 15%. Because iPfam does not cover functional relations and the input data for predictions (Pfam models, species range, PPI data) are highly incomplete this number is quite encouraging. At the same time, the figures clearly show the advantage of applying more than one method to the problem.

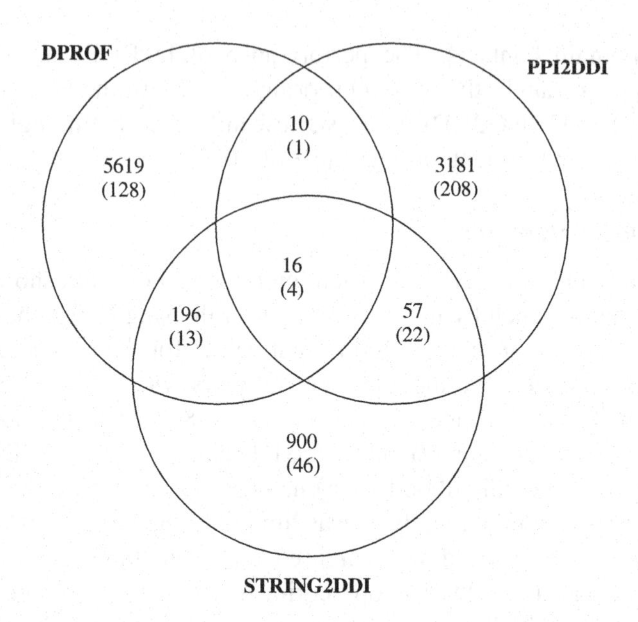

Fig. 3. Prediction overlap. Numbers represent the count in each intersection of the diagram. Values in parenthesis indicate how many of them are also found in iPfam. Clearly, the three sets of predictions overlap comparatively little with each other and the iPfam reference data.

3.5. Improved Prediction by Data Integration

As of today, domain contacts from known structural complexes (iPfam) represent the only reliable source of DDI and, thus, serve as our standard of truth for estimating the prediction performance in the following analysis. The small overlap among predictions with each other and with the iPfam data poses a major challenge for prediction assessment. Nevertheless, the numbers should suffice to give an approximation of the actual situation.

Combining different data sources can potentially improve prediction in two different ways, depending on the integration method. For better coverage, we use the union of predictions (i.e., all predictions supported by at least one method), whereas the intersection (i.e., all predictions supported by at least two or three different methods) will allow better predictive power at the expense of sensitivity. **Table 1** gives sensitivity and precision values for both modes of operation. Depending on the application both integration methods have their merits. For example, for a broad target selection for subsequent lab-testing it may be desirable to first search for possible targets in union-mode and then rank

Table 1
Combining Predictions

			union (∪)				intersection (∩)			
DPROF	PPI2DDI	STRING2DDI	total	in iPfam	PPV [%]	TPR [%]	total	in iPfam	PPV [%]	TPR [%]
+	+	+	9979	422	4.2	15.4	16	4	25.0	0.1
−	+	+	4360	294	6.7	10.8	73	26	35.6	1.0
+	−	+	6798	214	3.1	7.8	212	17	8.0	0.6
+	+	−	9079	376	4.1	13.8	26	5	19.2	0.2
−	−	+	1169	85	7.3	3.1	1169	85	7.3	3.1
−	+	−	3264	235	7.2	8.6	3263	235	7.2	8.6
+	−	−	5841	146	2.5	5.3	5841	146	2.5	5.3

The table contains positive predictive value (PPV) and sensitivity (true-positive rate; TPR) for union and intersection of different prediction data. The first three columns indicate which datasets were included. These are combined by either intersection or union of the predicted pairs. Numbers can be interpreted easily when comparing performance of the different datasets but absolute values mean little because of the incomplete standard of truth.

the predictions by intersection. This way the maximum number of predictions is recovered while the most promising targets are used for experimental work first.

Clearly, performance of individual methods will depend on the choice of thresholds but the data from the above survey already allows a rough assessment. Both PPI2DDI and STRING2DDI showed a precision of ≈7% under the chosen conditions, whereas the profiling method scores 2.5%. Sensitivities range from 3.1% (STRING2DDI) to 8.6% (PPI2DDI) when measured against iPfam. Of course, the absolute numbers should be interpreted with caution because they were derived from a highly incomplete and probably biased standard of truth and thus cannot be expected to be correct. Nevertheless they can safely be used for comparison of results of different methods.

4. Notes

Several conclusions can be drawn on the results obtained by us and other groups working on the problem of domain–domain interactions.

1. Overall, the PPI2DDI methods seems to produce the best results when measured against iPfam and DPROF holds a close second place. The fairly small overlap among them makes them valuable as complementary approaches.

2. STRING2PPI works on much less reliable input data and accordingly its performance is much less impressive than its counterpart PPI2DDI. Nevertheless, the possibility to boost coverage by introducing predicted PPI data may be a valuable endeavor in many cases.
3. No two methods we studied in this work showed much overlap. This gives strong emphasis to the approach of pooling results produced by different algorithms and/or datasets. Although the apparent precision of DDI predictions as evaluated using iPfam seems relatively low, they represent a valuable tool for producing a dataset strongly enriched in real interactions and one has to keep in mind that they were measured against a highly incomplete standard of truth, i.e., the obtained precisions represent lower bounds of the true values.
4. Even if a predicted domain pair may be found not to engage in physical binding in experimental tests many of them may turn out to be linked by functional associations such as binding a common target.
5. Researchers in bioinformatics may prefer to implement their own version of the described approaches to fit their specific needs. The DIMA website allows quick and easy access to these methods for a much wider audience.

Acknowledgments

This work was funded by a grant from the German Federal Ministry of Education and Research (BMBF) within the BFAM framework (031U112C). We would like to thank Robert Riley for helpful technical explanations regarding his algorithm, Thorsten Schmidt for helpful discussions, and Lousie Riley for careful reading of the manuscript.

References

1. Pawson, T. and Nash, P. (2003) Assembly of cell regulatory systems through protein interaction domains. *Science* **300,** 445–452.
2. Letunic, I., Goodstadt, L., Dickens, N. J., et al. (2002) Recent improvements to the SMART domain-based sequence annotation resource. *Nucleic Acids Res.* **30,** 242–244.
3. Henikoff, J. G., Henikoff, S., and Pietrokovski, S. (1999) New features of the Blocks Database servers. *Nucleic Acids Res.* **27,** 226–228.
4. Bateman, A., Coin, L., Durbin, R., et al. (2004) The Pfam protein families database. *Nucleic Acids Res.* **32,** D138–D141.
5. Mulder, N. J., Apweiler, R., Attwood T. K., et al. (2005) InterPro, progress and status in 2005. *Nucleic Acids Res.* **33,** D201–D205.
6. Xenarios, I., Salwínski, L., Duan, X. J., Higney, P., Kim, S. -M., and Eisenberg, D. (2002) DIP, the Database of Interacting Proteins: a research tool for studying cellular networks of protein interactions. *Nucleic Acids Res.* **30,** 303–305.
7. Bader, G. D., Betel, D., and Hogue, C. W. V. (2003) BIND: the Biomolecular Interaction Network Database. *Nucleic Acids Res.* **31,** 248–250.

8. Peri, S., Navarro, J. D., Kristiansen, T. Z., et al. (2004) Human protein reference database as a discovery resource for proteomics. *Nucleic Acids Res.* **32,** D497–D501.
9. Pagel, P., Kovac, S., Oesterheld, M., et al. (2005) The MIPS mammalian protein-protein interaction database. *Bioinformatics* **21,** 832–834.
10. von Mering, C., Jensen, L. J., Snel, B., et al. (2005) STRING: known and predicted protein-protein associations, integrated and transferred across organisms. *Nucleic Acids Res.* **33,** D433–D437.
11. Berman, H. M., Westbrook, J., Feng, Z., et al. (2000) The Protein Data Bank. *Nucleic Acids Res.* **28,** 235–242.
12. Finn, R. D., Marshall, M., and Bateman, A. (2005) iPfam: visualization of protein-protein interactions in PDB at domain and amino acid resolutions. *Bioinformatics* **21,** 410–412.
13. Stein, A., Russell, R. B., and Aloy, P. (2005) 3did: interacting protein domains of known three-dimensional structure. *Nucleic Acids Res.* **33,** D413–D417.
14. Riley, R., Lee, C., Sabatti, C., and Eisenberg, D. (2005) Inferring protein domain interactions from databases of interacting proteins. *Genome Biol.* **6,** R89.
15. Deng, M., Mehta, S., Sun, F., and Chen, T. (2002) Inferring domain-domain interactions from protein-protein interactions. *Genome Res.* **12,** 1540–1548.
16. Huang, C., Kanaan, S. P., Wuchty, S., Chen, D. Z., and Izaguirre, J. A. (2004) Predicting protein-protein interactions from protein domains using a set cover approach. Submitted manuscript.
17. Kim, W. K., Park, J., and Suh, J. K. (2002) Large scale statistical prediction of protein-protein interaction by potentially interacting domain (PID) pair. *Genome Inform. Ser. Workshop Genome Inform.* **13,** 42–50.
18. Pagel, P., Wong, P., and Frishman, D. (2004) A domain interaction map based on phylogenetic profiling. *J. Mol. Biol.* **344,** 1331–1346.
19. Pellegrini, M., Marcotte, E. M., Thompson, M. J., Eisenberg, D., and Yeates, T. O. (1999) Assigning protein functions by comparative genome analysis: protein phylogenetic profiles. *Proc. Natl. Acad. Sci. USA* **96,** 4285–4288.
20. Riley, M. L., Schmidt, T., Wagner, C., Mewes, H. -W., and Frishman, D. (2005) The PEDANT genome database in 2005. *Nucleic Acids Res.* **33,** D308–D310.
21. Sprinzak, E. and Margalit, H. (2001) Correlated sequence-signatures as markers of protein-protein interaction. *J. Mol. Biol.* **311,** 681–692.
22. Pagel, P., Oesterheld, M., Stümpflen, V., and Frishman, D. (2006) The DIMA web resource: exploring the protein domain network. *Bioinformatics* **22,** 997–998.

2

Domain Team

Synteny of Domains is a New Approach in Comparative Genomics

Sophie Pasek

Summary

We present here a method to identify microsyntenies across several genomes. This method adopts the innovative approach of deconstructing proteins into their domains. This allows the detection of strings of domains that are conserved in their content, but not necessarily in their order, that we refer to as *domain teams* or *syntenies of domains*. The prominent feature of the method is that it relaxes the rigidity of the orthology criterion and avoids many of the pitfalls of gene families identification methods, often hampered by multidomain proteins or low levels of sequence similarity. This approach, that allows both inter- and intrachromosomal comparisons, proves to be more sensitive than the classical methods based on pairwise sequence comparisons, particularly in the simultaneous treatment of many species. The automated and fast detection of domain teams is implemented in the DomainTeam software. In this chapter, we describe the procedure to run DomainTeam. After formatting the input and setting up the parameters, running the algorithm produces an output file comprising all the syntenies of domains shared by two or more (sometimes all) of the compared genomes.

Key Words: Protein domains; (micro)synteny; gene fusion; gene cluster; duplication; interaction.

1. Introduction

A synteny can be defined as a chromosomal region showing a local conservation of gene content and proximity—not necessarily contiguity—across several genomes. Because this definition includes the constraint of gene proximity on all of the compared genomes ***(1–3)***, one can speak of microsynteny

From: *Methods in Molecular Biology, vol. 396: Comparative Genomics, Volume 2*
Edited by: N. H. Bergman © Humana Press Inc., Totowa, NJ

rather than synteny. Microsyntenies are mainly encountered among prokaryotic genomes (operons are well known examples of microsyntenies in bacteria) and lower eukaryota such as yeast. This conservation probably points out, in many cases, to a selection pressure that tends to preserve the very proximity of the genes ***(4)***. As a consequence the detection, across several genomes, of these syntenic segments considerably helps the prediction of features of interest such as the physical interaction of proteins or their participation in common metabolic/regulatory networks ***(3,5–11)***. It also enables phylogenetic reconstructions through the identification of some of the numerous rearrangement events that can affect a genome: transpositions, deletions, insertions, inversions, fusions, and fissions ***(12,13)***.

The search for syntenies is commonly integrated in many studies of comparative genomics and a wide range of tools is available to search for microsyntenies ***(1,2,14–18)***. Some of them allow the simultaneous treatment of many species ***(1,2,16,17)***. Others perform both inter- and intraspecies comparisons but impose the number of species to be restricted to two to be tractable (to achieve polynomial time-complexity of their algorithm) ***(15)***. Our approach allows the simultaneous detection of intraspecies and interspecies microsyntenies in a number of species ***(18)***. Moreover the use of domains (instead of genes) as atomic units of the syntenies relaxes the rigidity of the orthology criterion by removing the constraint that each gene must have a unique occurrence in each species, and thus allowing the analysis of genomes with extensive paralogs. Therefore, the notion of synteny of domains allows (1) an extrasensitivity of the detection of syntenies thanks to the identification of domain-homologous genes ***(19)*** sometimes sharing a low level of sequence similarity, (2) the search for duplicated syntenic segments, and (3) to take into account the rearrangements of domains such as fusions ***(6,9,10,20–22)*** and circular permutations ***(23)***.

In this chapter, we describe the use of DomainTeam ***(18)***. We define chromosomal regions of conserved protein domains as *domain teams* or *synteny of domains*. A chromosome is defined as an *ordered* sequence of genes where a unique coding sequence is associated to the nucleic acid sequence of a gene. Each gene is divided into one or more consecutive *domains*, using the PfamA domains of the encoded proteins ***(24)***. Computation of the DomainTeam software produces an output comprising all the syntenies of domains shared by two or more (sometimes all) of the compared genomes. This output reports also the fusions detected inside a domain team. Here, we describe the preparation of the input file, the setup of the parameters, and the command-line of the DomainTeam software. Finally, we explain how to retrieve information from the output file.

2. Materials

2.1. The Chromosome Tables

For all the compared chromosomes (*see* **Note 1**), the chromosomal ordered lists (chromosome tables) of the genes under study and their products (together with their protein identifiers) are required. They can be retrieved from the EBI "proteome" site (http://www.ebi.ac.uk/integr8/EBI-Integr8-HomePage.do).

In the chromosome tables, the following information has to be extracted:

1. A unique gene name for each gene of a chromosome (*see* **Note 2**).
2. The orientation of the gene on the chromosome (forward or reverse, *see* **Note 3**).
3. The rank of the gene, that is, its order on the chromosome (*see* **Note 4**).
4. The physical position (bp) of the first base of the gene (optional, *see* **Note 5**).
5. The identifier or accession number of the protein encoded by the gene (*see* **Note 6**).
6. The size of the protein in amino acids (optional, *see* **Note 7**).
7. The annotation of the protein (optional, *see* **Note 8**).

2.2. The Domain Annotations

For all the proteins encoded by the genes mentioned in the chromosome tables, it is required to provide the ordered suite of domains that belong to the proteins (from the N- to the C-terminal extremities of the protein). Several domain databases can be used for this purpose but we prefer to use the Pfam database (*see* **Note 9**). The PfamA annotations pertaining to the compared proteomes can be downloaded from ftp://ftp.sanger.ac.uk/pub/databases/Pfam/database-files. Home-defined domains can also be used (for example, using HMMER, http://hmmer.wustl.edu/). Whatever the domain database chosen, the following information has to be extracted for each gene:

1. Identifiers of all the domains belonging to the protein encoded by the considered gene.
2. The rank of each domain, i.e., its ordered position in the amino-acids sequence of the protein (*see* **Note 10**).
3. The domain annotation corresponding to each domain identifier (optional, *see* **Note 8**).

3. Methods

3.1. Preparation of the Input File

An example of the input file format is given **Fig. 1**. Each entry begins with "A>" and corresponds to a domain in a specific position on a unique chromosome.

```
A>PF08254	0	0	Leader_Thr
b0001	LPT_ECOLI	0	1	Thr operon leader peptide (Thr operon attenuator)
190	21	F

A>PF00696	0	1	AA_kinase
b0002	AK1H_ECOLI	0	5	Bifunctional aspartokinase/homoserine dehydrogenase I (AKI-HDI)
337	820	F

A>PF01842	0	2	ACT
b0002	AK1H_ECOLI	1	5	Bifunctional aspartokinase/homoserine dehydrogenase I (AKI-HDI)
337	820	F

A>PF01842	0	3	ACT
b0002	AK1H_ECOLI	2	5	Bifunctional aspartokinase/homoserine dehydrogenase I (AKI-HDI)
337	820	F

A>PF03447	0	4	NAD_binding_3
b0002	AK1H_ECOLI	3	5	Bifunctional aspartokinase/homoserine dehydrogenase I (AKI-HDI)
337	820	F

A>PF00742	0	5	Homoserine_dh
b0002	AK1H_ECOLI	4	5	Bifunctional aspartokinase/homoserine dehydrogenase I (AKI-HDI)
337	820	F

A>PF00288	0	6	GHMP_kinases_N
b0003	KHSE_ECOLI	0	2	Homoserine kinase (EC 2.7.1.39) (HSK) (HK)
2801	310	F

A>PF08544	0	7	GHMP_kinases_C
b0003	KHSE_ECOLI	1	2	Homoserine kinase (EC 2.7.1.39) (HSK) (HK)
2801	310	F

A>PF00291	0	8	PALP
b0004	THRC_ECOLI	0	1	Threonine synthase (EC 4.2.3.1)
3734	428	F

/…/

A>PF00486	1	6026	Trans_reg_C
STM4598 Q7CP63	0	2	Response regulator (OmpR family) in two-component regulatory system with
ArcB (Or CpxA)
4855370 238	R

A>PF00072	1	6027	Response_reg
STM4598 Q7CP63	1	2	Response regulator (OmpR family) in two-component regulatory system with
ArcB (Or CpxA)
4855370 238	R

A>XXX	1	6028	XXX
STM4599 Q7CP62	0	0	Putative inner membrane protein
4856182 46	F

A>PF00588	1	6029	SpoU_methylase
STM4600 Q8ZJT6	0	1	Putative tRNA/tRNA methyltransferase (EC 2.1.1.-)
4856722 228	F

//
```

The first line (line 1 of the example) is formatted in the following way:

1. The accession number of the domain (e.g., PF00696).
2. The identifier of the chromosome on which it stands (beginning at 0, e.g., 0 for *Escherichia coli* and 1 for *Salmonella typhimurium*).
3. The position of the domain (rank) on the chromosome (beginning also at 0, e.g., 1).
4. The annotation of the domain (e.g., AA_kinase).

The next two lines (lines 2 and 3 in this example) are devoted to the gene that codes for the protein to which the domain belongs. The first line is composed of:

1. The gene identifier (e.g., b0002).
2. The protein identifier or accession number (e.g., AK1H_ECOLI) of the protein the gene codes for.
3. The position of the domain in the protein, starting from the N-terminus (e.g., 0).
4. The total number of domains in the protein (e.g., 5).
5. The annotation of the protein (e.g., Bifunctional aspartokinase/homoserine dehydrogenaseI).

The second line is composed of:

1. The physical position (bp) of the first base of the gene (e.g., 337).
2. The size of the protein in amino acids (e.g., 820).
3. Its orientation on the chromosome (F stands for forward, R for reverse).

One given gene may be associated to several domains (in this example, the protein AK1H_ECOLI is associated to five domains).

Special domains, labeled XXX (*see* **Fig. 1**), indicate that the corresponding protein is not annotated with any PfamA domain (e.g., Q7CP62, *see* **Note 11**).

3.2. Running DomainTeam

3.2.1. Some Precision Concerning the DomainTeam Source Code

The program is implemented with the C programming language. A compress archive is downloadable at http://stat.genopole.cnrs.fr/domainteams. Uncompress and untar this archive. This will create the DomainTeam directory containing the source files. In this directory, use gcc for the compilation of the source code:

gcc *.c –o domainteam

Fig. 1. Figure represents part of a DomainTeam input file. Several entries (beginning with A>) pertaining to the chromosomes of *Escherichia coli* and *Salmonella typhimurium* are represented. A box at the top of the figure indicates the example given in **Subheading 3**. The second box (bottom) indicates a protein devoid of domain annotation.

3.2.2. The DomainTeam Command-Line

The DomainTeam command-line is the following:
domainteam -a <input file> -d <delta> -e 'criterion'
The input parameters are the name of the previously described input file (<input file>, option –a), the maximum distance *delta* (<delta>, option -d), and the optional criterion to sort the output (option -e). The computation time depends on the data under study (*see* **Note 12**).

3.2.3. Choice of the Delta Parameter

The *distance* between two domains on the same chromosome is the difference between their positions where the position of a domain is defined using the order in which the domain appears on the chromosome (*see* **Note 13**). Then the *delta* value is the maximum distance allowed between two domains in a run of DomainTeam. For example, if the *delta* value is three, it allows gaps of two consecutive domains (*see* **Note 14**).

3.2.4. Using Sorting Criteria

The option –e allows to define a score that is used to sort the domain teams by decreasing score. This score is the result of a user-defined rational expression (comprising +, −,*, /, >, <, =,!) using global and local parameters computed for each domain team. For example, `-e 'NBCHR'` will sort the output as a function of the number of chromosomes the teams are found in and `-e 'NBFUS'` will sort the output as a function of the number of fusions that were found within the teams. Using `-e 'NBCHR * NBFUS'` will sort the output as a function of the product of the number of chromosomes and the number of fusions.

Global parameters are calculated from the information included in the input file:

1. TOTALDOM, the total number of domains.
2. TOTALPROT, the total number of proteins.
3. TOTALCHR, the number of compared chromosomes.
4. DELTA, the input distance *delta*.

The local parameters (specific to a domain team) are the following:

1. NBFUS the number of fusions detected in the team.
2. NBCHR, the number of chromosomes among which the domain team was found.
3. NBOCC, the number of occurrences of the team (*see* **Note 15**).
4. NBDOM, the number of distinct domains in the team.
5. NBPROT, the total number of proteins that belong to the domain team.

6. PRCOMPL, the number of complete proteins it contains (*see* **Note 16**).
7. POSTOT, the total number of domain positions, including insertions and partial proteins.
8. POSIN, the number of positions in the domain team.
9. POSOUT, the number of positions within the team but not belonging to it (insertions, partial proteins, proteins devoid of pfamA annotation [*see* **Note 17**]).
10. MAXDIS and MINDIS the maximal and minimal number of total positions of an occurrence.
11. AVDIS the average distance between two positions in the domain team.

3.3. Reading the Output File

3.3.1. Representation of a Domain Team

Each domain team is reported as in the example given **Fig. 2**.

The parameters pertaining to each domain team are listed in a table (*see* next paragraph), followed by the domain team itself. Each occurrence of the team begins with the identifier of the chromosome it belongs to, followed by a block of elements of the form "*Gene*[*Domain*][*e*/*t*]" where *Gene* is here the ordered locus name of the gene, *Domain* is the Pfam accession number of the domain, *e* is the position of the domain within the protein and *t* the total number of domains in the protein (e.g., STM3982[PF02803][1/2]).

An element enclosed between # (e.g., #b0993[PF00512][4/5]#) is an insertion that does not belong to the domain team (*see* **Note 18**, the *delta* parameter defines the maximum number of such consecutive insertions).

An element enclosed between * (ex., *[242]*) is a protein (the rank of which is 242 on the chromosome) devoid of Pfam domain. In this case, the protein is considered as an inserted domain within the team but the notation * allows to distinguish between this type of insertion and the previously mentioned case noted #. Two occurrences on the same chromosome and separated by a distance of 1981 domain positions is represented as <- 1981 -> (*see* **Fig. 2**).

When detected, a fusion is listed separately. The sum of the lengths of the unfused proteins and the length of the fused protein appear on the side of each occurrence enclosed between {} (*see* the fusion represented in **Fig. 2**).

3.3.2. The Set of Results

The resulting set of domain teams is summarized in a table sorted by decreasing score (the score defined by the user with the –e option) such as exemplified in **Fig. 3**.

```
------------------------------------------------------------------------------------------------------------------------------------
NUMBER |[  EVAL ]|  NBCHR   |  NBOCC   |  NBDOM   |  NBPROT | PRCOMPL |  POSTOT |  POSIN   |  POSOUT |  MAXDIS |  MINDIS |  AVDIS  |  NBFUS   |
   429 |  2       |  2       |  4       |  5       | 10      | 10      | 26      | 25       |  1      |  8      |  6      |  1.879 |  1       |
------------------------------------------------------------------------------------------------------------------------------------
*
0        b1393[PF00378][1/1]  b1394[PF00378][1/1]  b1395[PF02737][1/3]  b1395[PF00725][2/3]  b1395[PF00725][3/3]  #b1396[PF03061][1/1]#
b1397[PF00108][1/2]  b1397[PF02803][2/2]  <-1258->  b2341[PF00725][1/4]  b2
341[PF00725][2/4]  b2341[PF02737][3/4]  b2341[PF00378][4/4]  b2342[PF02803][1/2]  b2342[PF00108][2/2]  <-1977->  b3845[PF02803][1/2]
b3845[PF00108][2/2]  b3846[PF00725][1/4]  b3846[PF00725][2/4]  b3846[PF02737][
3/4]  b3846[PF00378][4/4]
1        STM3982[PF02803][1/2]  STM3982[PF00108][2/2]  STM3983[PF00725][1/4]  STM3983[PF00725][2/4]  STM3983[PF02737][3/4]  STM3983[PF00378][4/4]

!
// FUSION
*
0        b1394[PF00378][1/1]  b1395[PF02737][1/3]  b1395[PF00725][2/3]  b1395[PF00725][3/3]  {737}  <-1261->  b2341[PF00725][1/4]
b2341[PF00725][2/4]  b2341[PF02737][3/4]  b2341[PF00378][4/4]  {714}  <-1981->  b
3846[PF00725][1/4]  b3846[PF00725][2/4]  b3846[PF02737][3/4]  b3846[PF00378][4/4]  {729}
1        STM3983[PF00725][1/4]  STM3983[PF00725][2/4]  STM3983[PF02737][3/4]  STM3983[PF00378][4/4]  {729}
!
------------------------------------------------------------------------------------------------------------------------------------
NUMBER |[  EVAL ]|  NBCHR   |  NBOCC   |  NBDOM   |  NBPROT | PRCOMPL |  POSTOT |  POSIN   |  POSOUT |  MAXDIS |  MINDIS |  AVDIS  |  NBFUS   |
   430 |  2       |  2       |  3       |  6       | 12      |  6      | 30      | 20       | 10      | 13      |  8      |  2.349 |  0       |
------------------------------------------------------------------------------------------------------------------------------------
*
0        b0992[PF00037][1/2]  b0992[PF00037][2/2]  #b0993[PF01627][1/5]#  b0993[PF00072][2/5]  b0993[PF02518][3/5]  #b0993[PF00512][4/5]#
b0993[PF00672][5/5]  #b0994[PF00532][1/1]#  #b0995[PF00486][1/2]#  b0995[
PF00072][2/2]  #b0996[PF03264][1/1]#  b0997[PF00384][1/2]  b0997[PF01568][2/2]  <-274->  b1221[PF00072][1/2]  #b1221[PF00196][2/2]#
b1222[PF00672][1/3]  #b1222[PF07730][2/3]#  b1222[PF02518][3/3]  #b1223[PF07690
][1/1]#  b1224[PF00384][1/2]  b1224[PF01568][2/2]  b1225[PF00037][1/1]
1        STM1763[PF00037][1/1]  STM1764[PF01568][1/2]  STM1764[PF00384][2/2]  #STM1765[PF07690][1/1]#  STM1766[PF00672][1/3]
#STM1766[PF07730][2/3]#  STM1766[PF02518][3/3]  STM1767[PF00072][1/2]
!

/.../
```

Fig. 2. Representation of two domain teams (number 429 and 430). This is a part of the output file obtained by running DomainTeam on two genomes (*Escherichia coli* and *Salmonella typhimurium*). The domain team 429 contains also a fusion, which is reported in the output file such as represented here.

```
Input File: Data/FIC_ANNOT_EcoliK12_Styphimur

11849 Domains     (TOTALDOM)
8710 Proteins     (TOTALPROT)
2 Chromosomes     (TOTALCHR)
Delta: 3          (DELTA)

2138 Domain Teams sorted by: NBCHR
Total number of fusions: 65
```

NUMBER	EVAL	NBCHR	NBOCC	NBDOM	NBPROT	PRCOMPL	POSTOT	POSIN	POSOUT	MAXDIS	MINDIS	AVDIS	NBFUS
1	2	2	4	2	8	6	14	10	4	4	3	2	0
2	2	2	2	1	2	2	4	4	0	2	2	1.500	0
3	2	2	2	3	4	4	8	6	2	5	3	2	0
4	2	2	2	2	2	0	4	4	0	2	2	1.500	0
5	2	2	2	7	12	12	20	16	4	10	10	2.125	0
6	2	2	3	1	3	0	3	3	0	1	1	1	0
7	2	2	5	2	10	10	19	10	9	4	3	2.400	0
8	2	2	5	1	5	5	5	5	0	1	1	1	0
9	2	2	5	2	5	5	12	12	0	3	2	1.583	0
10	2	2	4	1	4	4	4	4	0	1	1	1	0
11	2	2	2	4	4	4	10	8	2	6	4	2	0
12	2	2	5	2	10	5	14	11	3	3	2	1.818	0
13	2	2	4	1	4	0	4	4	0	1	1	1	0
14	2	2	3	5	15	15	25	20	5	9	8	2.099	0
15	2	2	4	1	4	4	4	4	0	1	1	1	0
16	2	2	2	39	56	56	87	78	9	44	43	2.089	0
17	2	2	4	2	8	8	28	28	0	9	5	1.857	0
18	2	2	3	5	14	12	26	18	8	14	6	2.277	0

/.../

Fig. 3. Part of the table summarizing the resulting set of domain teams in the output file. Global and local parameters are indicated in this table. In this example, the teams were sorted (using the –e option) according to the number of chromosomes (NBCHR). Note that in this case the EVAL value is equal to the NBCHR one.

The parameters reported in the tables are:

1. The identifier of the team (NUMBER) attributed after sorting the domain teams according to the scoring option.
2. The value of the score (EVAL).
3. All the previously mentioned local parameters.

In the example given in **Fig. 3**, the teams were sorted according to the number of chromosomes (NBCHR).

4. Notes

1. One may speak about a species instead of a chromosome because most of the bacteria possess one unique chromosome. However some bacteria such as *Vibrio cholerae* have two chromosomes or one may want to treat both the chromosome(s) and the plasmid(s) of some species. Nevertheless, each of the chromosomes and/or plasmids are independently treated by the algorithm, which is not able distinguishing between two chromosomes of the same species or two chromosomes belonging to two distinct species. That is why we prefer speaking about a chromosome rather than a species. Note also that one could provide several contigs instead of the complete chromosomal sequence, each contig being considered as one chromosome by the algorithm.
2. To ensure the uniqueness of the gene name we advise using the ordered gene's name or locus tag nomenclature.
3. The domain teams are searched for irrespective of the conservation of the gene orientation. Nevertheless the gene orientation is an important information because the ordered list of domains of a protein (from the N- to the C-terminal) must be totally reversed for those genes in reverse orientation to attribute a position for the encoded domain on the chromosome.
4. The rank of the gene is simply the rank of the line containing the information about the considered gene in the chromosome table file. It can also be deduced from the physical position of the gene on the chromosome. In this case, the genes can be ranked using the start position for forward genes and the stop position for reverse genes. Overlapping genes *(25)* are considered and noted as adjacent.
5. The physical position (bp) of the first base of the gene is at this time only an informative value. In a first version, it was possible to set the *delta* parameter to a physical distance measured in basepairs but this option is deprecated because we prefer and advise using a number of genes for the value of delta (*see* **Subheading 3.** for an explanation concerning the *delta* parameter). This value is optional, put 0 if you do not provide this information.
6. The protein identifier or accession number of the chromosome tables has to be cross-referred with a domain database to extract the domains associated to the protein (*see* the domain annotation section).

7. The size of the protein is an optional informative value appearing in the output file only if this protein is related to a fusion event. In this case, both the size of the fused protein and the sum of the sizes of the unfused proteins is indicated (*see* **Subheading 3.3.**).
8. This value is optional. The user can put XXX instead of the requested information.
9. As the Pfam domain database is linked to the Uniprot protein database, the use of the Pfam files implies the use of the chromosome tables that contain the Uniprot identifiers (chromosome tables from the EBI "proteome" site).
10. In those few cases where a domain is inserted within another one *(24)*, the two domains are considered as adjacent.
11. Two different XXX domains are never considered as ortholog or paralog. You can use this notation if you want to force the algorithm to ignore a particular domain.
12. The number of domain teams is theoretically exponential but all the experiments we performed were done in a very reasonable time on a 1 Ghz Sun ultrasparc III+processor (5 min for a set of 16 Archaebacteria, 320 min for a set of 15 Gram-negative bacteria (containing very close species) and 29 min for a set of 13 Gram-positive bacteria). The computation time increases as a function of the number of chromosomes, the number of proteins in the set under study, the value of *delta* and the degree of conservation between the organisms under study. Thus, in case of comparing two strains of the same species or comparing tens of organisms, we advise using a low value of delta (for example, *delta*= 3).
13. The distance between two consecutive domains is 1 because the difference between domain in position $i+1$ and domain in position i equals 1.
14. Because *delta* is the maximum distance allowed between two consecutive domains it is distinct from the notion of gap which generally refers to the number of consecutive inserted domains between two domains belonging to the domain team. These two notions are linked but differ by 1: allowing no gap (0 inserted domain) correspond to running DomainTeam with a *delta* value of 1.
15. The number of occurrences (NBOCC) differs from the NBCHR because DomainTeam performs both inter- and intrachromosomal comparison. Thus, in the search for duplicated microsyntenies, the rational expression NBOCC/NBCHR can be used as a criterion to sort the results.
16. Domains being atomic units of the syntenies of domain, it may happen that one or more domains of a protein do not belong to the team the protein belongs to. PRCOMPL is the number of proteins, all the domains of which belong to the team.
17. Insertions are not counted, as well as those proteins that contain one or more domains that are not members of the team.
18. Such inserted elements (#) can be a domain of a protein or a complete protein if it is a one-domain protein.

Acknowledgments

Much of this research took place while the author was at the Laboratoire Génome et Informatique. At this time, implementation and data were publicly available on the Infobiogen website http://lgi.infobiogen.fr/DomainTeam whose team provided a substantial help for this. The author is grateful to the members of her present lab, Laboratoire Statistique et Génome, particularly to Mark Hoebeke who performed the migration of the website to http://stat.genopole.cnrs.fr/domainteams.

References

1. Bergeron, A., Corteel, S., and Raffinot, M. (2002) The algorithmic of gene teams. *Lecture Notes Comput. Sci.* **2452,** 464–476.
2. Luc, N., Risler, J. -L., Bergeron, A., and Raffinot, M. (2003) Gene teams: a new formalization of gene clusters for comparative genomics. *Comput. Biol. Chem.* **27,** 59–67.
3. von Mering, C., Huynen, M., Jaeggi, D., Schmidt, S., Bork, P., and Snel, B. (2003) STRING: a database of predicted functional associations between proteins. *Nucleic Acids Res.* **31,** 258–261.
4. Overbeek, R., Fonstein, M., D'Souza, M., Pusch, G. D., and Maltsev, N. (1999) The use of gene clusters to infer functional coupling. *Proc. Natl. Acad. Sci.* **96,** 2896–2901.
5. Sali, A. (1999) Functional links between proteins. *Nature* **402,** 23–26.
6. Marcotte, E. M., Pellegrini, M., Thompson, M. J., Yeates, T. O., and Eisenberg, D. (1999) A combined algorithm for genome-wide prediction of protein function. *Nature* **402,** 83–86.
7. Galperin, M. Y. and Koonin, E. V. (2000) Who's your neighbor? New computational approaches for functional genomics. *Nature Biotech.* **18,** 609–613.
8. Suyama, M. and Bork, P. (2001) Evolution of prokaryotic gene order: genome rearrangements in closely related species. *Trends Genet.* **17,** 10–13.
9. Enright, A. J. and Ouzounis, C. A. (2001) Functional associations of proteins in entire genomes by means of exhaustive detection of gene fusions. *Genome Biol.* **2,** research0034.1–0034.7.
10. Suhre, K. and Claverie, J. -M. (2004) FusionDB: a database for in-depth analysis of prokaryotic gene fusion events. *Nucleic Acids Res.* **32,** D273–D276.
11. Korbel, J. O., Jensen, L. J., von Mering, C., and Bork, P. (2004) Analysis of genomic context: prediction of functional associations from conserved bidirectionally transcribed gene pairs. *Nature Biotech.* **22,** 911–917.
12. Tang, J. and Moret, B. M. (2003) Scaling up accurate phylogenetic reconstruction from gene-order data. *Bioinformatics* **19,** i305–i312.
13. Sankoff, D. (2003) Rearrangements and genome evolution. *Curr. Opin. Gen. Dev.* **13,** 583–587.

14. Delcher, A. L., Phillippy, A., Carlton, J., and Salzberg, S. L. (2002) Fast algorithms for large-scale genome alignment and comparison. *Nucleic Acids Res.* **30,** 2478–2483.
15. He, X. and Goldwasser, M. H. (2005) Identifying conserved gene clusters in the presence of homology families. *J. Comput. Biol.* **12,** 638–656.
16. Fujibuchi, W., Ogata, H., Matsuda, H., and Kanehisa, M. (2000) A heuristic graph comparison algorithm and its application to detect functionally related enzyme clusters. *Nucleic Acids Res.* **28,** 4021–4028.
17. Kolesov, G., Mewes, H. W., and Frishman, D. (2001) SNAPping up functionally related genes based on context information: a colinearity-free approach. *J. Mol. Biol.* **311,** 639–656.
18. Pasek, S., Bergeron, A., Risler, J. -L., Louis, A., Ollivier, E., and Raffinot, M. (2005) Identification of genomic features using microsyntenies of domains: domain teams. *Genome Res.* **15,** 867–874.
19. Fitch, W. M. (2000) Homology a personal view on some of the problems. *Trends Genet.* **16,** 227–231.
20. Enright, A. J., Iliopoulos, I., Kyrpides, N. C., and Ouzounis, C. A. (1999) Protein interaction maps for complete genomes based on gene fusion events. *Nature* **402,** 86–90.
21. Yanai, I., Derti, A., and DeLisi, C. (2001) Genes linked by fusion events are generally of the same functional category: a systematic analysis of 30 microbial genomes. *Proc. Natl. Acad. Sci.* **98,** 7940–7945.
22. Yanai, I., Wolf, Y. I., and Koonin, E. V. (2002) Evolution of gene fusions: horizontal transfer versus independent events. *Genome Biol.* **3,** research 0024.1–0024.13.
23. Weiner, J., 3rd, Thomas, G., Bornberg-Bauer, E. (2005) Rapid motif-based prediction of circular permutations in multi-domain proteins. *Bioinformatics* **21,** 932–937.
24. Bateman, A., Coin, L., Durbin, R., et al. (2004) The Pfam protein families database. *Nucleic Acids Res.* **32,** D138–D141.
25. Fukuda, Y., Washio, T., and Tomita, M. (1999) Comparative study of overlapping genes in the genomes of Mycoplasma genitalium and Mycoplasma pneumoniae. *Nucleic Acids Res.* **27,** 1847–1853.

3

Inference of Gene Function Based on Gene Fusion Events

The Rosetta-Stone Method

Karsten Suhre

Summary

The method described in this chapter can be used to infer putative functional links between two proteins. The basic idea is based on the principle of "guilt by association." It is assumed that two proteins, which are found to be transcribed by a single transcript in one (or several) genomes are likely to be functionally linked, for example by acting in a same metabolic pathway or by forming a multiprotein complex. This method is of particular interest for studying genes that exhibit no, or only remote, homologies with already well-characterized proteins. Combined with other non-homology based methods, gene fusion events may yield valuable information for hypothesis building on protein function, and may guide experimental characterization of the target protein, for example by suggesting potential ligands or binding partners. This chapter uses the FusionDB database (http://www.igs.cnrs-mrs.fr/FusionDB/) as source of information. FusionDB provides a characterization of a large number of gene fusion events at hand of multiple sequence alignments. Orthologous genes are included to yield a comprehensive view of the structure of a gene fusion event. Phylogenetic tree reconstruction is provided to evaluate the history of a gene fusion event, and three-dimensional protein structure information is used, where available, to further characterize the nature of the gene fusion. For genes that are not comprised in FusionDB, some instructions are given as how to generate a similar type of information, based solely on publicly available web tools that are listed here.

Key Words: Gene fusion; Rosetta stone method; nonhomology-based function prediction; functional networks; metabolic pathways; protein–protein interaction; protein complexes.

From: *Methods in Molecular Biology, vol. 396: Comparative Genomics, Volume 2*
Edited by: N. H. Bergman © Humana Press Inc., Totowa, NJ

1. Introduction

Gene fusion events represent valuable "Rosetta stone" information for the identification of potential protein–protein interactions and metabolic or regulatory networks *(1,2)*. More generally, information on gene fusion events can be complementary to results from other, nonhomology-based approaches, such as phylogenomic profiling and identification of conserved chromosomal localization, to provide testable hypotheses for the characterization of proteins of unknown function *(3–5)*. A number of web-based databases, such as AllFuse *(3)*, STRING *(6)*, and Predictome *(7)*, implement this idea already. However, most of the available databases limit the definition of a gene fusion event to simple nonoverlapping side-by-side BLAST *(8)* matches of two genes from a reference genome to a single open reading frame (ORF) in a target genome, and without providing much information for further in-depth analysis. Searches based on these databases give good starting points for hypothesis building, but the false-positive rate may be quite high (in particular in cases where genes evolve through gene duplication and where the identification of gene orthology is hence difficult). The user is then left with the task of assembling the data required for more extensive case analysis.

This chapter presents FusionDB *(9)* (http://www.igs.cnrs-mrs.fr/FusionDB/), a database that is based on a more strict definition of a gene fusion event, applying a mutual best match criteria *(10)*. It drastically reduces the number of false-positives, at the expense of a potentially similarly high number of false-negatives. To recover from this drawback, gene fusion events between genes from different genomes that belong to the same Cluster of Orthologous Groups (COG) *(11)* are pulled together in what is termed a "COG fusion event." Analysis of COG fusion events allows to investigate gene fusion events in their phylogenomic context by using multiple alignments and phylogenetic tree reconstruction. Questions on the history of individual gene fusion events, such as whether a particular event occurred only once or several times during evolution, or whether more complex processes, such as gene insertion though horizontal gene transfer, gene fission, or gene decay are involved may be addressed using the information provided by FusionDB. The extension to "COG fusion events" also provides information on general tendencies in a whole bacterial genomic context, that is to address questions such as "Which type of genes are most likely to fuse?". FusionDB thereby complements the phylogenetic profiling web server PhydBac *(12–14)*, which is based on the same philosophy: providing detailed non-homology-based information for in-depth analysis of putative protein–protein interactions. FusionDB is thus a complementary tool to other nonhomology based function predictors like the databases previously cited *(3,6,7)*.

2. Materials

All methods described in this chapter may be applied using a standard computer that is equipped with a web browser and an internet connection. For proteins that are included in FusionDB, all required information can be accessed through the FusionDB web interface. For proteins that are not included in FusionDB, a list of freely available web servers that can be used in the context of this method are summarized next.

2.1. Basic Local Alignment Search Tool

The Basic Local Alignment Search Tool (BLAST) *(8)* is used to identify homologies between different proteins. A BLAST web server is available at the NCBI at http://www.ncbi.nlm.nih.gov/BLAST/.

2.2. NCBI RefSeq

All bacterial genes and annotations used in FusionDB are derived from the NCBI taxonomy-based reference sequence (RefSeq) ***(15)*** records for complete bacterial genomes that can be downloaded from ftp://ftp.ncbi.nih.gov/genomes/Bacteria/.

2.3. Cluster of Orthologous Groups

Genes from different genomes are identified to belong to a COG ***(11)*** using an algorithm described in detail by Tatusov et al. ***(10)***. Genes that are member of a same COG can be considered as orthologues within the specificity of the COG annotation. In FusionDB, a COG family is viewed as a generalization of an individual gene. Therefore, the term COG-fusion event is used here. Functional annotation of the different COGs is available at http://www.ncbi.nlm.nih.gov/COG/. FusionDB provides out-links to this database. The COG annotation for some bacterial genomes is also included in the .ptt files provided by RefSeq.

2.4. Protein Data Bank

The Protein Data Bank (PDB) ***(16)*** provides three-dimensional models for protein structures. Of particular interest for the analysis of gene fusion events are heterologuous protein complexes of two proteins that are also found as a fusion protein in some organism. Such cases are included and flagged in FusionDB. The PDB is accessible via the link http://www.rcsb.org.

2.5. Multiple Alignments

Multiple alignments between the N- and C-terminal genes and the corresponding fusion ORFs in FusionDB are computed using the T-Coffee software *(17)*. A T-Coffee web-server *(18)* is available at http://www.igs.cnrs-mrs.fr/Tcoffee/.

2.6. Phylogenetic Tree Reconstruction

Phylogenetic trees are reconstructed using the programs protdist and neighbor from the Phylip package *(19)*, based on the nongapped part of the T-Coffee alignments. A Phylip web server is available at the Pasteur Institute in Paris at http://bioweb.pasteur.fr/seqanal/phylogeny/phylip-fr.html.

3. Methods

1. It is assumed that the basic question to be addressed using the method described in this chapter is the following: "Given a prokaryotic gene – what is its putative role in the cell and what are potential interacting proteins?". Methods complementary to the Rosetta stone method exist and should be used in parallel (*see* **Note 1**). The query gene is supposed to be provided in FASTA format and translated to amino acids. The general method using FusionDB is then as follows:

1. Submit a query to FusionDB.
2. View the individual fusion events: analyze the scores and verify if these are "real" fusion events.
3. If the fused genes are both members of a COG family, view the COG-Analysis page.
4. If the fused genes have their protein structure determined as a multiprotein complex, view the PDB-dimer analysis page.

These steps will now be explained in detail.

3.1. Querying FusionDB

Submit a query to FusionDB using the sequence search option http://www.igs.cnrs-mrs.fr/FusionDB/gene_search.html (*see* **Note 2**). This will run a BLAST search against all genes that are listed in FusionDB. A list of BLAST hits is returned on a page named "Sequence Search Result." If the query gene (or a close homologue) has been found in the FusionDB database, a hit with close to 100% sequence identity will be returned. If this is not the case, a hit with lower sequence identity can be used, but the user should then verify that this gene is a true ortholog of the query gene (*see* **Note 3**).

A click on the top scoring BLAST hit will return all gene fusion events that are available in FusionDB for this selected gene (which may not necessarily be the query gene; indicated by sequence identity below 100%). If a large number of hits is returned, switch to tabular mode. The resulting hits are ordered by the "quality" of the gene fusion event, where a different number of stars is attributed to each hit as a function of the different fusion scores (*see* **subheading 3.2**). In addition, a "#" symbol may be present to indicate that this gene fusion event involves two genes that are member of a COG. A "%" symbol is displayed to indicate cases where the two genes are homologue to a dimer of proteins for which a structure is deposited in the PDB.

3.2. Inspection of Individual Gene Fusion Events

A putative fusion event (PFE) between two genes from a given reference genome in a given target genome is subject to three criteria (**Fig. 1**):

1. Each of the two reference genes must match the same ORF in the target genome as its highest scoring BLAST hit. The overlap between the BLAST hits of both genes must not exceed 10% of the size of the smaller of the two target genes.
2. When split between the two BLAST hits, the two halves of the target ORF must match back to the original two reference genes as their best BLAST hit to the reference genome.
3. Both reference genes must not be homologous to each other.

For every PFE FusionDB provides a series of information, including the sequence data itself. Related gene fusion events in other genomes and homologous gene pairs can be obtained. For a detailed analysis of a PFE, pair-wise alignments between the individual fused N- and C-terminal genes and the fusion ORFs are computed (**Fig. 2**). These pair-wise alignments give valuable information about which parts of the individual genes are also present in the fusion ORF, that is information of domain gains and losses. Ideally one would expect a nonoverlapping one-to-one correspondence between the two fused genes and the fusion ORF.

Every PFE is subjected to a scoring scheme based on different evaluations of its pair-wise and triple alignments by calculating the following five scores:

1. The separation index (*sep*) is a measure for the mix between the domains from the two reference genes when they are placed in a triple alignment with the target ORF. It is calculated as follows as

$$sep = \max_{n=1..L^{ali}} \sum_{i=1}^{n} \left(\frac{\delta_i^N}{L^N} - \frac{\delta_i^C}{L^C} \right)$$

where L^{ali} is the alignment length, L^N and L^C are the lengths of the N- and C-terminal genes, respectively. δ_i^N equals one if a residue of the N-terminal gene

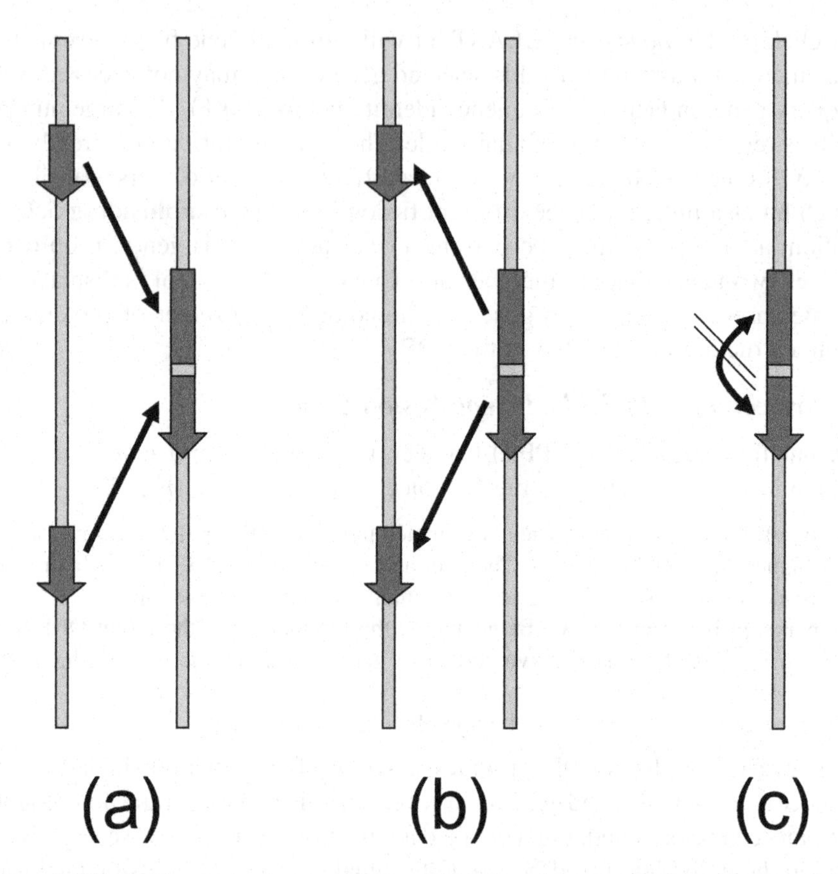

Fig. 1. Criteria for a putative gene fusion event based on a mutual best match criteria. (**A**) Best BLAST hit of two genes in the reference genome to the same open reading frame (ORF) in the target genome, (**B**) the reverse best Basic Local Alignment Search Tool hit of split ORF must correspond to the initial genes in the reference genome, (**C**) no homology between the two ORF segments is required.

appears at position i of the alignment and it is zero otherwise, similarly for δ_i^C. This index *sep* varies between 0 (total mix) and 1 (complete separation).

2. The fusion index (*fus*) is the fraction of residues in the concatenated reference genes that have similar properties as their aligned counterparts in the target ORF. This index may vary between 0 (virtually no homology between the reference genes and the target ORF) and 1 (strong homology).
3. The gene coverage (*cov*) is the fraction of the two reference genes that are alignable to the target ORF in a triple alignment. This index varies between 0 (no relationship

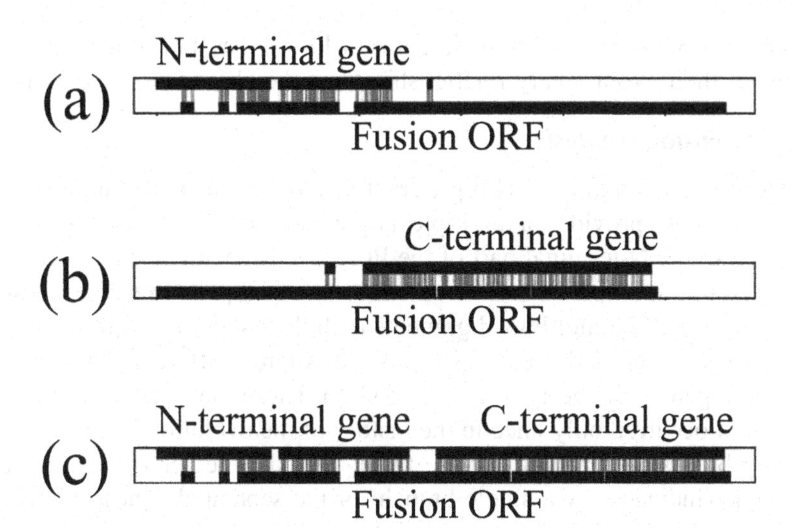

Fig. 2. Visualization of pair-wise alignments generated by FusionDB. (**A**) N-terminal gene from reference genome with fusion open reading frame (ORF) in target genome, (**B**) C-terminal gene with fusion ORF, (**C**) concatenated N- and C-terminal genes with fusion ORF. The degree of conservation of the aligned residues is represented by differently colored vertical lines. Gaps in the alignment can be spotted easily by breaks in the thick black lines that represent the different genes.

at all between the reference genes and the target ORF) and 1 (all domains of the reference genes have a counterpart in the target ORF).

4. The size ratio (*ratio*) between the size of the merged reference genes and the target ORF indicates possible domain gain or losses after the gene fusion event occurred.
5. The "baditude" (*bad*) is the fraction of residues that are aligned between both reference genes when placed in a triple alignment to the target ORF. This index varies between 0 (both reference genes are evolutionary unrelated) and 1 (both reference genes are homologues). A high "baditude" is an indicator for genes with paraloguous domains (domain duplication).

A central question is how to identify "real" gene fusion event and how to discriminate real events from other cases of gene and genome reorganization. Despite the application of the mutual best match criteria, a number of "dubious cases" still remains as can be seen when inspecting the alignments between the fused genes and the fusion ORF. These "dubious cases" correspond to situations where significant domains from two different genes are found in one single ORF, but where the overall architecture of these genes and the fusion ORF has changed so strongly through domain gains or losses that it becomes impossible to decipher their evolutionary relationship from the alignments (*see* **Note 4**).

3.3. COG Fusion Analysis

If two genes belonging to two different COGs are subject to a gene fusion event, FusionDB provides a multiple alignment of all nonfused genes that correspond to the N-terminal part of the PFE together with all fusion ORFs, as well as a similar multiple alignment for the C-terminal part (**Fig. 3**; *see* **Note 5**). From each of the two multiple alignments a phylogenetic tree is reconstructed. On both trees, fused and nonfused genes are colored differently. Comparison of the phylogenies of the two trees allows to determine whether a particular fusion event occurred only once in the history of the evolution of this gene pair (indicated by a single branch for the fused genes), whether it corresponds to a split gene (indicated by a single branch for the separated genes), or whether multiple events of gene fusion occurred. Inconsistent phylogenies between the N- and the C-terminal tree (e.g., a Gram-positive bacteria localized within a

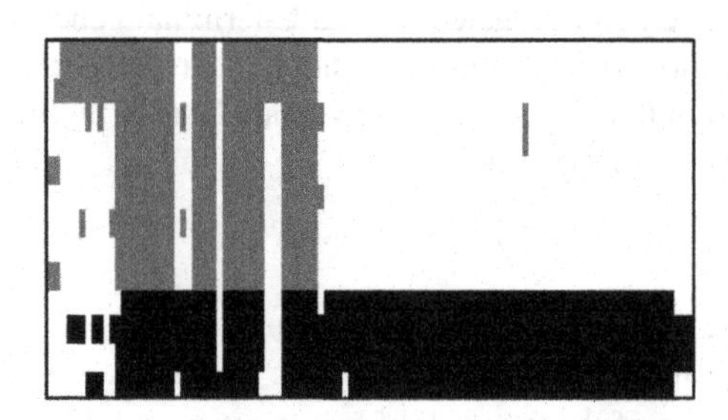

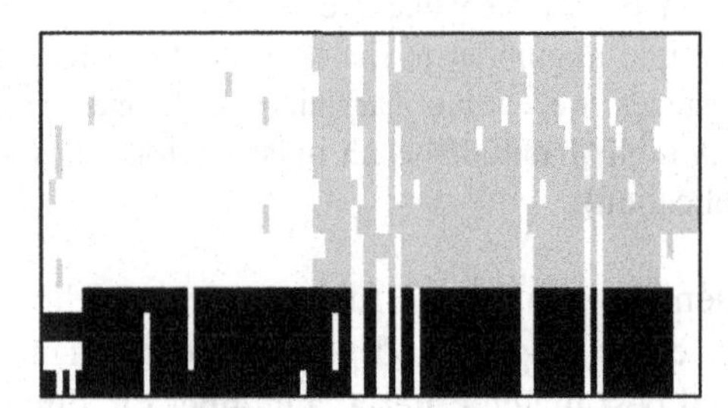

Fig. 3. Visualization of a multiple alignment generated by FusionDB between all N-terminal genes in the Cluster of Orthologous Groups family to which the query gene belongs and all fusion open reading frames (ORFs) that match this fusion event (top), between all C-terminal genes and all fusion ORFs (bottom).

branch of archeae bacteria) may be taken as an indicator of horizontal gene transfer, which may have led to the insertion of a horizontally transferred gene within another host gene.

If one or both of the COGs of a COG fusion event occur in a number of different fusion pairs, information on putative networks of interaction can be obtained. Up to 50 such network nodes following several levels are provided (*see* **Note 6**). For example, COG1290 (cytochrome b subunit of the bc complex) is found fused with COG2010 (cytochrome c, mono-, and diheme variants). On that COG fusion page one can find the information COG2010 is also found fused with COG1622 (Heme/copper-type cytochrome/quinol oxidases, subunit 2). This would then hint at a potential interaction between COG1290 and COG1622 (which is evident in this example, but may be informative when poorly characterized genes are involved).

3.4. PDB-Dimer Analysis

If two genes, that are found to have fused in some genome, have also their protein structure determined through X-ray crystallography (or other techniques), this represents experimental evidence for that these two proteins interact physically. It thus represents a validation of the gene fusion event. However, in most cases, only homologues of the genes under investigation are available as complexes in the PDB. Therefore, a triple alignment between the concatenated fusion genes, the concatenated sequences of the protein complex and the fusion ORF is required to verify the "reality" of such a case. Moreover, a multiple alignment between the fusion ORF and the sequences of the protein complex may give an idea of what part of the sequence has been deleted or added to produce a sequence that can code for a single protein that has—presumably—a similar structure as the protein complex. This kind of triple alignment is available in FusionDB for all such cases.

4. Notes

1. A very good server that combines several nonhomology-based methods for gene function prediction, including gene fusion, but which provides only little detail information on individual cases, is STRING ***(6)*** (http://string.embl.de/). Detailed analysis of phylogenetic profiles and gene coevolution can be done using PhydBac ***(12–14)*** (http://www.igs.cnrs-mrs.fr/phydbac/).
2. In addition to the sequence search option, gene fusion events can also be searched using gene names, COG numbers, and free text annotations, optionally limiting the query to selected genomes. Using the expert search option, it is possible to restrict gene fusion events based on the different fusion score cut-offs or to request

only cases for which COG and/or PDB fusion pages are available. If a large list of hits is expected, it is wise to use the option "tabular output format" to avoid huge downloads.

3. A simple and quick method to test for orthology between a query gene and a gene that is found in a gene fusion event in FusionDB is to retrieve the corresponding sequence data from the FusionDB server (if available all sequences from a COG fusion event) and to submit these sequences together with the query gene to the T-Coffee multiple alignment web server at http://www.igs.cnrs-mrs.fr/Tcoffee/. Check if the query sequence covers the entire length of the fusion gene. If this is the case, and if in addition the alignment core index of the query gene and the fusion gene are comparable and in a high range (indicated by warm colors), it is save to assume that both genes are orthologs (as far as such a conclusion can be drawn from sequence comparison alone).
4. Analysis of the statistics of the different PFE scores suggests that a separation index (*sep*) higher than 0.6 is a save indicator for "clear cases." Details can be found at http://www.igs.cnrs-mrs.fr/FusionDB/sepclass.html.
5. Compare the annotation of the different genes in the set of N- and C-terminal genes. These are supposed to be orthologs, but owing to historical reasons annotations can be quite different (and thus informative on how certain some annotations really are).
6. A list of all COG fusion events in FusionDB, classified by gene function can be found at http://www.igs.cnrs-mrs.fr/FusionDB/fmatrix.html. Interestingly, some genes are also found fused in reverse order. These gene pairs are listed at http://www.igs.cnrs-mrs.fr/FusionDB/symCOG.html (or follow the "results" link of FusionDB).

Acknowledgments

This work was partially supported by Marseille-Nice Génopole and the French Genomic Network (RNG). The author thanks Prof. Jean-Michel Claverie (head of IGS) for laboratory space and support.

References

1. Galperin, M. Y. and Koonin, E. V. (2000) Who's your neighbor? New computational approaches for functional genomics. *Nat. Biotechnol.* **18,** 609–613.
2. Sali, A. (1999) Functional links between proteins. *Nature* **402,** 23–26.
3. Enright, A. J. and Ouzounis, C. A. (2001) Functional associations of proteins in entire genomes by means of exhaustive detection of gene fusions. *Genome Biol.* **2,** RESEARCH0034.
4. Marcotte, E. M. (2000) Computational genetics: finding protein function by nonhomology methods. *Curr. Opin. Struct. Biol.* **10,** 359–365.

5. Marcotte, E. M., Pellegrini, M., Thompson, M. J., Yeates, T. O., and Eisenberg, D. (1999) A combined algorithm for genome-wide prediction of protein function. *Nature* **402,** 83–86.
6. von Mering, C., Huynen, M., Jaeggi, D., Schmidt, S., Bork, P., and Snel, B. (2003) STRING: a database of predicted functional associations between proteins. *Nucleic Acids Res.* **31,** 258–261.
7. Mellor, J. C., Yanai, I., Clodfelter, K. H., Mintseris, J., and DeLisi, C. (2002) Predictome: a database of putative functional links between proteins. *Nucleic Acids Res.* **30,** 306–309.
8. Altschul, S. F., Madden, T. L., Schaffer, A. A., et al. (1997) Gapped BLAST and PSI-BLAST: a new generation of protein database search programs. *Nucleic Acids Res.* **25,** 3389–3402.
9. Suhre, K. and Claverie, J. M. (2004) FusionDB: a database for in-depth analysis of prokaryotic gene fusion events. *Nucleic Acids Res.* **32,** D273–D276.
10. Tatusov, R. L., Koonin, E. V., and Lipman, D. J. (1997) A genomic perspective on protein families. *Science* **278,** 631–637.
11. Tatusov, R. L., Natale, D. A., Garkavtsev, I. V., et al. (2001) The COG database: new developments in phylogenetic classification of proteins from complete genomes. *Nucleic Acids Res.* **29,** 22–28.
12. Enault, F., Suhre, K., Abergel, C., Poirot, O., and Claverie, J. M. (2003) Annotation of bacterial genomes using improved phylogenomic profiles. *Bioinformatics* **19,** i105–i107.
13. Enault, F., Suhre, K., Poirot, O., Abergel, C., and Claverie, J. M. (2003) Phydbac (phylogenomic display of bacterial genes): an interactive resource for the annotation of bacterial genomes. *Nucleic Acids Res.* **31,** 3720–3722.
14. Enault, F., Suhre, K., Poirot, O., Abergel, C., and Claverie, J. M. (2004) Phydbac2: improved inference of gene function using interactive phylogenomic profiling and chromosomal location analysis. *Nucleic Acids Res.* **32,** W336–W339.
15. Pruitt, K. D., Tatusova, T., and Maglott, D. R. (2005) NCBI Reference Sequence (RefSeq): a curated non-redundant sequence database of genomes, transcripts and proteins. *Nucleic Acids Res.* **33,** D501–D504.
16. Berman, H. M., Westbrook, J., Feng, Z., et al. (2000) The Protein Data Bank. *Nucleic Acids Res.* **28,** 235–242.
17. Notredame, C., Higgins, D. G., and Heringa, J. (2000) T-Coffee: A novel method for fast and accurate multiple sequence alignment. *J. Mol. Biol.* **302,** 205–217.
18. Poirot, O., Suhre, K., Abergel, C., O'Toole, E., and Notredame, C. (2004) 3DCoffee@igs: a web server for combining sequences and structures into a multiple sequence alignment. *Nucleic Acids Res.* **32,** W37–W40.
19. Felsenstein, J. (1989) PHYLIP: Phylogeny Inference Package (Version 3.2). *Cladistics* **5,** 164–166.

4

Pfam

A Domain-Centric Method for Analyzing Proteins and Proteomes

Jaina Mistry and Robert Finn

Summary

The constant deluge of genome sequencing data means that annotating, classifying, and comparing proteins or proteomes can seam like an endless task. Furthermore, discovering and accessing such data is fundamental to biologists. There are, however, databases that perform these tasks. Pfam, a protein families database, is one such database. In this chapter, the use of the web interface to Pfam and the resources provided (annotation, sequence alignments, phylogenetic trees, profile hidden Markov models [HMMs]) are described. The exploitation of tools for searching sequences against the library of Pfam HMMs, searching for domain combinations, searching by taxonomy, browsing proteomes, and comparing proteomes are outlined in detail.

Key Words: Pfam; protein families; domains; profile hidden Markov model; HMM; annotation; sequence; alignments; phylogeny; taxonomic distribution.

1. Introduction

Pfam *(1)*, the protein families database, is a collection of multiple sequence alignments and profile hidden Markov models (HMMs) *(2)*. Each HMM is a probablistic model that represents a protein family or domain, where a domain can be thought of as a discrete structural unit that is often found in different combinations or "architectures". By searching a protein sequence against the Pfam library of HMMs, the domain architecture of a protein can be deduced. Each Pfam family has associated annotation, literature links, multiple sequence alignments, and links to other databases.

From: *Methods in Molecular Biology, vol. 396: Comparative Genomics, Volume 2*
Edited by: N. H. Bergman © Humana Press Inc., Totowa, NJ

Pfam contains two types of families, Pfam-A and Pfam-B. Pfam-A families are manually curated HMM-based families. The second type of family are Pfam-B families that are automatically generated from the PRODOM database *(3)*. Pfam-B families are alignments of sequence segments taken from PRODOM with any Pfam-A residues removed *(4)*. All families in Pfam are non-overlapping such that no amino acid belongs to more that one family/domain.

The coverage of Pfam is high. Approximately 75% of all proteins in UniProt *(5)* have a match to at least one Pfam-A family, and approx 94% have a match to at least one Pfam-A or Pfam-B family (release 19.0). Approximately 50% of residues in UniProt fall within a Pfam-A family. An alternative measure of coverage is to consider the proportion of a proteome that is covered. 68 percent of all human sequences and 85% of all *Escherichia coli* sequences contain at least one match to a Pfam-A family (release 19.0).

It is thought that the majority of proteins can be clustered into a few thousand protein families *(6)*. Functional information about a protein can usefully be transferred between related proteins and the high coverage of Pfam makes it an invaluable tool in genome and proteome annotation. This chapter describes how to use and access the different types of information contained within Pfam. It explains how to search protein and DNA sequences against the library of Pfam HMMs, and describes how to utilize Pfam in the analysis of proteomes and domain architectures. This chapter should be read alongside the help pages on the Pfam website and the Pfam publications.

Pfam data are freely available and can accessed from the following locations:
http://www.sanger.ac.uk/Software/Pfam (UK)
http://www.cgr.ki.se/Pfam (Sweden)
http://pfam.wustl.edu (US)
http://pfam.jouy.inra.fr/ (France)

2. Materials

2.1. Hardware

1. Computer with network connection.

2.2. Software

1. Javascript-capable browser (e.g., Internet Explorer 4.0, Mozilla 5.0).
2. Java.
3. HMMER2 software package *(7)*.
4. pfam_scan.pl (available from ftp://ftp.sanger.ac.uk/pub/databases/Pfam/Tools/pfam_scan.pl).

5. Perl (Windows users can use a precompiled binary for Perl, for example active Perl from www.activestate.com).

2.3. Other

1. Email account.
2. Sequence data in fasta format.
3. The following Pfam files, which are available from the Pfam ftp site (ftp://ftp.sanger.ac.uk/pub/databases/Pfam/current_release/): Pfam_ls, Pfam_fs, Pfam-A.seed, and Pfam-C.

3. Analyzing a Proteome

The following sections describe how to use the Pfam database. **Subheading 3.1.** describes the features of the family website pages that form the focal point for information relating to a particular family. **Subheading 3.2.** explains how to perform searches against the library of Pfam HMMs and **Subheading 3.3.** onward describes how to analyze proteomes and domain architectures.

3.1. Description of the Family Page and General Background to Pfam

The Pfam website family pages, an example of which is shown in **Fig. 1,** can be accessed by following links for a particular family/domain during the next sections, clicking on a graphical image of a domain or by searching for a particular family using the search box on the Pfam website. A list of families can also be accessed by clicking on "Pfam family ID" under the "browse by" menu on the Pfam homepage.

The family pages are the center point for viewing information about a particular family. They contain alignments, annotation (including annotation from the InterPro database *[8]* where available), and cross-links to other databases in addition to other tools for protein analysis. Structural information is displayed for each family where available. This section describes how to access the main information that Pfam keeps for each family.

Pfam has two types of alignments for each family that are called the "seed" and the "full" alignment. The seed alignment contains a set of representative sequences for a family and is the alignment that Pfam used to construct the HMMs for that family. Pfam uses the HMMER2 suite of programs *(7)* to build and search profile HMMs. Two HMMs are built for each family, one to represent fragment matches and one to represent full length matches. Each HMM has manually set threshold values that determine the minimum score a sequence must attain to belong to a family. All UniProt sequences are searched against each of the Pfam HMMs, and sequences that score above the cut-off value for a particular family are included in the family's full alignment.

Pfam annotation and associated literature references, and InterPro annotation where available, is found on every family page (*see* **Fig. 1A–C**).

3.1.1. Viewing Family Sequence Alignments

The seed and full alignment is available in a number of formats including Stockholm, FASTA, and plain Pfam format (*see* **Fig. 1D**).

1. Choose an alignment (either seed or full).
2. Choose a format from the drop down menu. Further alignment options can be viewed and selected by clicking on the "further alignment options" link and selecting from the drop down menus.
3. Click "get alignment."

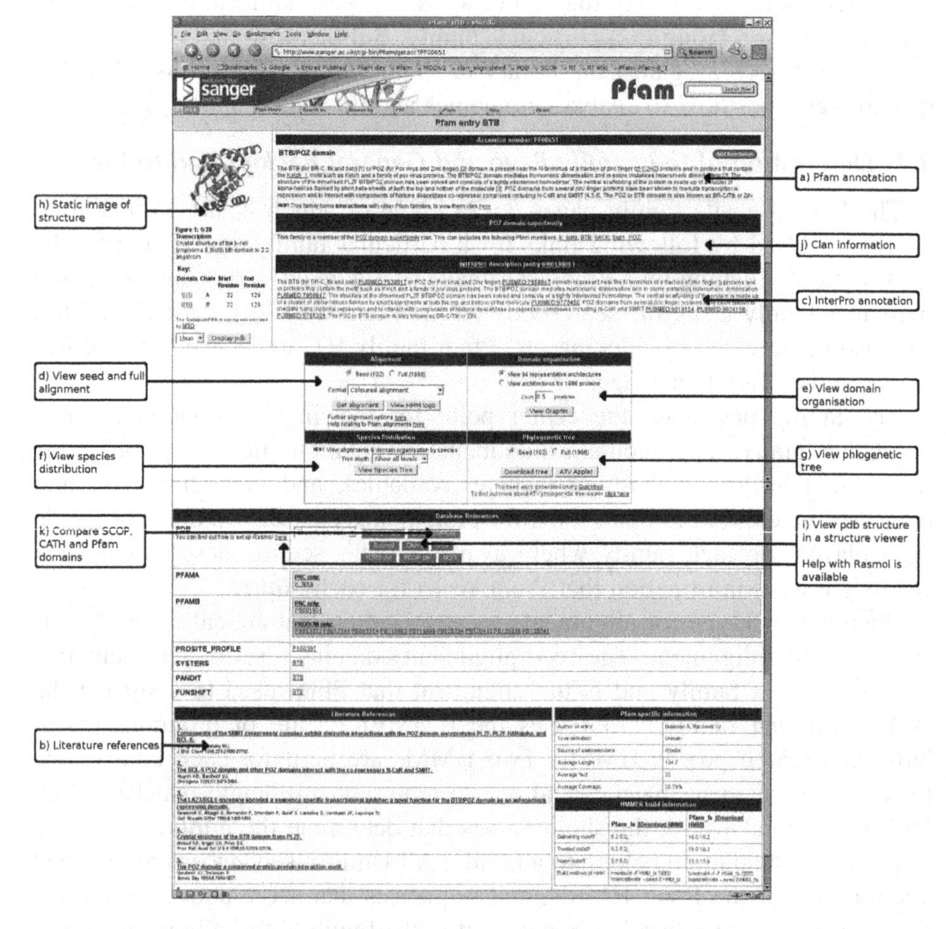

Fig. 1. Family page for the BTB domain with the main features highlighted.

3.1.2. Viewing Domain Organization

A graphical view of the domain organization is a practical way of seeing all the different types of proteins architectures that contain a particular domain. The user has the option of selecting all domain architectures for a given family, or viewing a smaller number of representative domain architectures (*see* **Fig. 1E**).

1. Choose the type of architectures to view (either all or a representative set).
2. Click "view graphic."

The Pfam domain definitions can also be compared that of the structural databases SCOP and CATH (*see* **Note 1**).

3.1.3. Viewing Species Distribution

The species distribution shows which phyla contain the domain. The user is able to select species from the tree and view an alignment containing only the species of interest (*see* **Fig. 1F**).

1. Click on the "View species tree" under the species distribution heading. This gives a phylogenetic tree of all the species that contain the family/domain.
2. Click on the species to see the domain organization of the proteins from a particular species that contain the domain. The alignment for a specific set of species can be viewed by clicking on the tick boxes next to the species of interest and clicking on the "view selected species alignment" box near the top of the page.

3.1.4. Viewing the Phylogenetic Tree

The phylogenetic tree for a Pfam family can be viewed both in Newick format and graphically via the ATV phylogenetic tree viewer. (More information on how to use the ATV viewer can be accessed through the link under the phylogenetic tree heading [*see* **Fig. 1G**]).

1. Select an alignment (either seed or full).
2. Click on "download tree" to see the tree in Newick format, or on the "ATV applet" button view the tree graphically in ATV phylogenetic tree-viewer.
3. If the tree has been downloaded, it can also be viewed using other tree viewing software of the users choice.

3.1.5. Viewing Three-Dimensional Structures

If a known three-dimensional structure is available for a member of the family, a static image highlighting the Pfam family/families will be shown on the family page. Where there is more than one structure, the user can select different static images as follows:

1. Choose a pdb code from the drop down menu (*see* **Fig. 1H**).
2. Click on "display pdb."

Structures can also be explored using one of the structure viewers Jmol, Chime, and Rasmol. Jmol is an applet whereas Chime and Rasmol will need to be installed (if the user does not already have them) and the browser configured. For help with the browse configuration for Rasmol and Chime, follow the help link (*see* **Fig. 1I**). The Pfam family/families along with any active site residues are highlighted on the three-dimensional structure.

Viewing structures in a structure viewer:

1. Under PDB in the database section of the family page, choose a structure from the drop down menu.
2. Click on one of the structure viewer buttons (Jmol, Chime, or Rasmol) (*see* **Fig. 1I**).

3.1.6. Viewing Clan Information

Some Pfam families are grouped into "clans". Pfam defines a clan as a collection of families that have arisen from a single evolutionary origin. Evidence of their evolutionary relationship can be in the form of similarity in tertiary structures, or when structures are not available, by common sequence motifs. The seed alignments for all families within a clan are aligned and the resulting alignment (called the clan alignment) can be accessed from a link on the clan page. Each clan page includes a clan alignment, a description of the clan, and database links where appropriate. The clan pages can be accessed by following a link from the family page, or alternatively they can be accessed by clicking on "clans" under the "browse" by menu on tab on any Pfam page.

If a family is a member of a clan, the clan name and other members of the clan are shown on the family page.

1. Clicking on the clan name to access the clan index page (*see* **Fig. 1J**).
2. From here, click on the clan of interest. This will bring up the clan page where the clan alignment, description, and database links can be accessed.

3.2. Searching Against Pfam

The next section explains how to search protein sequences against the library of Pfam HMMs. Pfam also has the facility to search single DNA sequences against Pfam by utilizing the Wise2 software package *(9)*, which allows the comparison of protein HMMs to genomic DNA.

For more information on scores and *e*-values see the help pages on the Pfam website.

3.2.1. Single Sequence Protein Searches

This section describes how to perform single protein sequences against the library of Pfam HMMs via the website.

If the sequence of interest is already in UniProt, Pfam may have already calculated the domain architecture for it. Enter the accession/id on the Pfam homepage in the "Enter a SWISS-PROT or TrEMBL name or accession number" box and press "go" (*see* **Fig. 2A**). If present in Pfam, the precalculated domain architecture will be retrieved. If the sequence is not present in Pfam, use the following instructions to perform the searches.

1. Click on "Protein name or sequence" under the "search by" menu on the Pfam homepage.
2. Paste the protein sequence of interest in FASTA format into the box under "cut and paste your sequence here" (*see* **Fig. 2B**).
3. Click on the "search Pfam" button. The results should be displayed within 2 min (*see* **Note 2**).

3.2.2. Medium-Scale Protein Searches (< 5000 Sequences)

The batch upload facility allows medium scale searching (up to 5000 sequences) of the Pfam HMM libraries and is useful if local searching of Pfam is not a feasible option. The user uploads a FASTA file of protein sequences to the Pfam site and the results are emailed back to them.

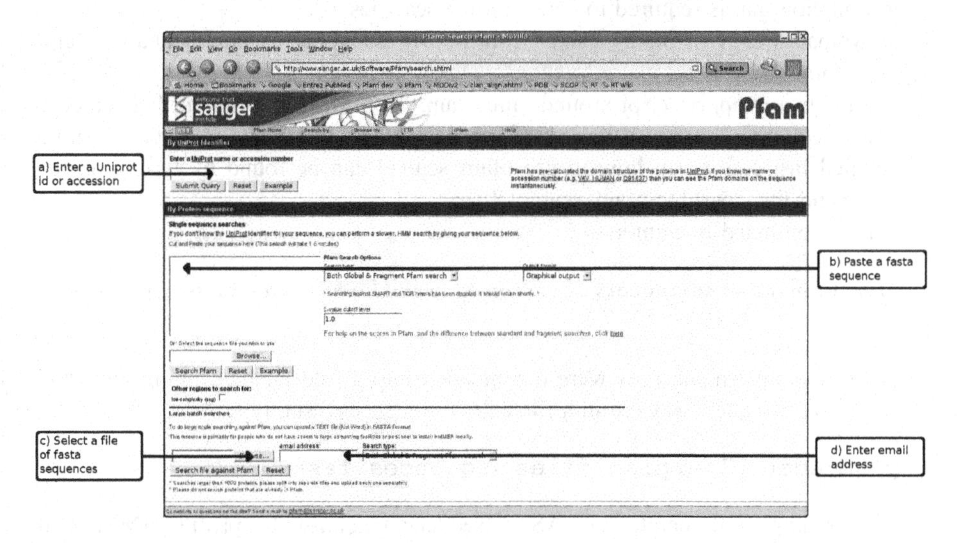

Fig. 2. The Pfam search page. This page is used for selecting protein sequence(s) to search the Pfam library of hidden Markov models.

1. Save the FASTA sequences as text files, each containing a maximum of 1000 sequences (*see* **Note 3**).
2. Go to the "Search by" menu on the Pfam homepage and select "Protein name or sequence."
3. At the bottom of the page use the browse button to select a file of sequences, enter the email address where the results should be mailed, and choose the search type, local (to searches against only fragment models and retrieve fragment matches), global (to search against full length model and retrieve only full length matches), or both. Click on "search against Pfam" (*see* **Fig. 2C,D**).
4. The search results will be emailed to the address specified once they are complete. The results are usually ready within 48 h.

3.2.3. Large-Scale Protein Searches

For large-scale searches the most favorable solution is to run Pfam locally. This is a more arduous task then the web-based searches and requires the user to download the HMMER2 software *(7)*, in addition to a Perl script from the Pfam website. The user will also need a copy of the Pfam files that contain the HMM libraries, these are available for download from the Pfam ftp site. Further information about each of the files can be found on the Pfam website.

1. Download the HMMER2 software *(7)*.
2. Download the Pfam files Pfam_ls, Pfam_fs, Pfam-A.seed, and Pfam-C from the ftp site linked from the Pfam homepage. These files contain the HMMs and additional information that is required to carry out the searches.
3. Download a copy of pfam_scan.pl from ftp://ftp.sanger.ac.uk/pub/databases/Pfam/Tools/pfam_scan.pl.

   This is a wrapper script around hmmpfam, the HMMER program that searches query sequences against a library of profile HMMs. If perldoc is installed, more detailed instructions on how to use pfam_scan.pl can be found by typing on the command-line "perldoc pfam_scan.pl."
4. On the command-line enter:

```
pfam_scan.pl -d <directory location of Pfam files><fasta file of proteome>
```

For example if the files were downloaded into a folder called pfam_files, and the FASTA sequences were in a file called sequences.fasta, type:

```
pfam_scan.pl -d pfam_files sequences.fasta
```

pfam_scan.pl will search the FASTA sequences against the profile HMMs and report all matches to families that score higher than the manually set thresholds for each of the Pfam families (*see* **Note 4**). The output is in the following format:

```
seq_id seq_start seq_end hmm_acc hmm_start hmm_end bit_score evalue hmm_name
000519 95 562 PF01425.9 1 513 526.5 2.6e-155 Amidase
001636 22 139 PF02408.8 1 137 157.3 3.6e-44 DUF141
003046 12 137 PF02788.5 1 132 295.2 1.1e-85 RuBisCO_large_N
```

3.2.4. Single Sequence DNA Searches

This section describes how to search single DNA sequences against the library of Pfam HMMs.

1. Click on "DNA sequence" under the "search by" menu on the Pfam homepage.
2. Paste the DNA sequence of interest in FASTA format into the box under "cut and paste your sequence here."
3. Select the type of genomic DNA from the drop down menu, and click on the "submit query" button. The results will take approx 2 min for a 1-kb sequence, and approx 1 h for a 80-kb sequence.

3.3. Finding the Families/Domains Contained Within a Proteome

Pfam precalculates the domain architecture for the proteomes defined in Integr8 *(10)* (*see* **Note 5**). This section describes how retrieve domain information for a particular proteome. **Figure 3** shows a sample results page which shows the Pfam domains and families that are present in the proteome of *E. coli*.

1. Go to the "genomes" page under the Browse by menu.
2. Click on the "+" buttons to see a list of all the proteomes for which Pfam has precalculated the domain structure for, and click on the species name of interest. The resulting table shows the Pfam families that are found in the species, a short description of each family, the number of proteins that match each family, and the number of regions found in the species.
3. The user can find out which proteins contain a particular family by selecting the family/families of interest by clicking in the "view regions" column (*see* **Fig. 3A**). Both the FASTA sequence of a family and the complete sequence are available by choosing from the drop down menu at the top of the page. The different domain architectures for a particular domain can be viewed by selecting graphical output from the drop down menu, or the text output of this information can be viewed by selecting "text output". After choosing from the drop down menu, the user can either retrieve the information for the selected domain(s) by clicking on "selected families," or retrieve the information for all families by clicking on "all families".
4. Information about a particular family can be accessed by clicking on a family in the first column of the table. This will take the user to the family page for that particular Pfam entry (*see* **Fig. 3B**).

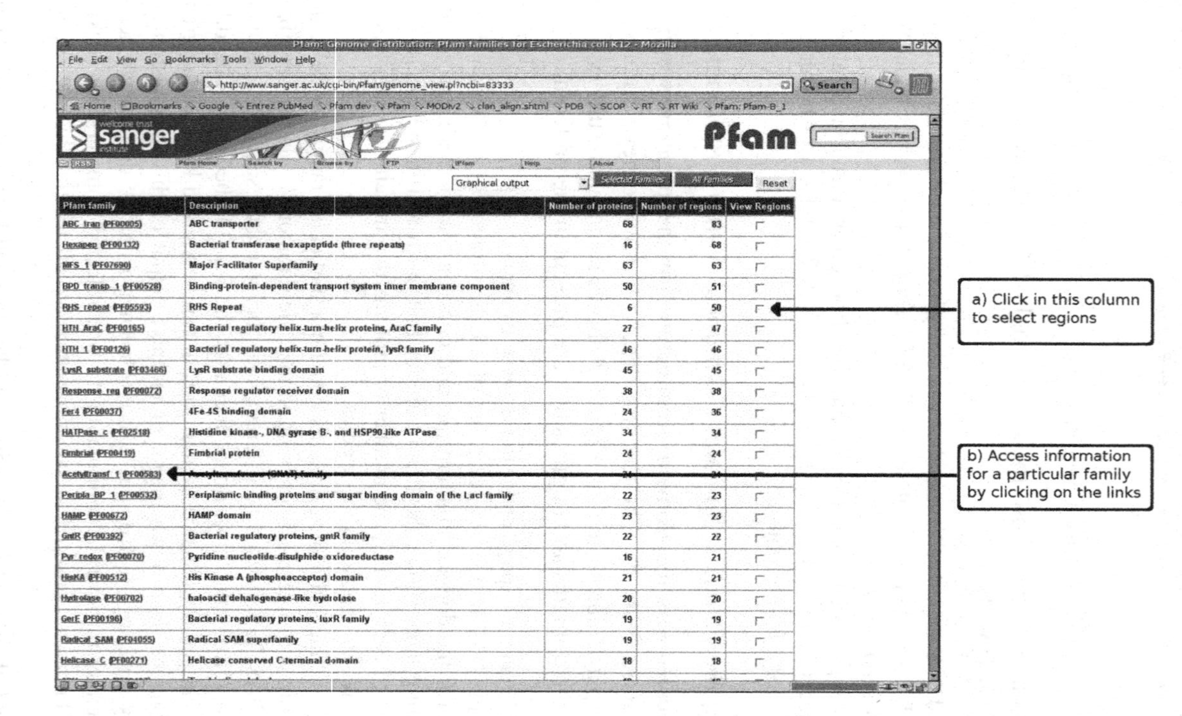

Fig. 3. Results page showing the Pfam domains and families present in the proteome of *Escherichia coli*.

3.4. Comparing Families/Domains Shared Between Proteomes

The proteome pages allow the user to compare which domains are shared between proteomes. It can also highlight the domains that are found in one proteome and not in another. This extracts information about the repertoire of domains contained within the proteomes being investigated.

1. Go to the "genomes" page under the Browse by menu.
2. Click on the "+" buttons to access the list of proteomes and click on the "compare genomes" boxes beside the proteomes of interest. At the top of the page click on the "compare selected genomes" button. For the large eukaryote species this can take several minutes.
3. The results are displayed in a table that shows for each species, the number of domains and in brackets, the number of proteins that match a Pfam family (*see* **Fig. 4A**). Domains that are present only in one species are highlighted in gray (*see* **Fig. 4B**).
4. Clicking on the numbers generates a page that displays a graphical view of the proteins for that specific Pfam domain. This view can be toggled to the FASTA sequence or text output view using the drop down menu at the top of the page.

3.5. Comparing Families/Domains Between Species

The taxonomy query allows quick identification of families/domains, which are present in one species but are absent from another.

1. Click "Taxonomy query" under the "search by" menu on the Pfam homepage.
2. Enter the query into the search box. Taxonomic names can either be picked from the drop down menu or typed in the search box. For example:
   Nematoda AND NOT Caenorhabditis elegans

This query will return the list of Pfam families present in Nematoda but not in *C. elegans*.

3.6. Finding Families/Domains Unique to a Particular Species

The taxonomy query can also be used to identify families/domains that are unique to a particular species.

1. Click "Taxonomy query" under the "search by" menu on the Pfam homepage.
2. To find families/domains unique to a particular species tick the box "find domain unique to a specific family", enter the species in the search box and click "search". Note that this can be very slow.

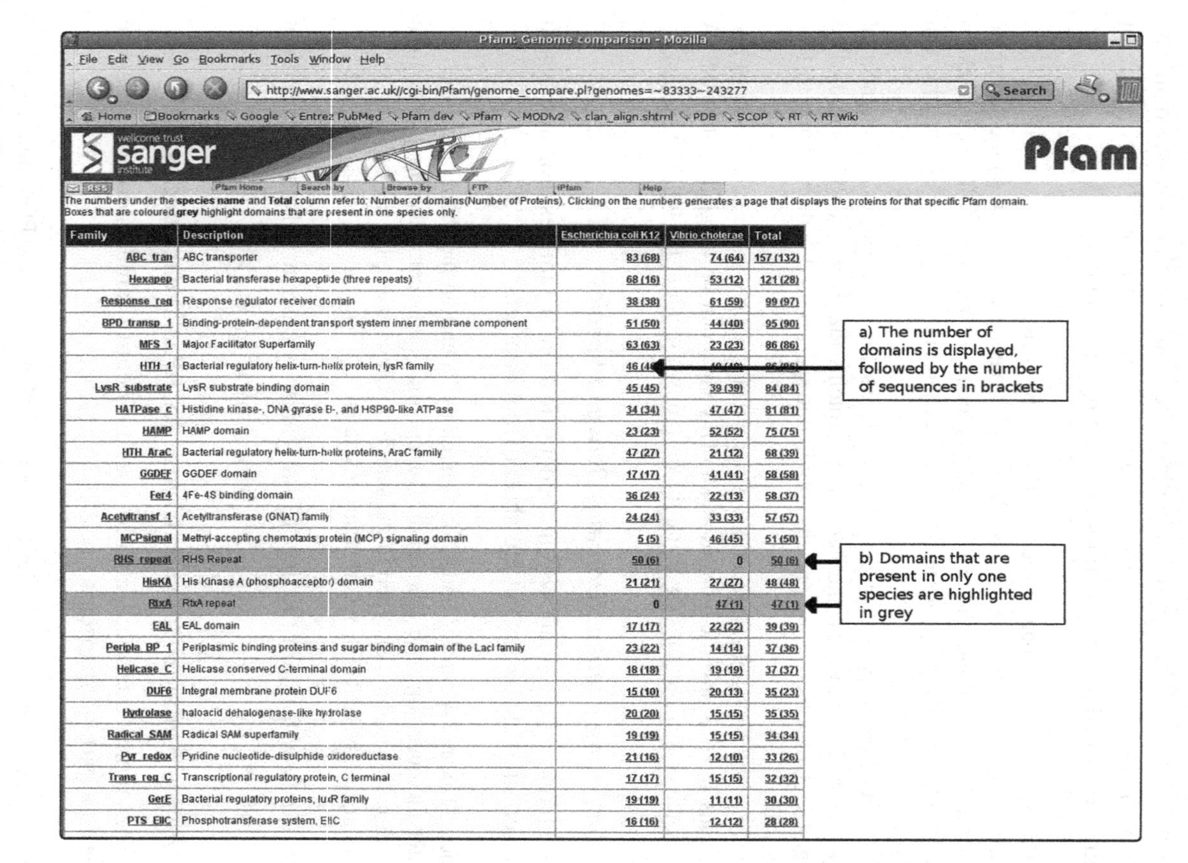

Fig. 4. Using Pfam to compare the domains present in the proteomes of *Escherichia coli* and *Vibrio cholerae*.

3.7. Finding Proteins That Contain a Specific Set of Domain Combinations

The domain query is a simple way to retrieve proteins which contain certain combinations of domains (e.g., proteins containing both a CBS domain and an IMPDH domain).

1. Click on "domain query" under the "search by" menu on the Pfam homepage.
2. In the search box enter the domain query. Some example domain queries are given:

   a. To get all proteins with both a RhoGEF domain and a PH domain type: RhoGEF and PH
   b. To get all proteins with a RhoGEF domain that do no contain a PH domain type: RhoGEF not PH

3. Click "Go" (*see* **Note 6**).

3.8. Finding Proteins That Contain Similar Domain Organization

PfamAlyzer is a java applet accessible from the Swedish Pfam site that allows a more detailed study of domain architectures. It allows the user to find proteins and organisms that contain a specific combination of domains and allows the user to specify the distances allowed between domains.

1. Point the web browser at http://pfam.cgb.ki.se/pfamalyzer/index.html and click "Start Pfam::Alyzer."
2. Choose "Domain query" from the "View menu."
3. Select a combination of domains of from the list on the left and drag and drop them onto the window in the upper right. (The first letter of the domain must first be selected from the drop down menu at the top [*see* **Fig. 5A**].)
4. Select the "species" tab on the upper right window and select the species of interest. Species with a circle next to them can be expanded and press "okay" in the bottom left corner of the window (*see* **Fig. 5B**).
5. If desired, amino acid distances can be specified for region before, after, and between domains. To specify distances click on gap range unit (*see* **Fig. 5C**), and click underneath the stars on the upper right window to specify the "from" and "to" distances.
6. All the proteins that contain the combination of domains within the species specified are displayed in the lower right window (*see* **Fig. 5D**). Both Pfam-A and Pfam-B domains are displayed, Pfam-B families being displayed as rectangles with the accession number displayed within the rectangle. The species distribution can be displayed by clicking on the species distribution tab. Species can be expanded by clicking on the circles next to them (*see* **Fig. 5E**).

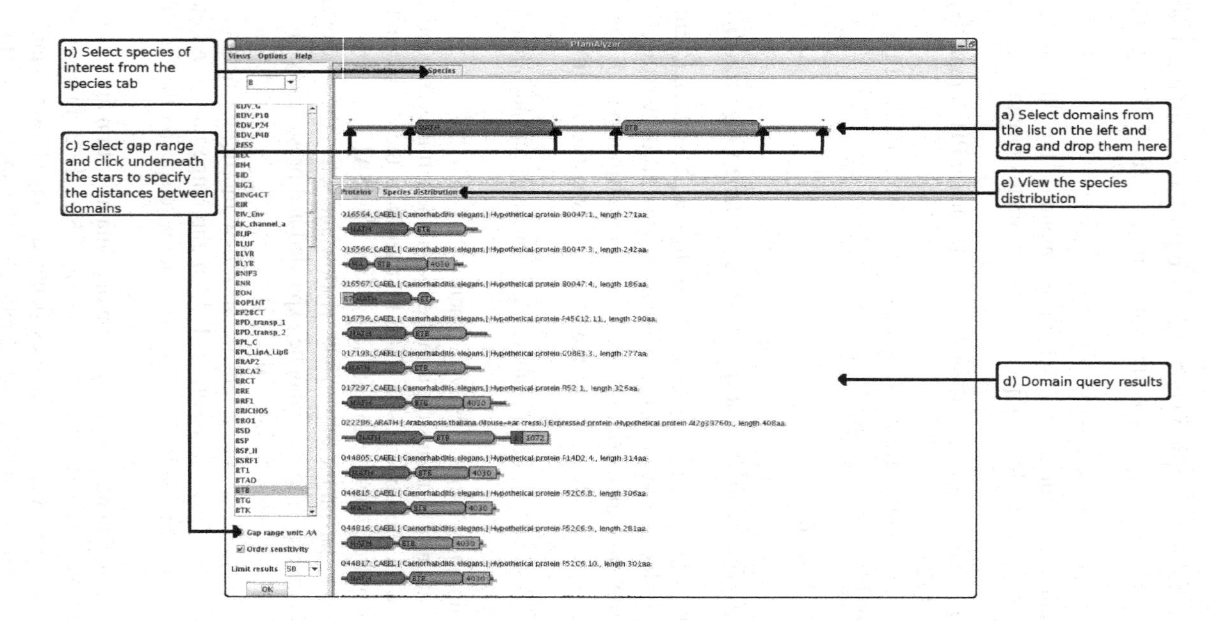

Fig. 5. PfamAlyzer results for a query for proteins containing a MATH domain followed by a BTB domain.

4. Notes

1. On the family page, try looking at the Pfam/SCOP/CATH comparisons. This will show the different domain boundaries from each of the three protein databases (*see* **Fig. 1K**).
2. When searching a sequence on the Pfam website and no families/domains are detected, try raising the *e*-value cutoff level on the "search by protein name or sequence" page under the search menu. This may give additional families/domains which must be treated cautiously but can be useful information.
3. When uploading sequence files, make sure the sequence data is in FASTA format and the file is a text file (e.g., .txt) and not in the format of a specific application such as a Word (.doc) document format as this will result in the searches failing.
4. If local searching using pfam_scan.pl is proving to be slow, try using the "–fast" option on the command-line. This will use BLAST to preprocess the input sequences and results in only having to search a subset of sequences against a subset of HMMs. Another way to speed up the searches is to use the "–mode ls" option on the command-line, which means only full length models are searched resulting in only half as many searches take place.
5. If the proteomes of interest are not in Pfam/Integr8, single linkage clustering might be a useful technique to employ in comparing domain structure between proteomes. This technique clusters together proteins containing similar domain architectures and will help to identify domains/families that are present in one species and not in another.
6. The Domain Query tool can be slow when identifying domains that are unique to a particular proteome. This query is computationally intense and the user will need to be patient to retrieve the results.

References

1. Finn, R. D., Mistry, J., Schuster-Bockler, B., et al. (2006) Pfam: clans, web tools and services. *Nucleic Acids Res.* **34,** D247–D251.
2. Eddy, S. R. (2004) What is a hidden Markov model? *Nat. Biotechnol.* **22,** 1315–1316.
3. Servant, F., Bru, C., Carrere, S., et al. (2002) ProDom: automated clustering of homologous domains. *Brief Bioinform.* **3,** 246–251.
4. Bateman, A., Birney, E., Durbin, R., Eddy, S. R., Howe, K. L., and Sonnhammer, E. L. (2000) The Pfam protein families database. *Nucleic Acids Res.* **28,** 263–266.
5. Wu, C. H., Apweiler, R., Bairoch, A., et al. (2006) The Universal Protein Resource (UniProt): an expanding universe of protein information. *Nucleic Acids Res.* **34,** D187–D191.
6. Chothia, C. (1992) Proteins. One thousand families for the molecular biologist. *Nature* **357,** 543–544.
7. http://hmmer.janelia.org/.

8. Mulder, N. J., Apweiler, R., Attwood, T. K., et al. (2005) InterPro, progress and status in 2005. *Nucleic Acids Res.* **33,** D201–D205.
9. Birney, E., Clamp, M., and Durbin, R. (2004) GeneWise and Genomewise. *Genome Research* **14,** 988–995.
10. Kersey, P., Bower, L., Morris, L., et al. (2005) Integr8 and Genome Reviews: integrated views of complete genomes and proteomes. *Nucleic Acids Res.* **33,** D297–D302.

5

InterPro and InterProScan

Tools for Protein Sequence Classification and Comparison

Nicola Mulder and Rolf Apweiler

Summary

Protein sequence classification and comparison has become increasingly important in the current "omics" revolution, where scientists are working on functional genomics and proteomics technologies for large-scale protein function prediction. However, functional classification is also important for the bench scientist wanting to analyze single or small sets of proteins, or even a single genome. A number of tools are available for sequence classification, such as sequence similarity searches, motif- or pattern-finding software, and protein signatures for identifying protein families and domains. One such tool, InterPro, is a documentation resource that integrates the major players in the protein signature field to provide a valuable tool for annotation of proteins. Protein sequences are searched using the InterProScan software to identify signatures from the InterPro member databases; Pfam, PROSITE, PRINTS, ProDom, SMART, TIGRFAMs, PIRSF, SUPERFAMILY, Gene3D, and PANTHER. The InterPro database can be searched to retrieve precalculated matches for UniProtKB proteins, or to find additional information on protein families and domains. For completely sequenced genomes, the user can retrieve InterPro-based analyses on all nonredundant proteins in the proteome, and can execute user-selected proteome comparisons. This chapter will describe how to use InterPro and InterProScan for protein sequence classification and comparative proteomics.

Key Words: Functional classification; domain; protein family; comparative proteomics; InterPro.

1. Introduction

The annotation of protein sequences using computational tools has always been important, particularly for scientists working on novel proteins. This importance is growing as new technologies for DNA sequencing and functional

From: *Methods in Molecular Biology, vol. 396: Comparative Genomics, Volume 2*
Edited by: N. H. Bergman © Humana Press Inc., Totowa, NJ

genomics are developed and raw scientific data is generated more rapidly. Protein functional classification can be achieved through sequence similarity searches and identification of the closest characterized homologue. However, the homologues are not always annotated themselves, and the method has other inherent limitations. More reliable methods of functional classification use protein signatures, which are descriptions of protein families or domains derived from multiple sequence alignments of proteins that are known members of a family or related proteins with a known function. Protein signatures can be produced using different techniques such as regular expressions, profiles, or hidden Markov models, and are more sensitive and effective than sequence similarity searches. There are a number of public protein signature databases, including Pfam ***(1)***, PROSITE ***(2)***, PRINTS ***(3)***, SMART ***(4)***, TIGRFAMs ***(5)***, and others, that developed independently of each other. Each database has a different focus, but many have overlapping signatures. To increase their potential, these and a handful of other such databases, including ProDom ***(6)***, PIRSF ***(7)***, SUPERFAMILY ***(8)***, Gene3D ***(9)***, and PANTHER ***(10)***, have been integrated into InterPro ***(11)***, which rationalizes their overlaps and adds biological annotation and cross-references to diverse data sources, for example the Gene Ontology (GO) ***(12)***. All signatures representing the equivalent domain or family are merged into single InterPro entries with annotation describing the domain/family. If a signatures describes a subset of proteins or a region on a protein sequence falling within another signature, then relationships are inserted between the corresponding InterPro entries to represent parent/child (family) or contains/found in (domain composition) relationships.

A major component of the InterPro database is the software used to run protein sequences through all the protein signatures. InterProScan ***(13)*** integrates the algorithms from the member databases into a single software package, generating results in a unified format. All proteins in the UniProt Knowledgebase (UniProtKB) ***(14)*** have been run through InterProScan and the results can be viewed in the InterPro database. For novel sequences not yet in UniProtKB, the software package is available through a web interface, and as a stand-alone tool for installing locally. The latter is only necessary if bulk searches or confidentiality are required. The InterProScan output provides the sequence search results in a number of formats, including a graphical view and a table view. The table view includes GO annotations and information on InterPro relationships where applicable. Matches of protein signatures on the query sequence are grouped by InterPro entry, which, together with information on which entries are related to each other, helps to rationalize multiple protein signature hits.

The InterPro data is available for searching via a web interface. For completely sequenced genomes, the Integr8 resource ***(15)*** also provides data on InterPro matches for all proteins in a proteome. InterPro analyses for each proteome include tables of the top 30 or top 200 InterPro entries, 15 most common families, 15 most common domains, and so on. A complete table of InterPro matches for the proteome is also available for viewing or for downloading. Integr8 also provides a useful interface for comparative genomics. Each proteome has a precomputed comparison of InterPro matches between the reference proteome and one or more preselected proteomes. If the user has a different set of proteomes to compare, an interactive comparative proteome analysis tool is available, where the user can select a reference proteome and one or more other proteomes to compare it to, and can select the type of InterPro statistics to output. This chapter will describe how to browse, search, and analyze protein sequences and protein families and domains using InterPro and InterProScan. It will discuss how to use the Integr8 Proteome Analysis tools for InterPro comparisons across proteomes.

2. Materials

The methods described in this chapter are *in silico* and require a workstation, personal computer, or terminal connected to the Internet. If the user prefers to submit sequences via email, then an email program for the mail server is also required. For searching or browsing InterPro and Integr8, any type of web browser will suffice, although Internet Explorer is faster. Input sequences for InterProScan can be copied and pasted into the appropriate part of the form, or they can be stored in separate files in FASTA format and uploaded to the form.

3. Methods

The InterPro database is available via a web server at: http://www.ebi.ac.uk/interpro. The homepage provides basic information about the database, a search box, and links to useful files and documentation. A user manual provides all the information required by a user to understand the database, and additional support is provided by the InterPro tutorial (http://www.ebi.ac.uk/interpro/tutorial.html) and FAQ (frequently asked questions) pages. The database can be searched using a query sequence, or through a text search facility. For complete proteomes the InterPro results are also available in the Integr8 Proteome Analysis pages at: http://www.ebi.ac.uk/integr8. The methods will be broken down into three parts to describe the sequence search tool, searching the InterPro data, and using Integr8 for InterPro comparative proteome analyses.

3.1. Searching InterPro With a Query Sequence

InterProScan is the software package used to search query sequences for matches to the protein signatures in InterPro. The input sequence may be either a nucleotide or protein sequence.

1. Go to the InterProScan webpage (http://www.ebi.ac.uk/InterProScan/).
2. Enter an email address in the appropriate box (this is required in case the search takes too long and times out). Paste the sequence into the sequence box or upload the sequence from a file using the "upload a file" section.
3. If a nucleotide sequence is submitted, change "PROTEIN" to "DNA," and choose the translation table to be used. The default minimum open reading frame size is 100.
4. Press the "Submit Job" button to proceed with the search.

(A)

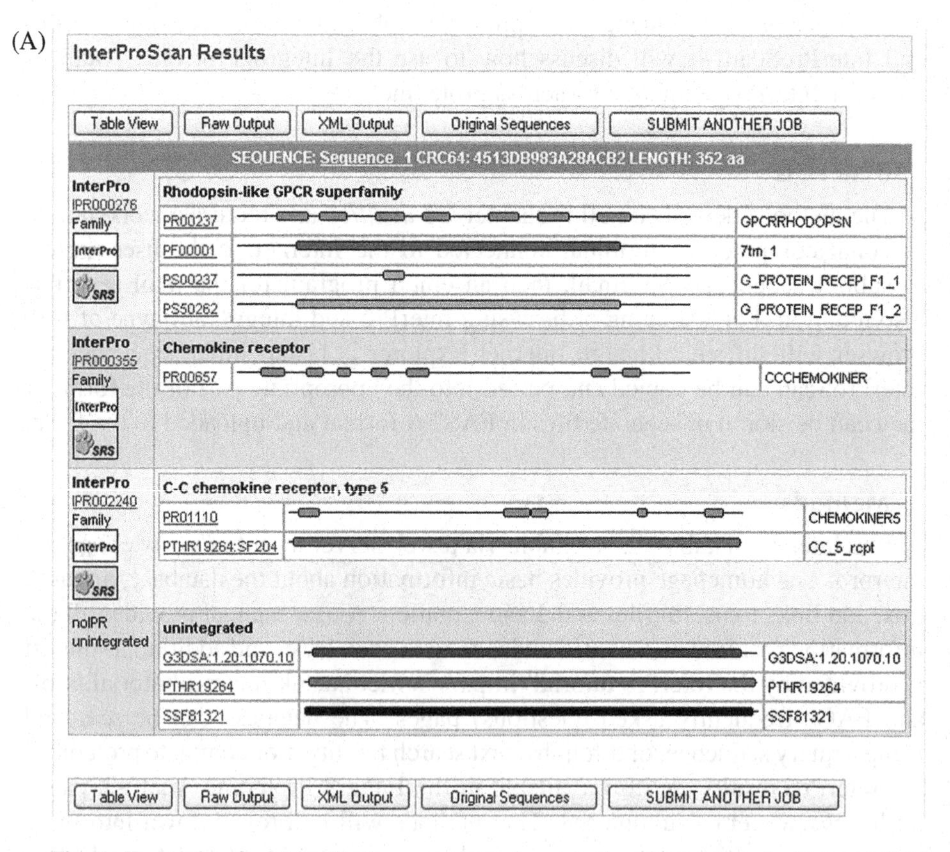

Fig. 1. Example of an InterProScan output, showing the **(A)** graphical view and **(B)** table view.

(B)

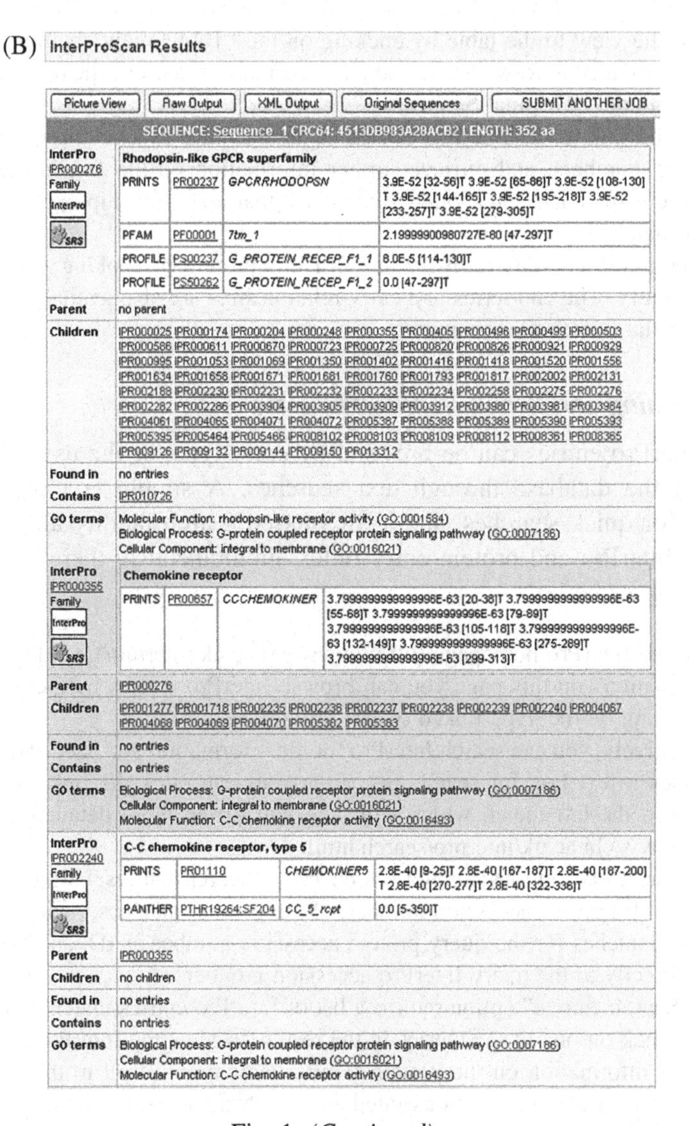

Fig. 1. *(Continued)*

5. View the results as they appear in the default graphical format (*see* example output in **Fig. 1**) (*see* **Note 1**).
6. Mouse-over the matching methods to see a pop-up window that provides the sequence positions of the method, its match status, the *e*-value of the match and the name of the method.

7. Change the view to the table by clicking on the "Table View" button. The results are also provided in Raw Output and XML format. To retrieve the original sequence submitted click "Original Sequences."
8. Follow the link to the InterPro entries in the database by clicking on the InterPro accession numbers, and retrieve further information on the families in question. From the results and entry views we can see that a relationship exists between the entries matched.
9. Find the associated GO terms, if any, for this protein by looking at the GO links in each entry. The GO terms and entry relationships are also available in the Table View of the InterProScan output.

3.2. Searching and Browsing InterPro Data

The InterPro entries can be browsed by entry type, or the user can retrieve data from the database through text searches. A simple text search box is available for quick searches, and a SRS-based search facility allows users to search in InterPro and protein entry fields simultaneously thereby facilitating more complex queries.

1. Go to the InterPro homepage (http://www.ebi.ac.uk/interpro/). Under the heading "Information" on this page you can browse InterPro entries by their entry types, by clicking on the hyperlinked types.
2. Alternatively, you can search InterPro for the information you need. The homepage provides a text box for searching, or the user can click the link "Text Search" option in the left menu, which will take you to the more detailed search page: http://www.ebi.ac.uk/interpro/search.html.
3. There are three options for the search box, "Search entries," which will accept any search term or accession number, "Find protein matches," which returns the protein matches for the query protein accession number or ID, and "IPR," which goes directly to the query InterPro accession number.
4. The "Search entries" option returns a list of InterPro entries matched by the query term. Click on one of the entries in the results to view the InterPro entry.
5. To find information on the protein family or domain, scroll to the GO terms or the abstract field. Publications cited in the abstract are listed at the bottom of the entry.
6. To view the matches of the entry to UniProtKB proteins, select one of the Match views at the top of the entry. The proteins can be viewed in an Overview, Detailed view, Table view, or Architectures view. For the first three views, the proteins can be ordered by accession number or ID, or only proteins of known structure or with splice variants can be selected.
7. Click "Overview" to see a graphical view of the protein matches displaying only the individual InterPro entries hit.

8. Click "Detailed view" to see a graphical view displaying hits to all signatures from each InterPro entry hit.
9. Click "Table view" to get the matches from this entry in the table format.
10. Click "Architectures" to retrieve the InterPro Domain Architecture view, which displays the different domain architectures present in the entry (*see* **Note 2**).
11. To use the SRS-based text search from the InterPro web server, go to the search page (http://www.ebi.ac.uk/interpro/search.html).
12. Perform a query using up to two data fields, one from InterPro and the second from the protein matches. The two drop-down menus contain the names of all the currently relevant data fields from which you can choose a data field of interest. The first menu lists InterPro data fields, the second lists data fields from UniProtKB entries.
13. By default the wildcards box is ticked and wildcards are appended to all search words automatically. If you do not wish wildcards to be appended, uncheck the checkbox.
14. Select a predefined view using the drop-down menu, which contains the built-in views, to specify how to display the results of the search.
15. When the user has defined the query, click the "Query" button. The results are returned in the selected view and the user can again go to the relevant InterPro entries for more information.

3.3. Comparative Proteomics Using Integr8

Integr8 is a resource that integrates data for the genes and proteins from completely sequenced genomes. Data is integrated from databases such as UniProtKB, Genome Reviews, InterPro, GO, and so on. Here, we will focus on the Proteome Analysis part of Integr8 that provides InterPro analyses. The user can either browse the precomputed analyses, or perform an interactive comparative analysis.

3.3.1. Browsing Precomputed Analyses

1. Go to the Integr8 web server (http://www.ebi.ac.uk/integr8).
2. In the "Search for species" box, enter a search term for the organism of interest, and then click "Go." As an example, enter "human."
3. Either on the side menu or the menu at the bottom of the resulting page, click the "Proteome Analysis" option. This will automatically provide a table of the top 30 InterPro entries for the organism.
4. Use the scroll-down bar to select a different analysis, the choices include: top 200 hits, 15 most common families, 15 most common domains, 15 most common repeats, top 30 proteins with different InterPro hits, and general InterPro statistics. The latter lists all InterPro hits to the proteome in numerical order of InterPro accession numbers (*see* **Note 3**).

5. To view precomputed comparative analyses with other proteomes, click on the "Comparative" tab under the "Proteome Analysis" heading.
6. Choose the analysis type and the organism sets to compare against from the choices in the scroll-down bars. The choice of organism sets is limited to those predicted to be of interest to most users. If the user is interested in selecting different organism sets, proceed to **Subheading 3.3.2.**

3.3.2. Interactive Comparative Analysis

1. Go to the Integr8 web server: http://www.ebi.ac.uk/integr8.
2. On the left menu, click "Configure your own InterPro analysis."
3. Choose a reference proteome from the organisms listed on the left.
4. Select one or more proteomes to compare this to from the list in the middle. To select more than one organism hold the "Ctrl" button on the keypad and select the organisms.
5. Select the type of analysis on the right.
6. Click the "Compare" button at the bottom of the page (*see* **Note 4**).

4. Notes

1. Searching InterPro with a query sequence. When submitting a protein sequence, you can use free text/raw, FASTA, or UniProtKB formats, and for a nucleotide sequence free text/raw, FASTA, EMBL, or GenBank formats are accepted. Partially formatted sequences are not accepted, and it is useful to note that copying and pasting directly from word processors may yield unpredictable results as hidden characters may be present. Adding a return to the end of the sequence may help certain applications understand the input. When submitting a nucleotide sequence, it is translated in all six frames, where the length of the reading frame can be preset, and the translation products are queried against InterPro. Only one protein or nucleotide sequence is accepted by InterProScan at a time, however, it is also possible to run a sequence through InterProScan using web services, in which multiple sequences can be submitted to the service by means of scripts. For more information on the web services see: http://www.ebi.ac.uk/Tools/webservices/WSInterProScan.html.

   When a query sequence is submitted, InterProScan initially calculates its CRC64 (cyclic redundancy checksum—a unique number generated by each sequence) and compares it to the InterPro match XML file. If a match to the CRC64 is found, InterProScan returns the precomputed matches for that sequence, if not, it takes the sequence and analyses it against the signatures in the member databases. The advantage of retrieval of results from the XML file if the CRC64 is recognized is that UniProtKB/Swiss-Prot and InterPro curators manually curate the status of some InterPro matches in the database so false-positive and false-negative results may also be detected. Following analysis, each result is returned and combined, and the InterPro families and protein signatures are returned to the submitter. The results, which are kept for 1 wk on the EBI server, are presented as a graphical view.

The InterProScan results display matches to the member database signatures grouped by the corresponding InterPro entries, providing the positions of the signatures within the sequence, in a graphical format **(Fig. 1A)**. The protein sequence is split into several lines, one for each different signature hit. The table view **(Fig. 1B)** displays the results in a table format showing the positions in the sequence and the corresponding scores, and also displays entry relationships and GO terms. In the views, the InterPro accession number is hyperlinked to take the user directly to the InterPro entry from the InterPro database, and this can also be achieved by clicking on the InterPro image. Clicking on the SRS image takes the user to an SRS-view of the entry. It is important to note that the InterPro data in SRS is based on the InterPro XML file, and the most up to date view is that accessing the database. The results should give the user an idea of the family and domain composition of the query protein sequence. In some cases, there may be matches to signatures that have no corresponding InterPro entry, this can occur for two reasons; the signatures may be new and not yet integrated or they may be low complexity regions or ProDom domains, which will not be integrated. They may however provide additional information for the user, so are maintained in the InterProScan data files.

Rationalizing InterProScan results may seem daunting if multiple hits are detected. If multiple hits are recorded within overlapping positions on the sequence it is important to look further for relationships between the entries that are hit. It may be that they are related through parent child or contains/found in relationships. The former case facilitates characterization of the protein on different levels of family and superfamily, whereas the latter provides domain composition information. It is then possible to find other proteins that are similar on the superfamily level or more specifically at the family or subfamily levels, or alternatively those that share a domain or domain composition in common with the query protein. In the example result, IPR002240 is a child of IPR000355, which is a child of IPR000276, with each of these entries becoming less specific as you move up the hierarchy from the chemokine receptor type 5 subfamily, to the chemokine receptor family, to the general rhodopsin-type G-protein coupled receptor superfamily, respectively. If hits to multiple InterPro entries that are not related to each other are observed, and the hits are not overlapping on the sequence, this may be indicative of multifunctional proteins.

The question often asked by users is how to know whether the matches are true, and the answer is that the threshold scores for profiles or HMMs are provided by the member databases and are considered to be trustworthy for displaying only true hits. Some postprocessing of data is linked to the software package, for example the Pfam, SMART, and TIGRFAMs outputs are filtered through family-specific thresholds for increased accuracy of the results, and PRINTS results are filtered through an algorithm that considers the PRINTS family hierarchy and rationalizes hits to multiple siblings of the same parent. Therefore, the reported matches should be correct. Some false-positive hits may, however, still occur particularly for PROSITE patterns, so it is important to check if a hit may be spurious. This is done by looking at the InterPro entry and seeing if any other signatures from the same entry also hit the sequence. If so, the hit is probably true; but if not, it is better to consider the match with caution.

2. Searching and browsing InterPro. When searching InterPro, the common query syntax and operators apply. The general query syntax is name:value pairs, e.g., name:kinase, where the field "name" will be searched for value "kinase." Queries can be combined with operators & (AND), | (OR), – (MUST NOT), + (MUST), and (subquery). The "Search entries" option searches for the query in the following entry fields: name, abstract, method name, method accession, and InterPro entry accession. The query can be the entry accession number, e.g., IPR000001, the method accession number, e.g., PR00480, or a general text query looking for keywords in the entry. A simple text query with one or more words will return all the entries containing the search term(s). A list of the entries will be returned with the relevance of the search (R). If more than one search word is used, R indicates how many of the search terms were found in the entry. When the SRS-based query form is used, the results are returned within the SRS interface. To view the resulting entries in the InterPro interface, click on the "InterPro Oracle" image next to the accession number.

   Once the relevant InterPro entry is retrieved, opening the entry provides useful information on the protein family or domain. When viewing protein match lists from a resulting InterPro entry, it is important to know what each view is representing. The overview displays different colored bars for each InterPro entry hit, where the positions on the sequence are condensed to the first position of the first signature to the last position of the last signature in the entry. In the detailed view, on the other hand, the sequence is split into several lines, one for each hit by a unique signature, and the position of the signature hit reflects its position on the sequence. This view includes the hits by all signatures from the same and other InterPro entries. The signatures of the member databases are given specific colors: Gene3D –purple, PANTHER –brown, Pfam -dark blue, PIRSF –pink, PRINTS –green, ProDom –light blue, PROSITE patterns –yellow, PROSITE profiles –orange, SMART –red, SUPERFAMILY –black, and TIGRFAMs –teal. The table view lists the protein accession numbers, the positions in the amino acid sequence where each signature from that InterPro entry hits and the status of the match, i.e., true, false-positive, false-negative, and partial. This view only shows matches to signatures in the InterPro entry in question.

3. Browsing precomputed analyses in Integr8. When searching for an organism, if the exact name is entered, the homepage for that organism appears directly. However, if the search term matches more than one organism, a table is returned listing

the set of organisms matched and the user can select the correct organism and/or strain. It is also possible to find an organism by browsing the list of species using the "Browse species" button in the left menu. The user can choose to browse the list of "Bacteria," "Archaea," or "Eukaryota." The organism homepage provides information on the taxonomy and/or characteristics of the organism. Links are provided to "Literature," "Genome statistics," Proteome Analysis," and "Taxonomy" information related to the organism, and relevant data files can be retrieved from the "Downloads" page.

For the Proteome Analysis part of the resource, the resulting tables are based on InterPro matches for the proteome. In the top 30 entries table, for example, the InterPro accession is listed in the first column (these are hyperlinked to the InterPro database), and the number of proteins matched in the second column, with the percentage coverage of the proteome for that entry in brackets. Clicking on the protein number returns a table listing the protein accession numbers, IDs, descriptions, and number of InterPro hits. Clicking on the image in the fifth column takes the user to a view of the InterPro matches for the protein in different formats.

4. Interactive comparative analysis. The interactive analysis allows the user to do a comparison of InterPro hits between two or more different proteomes. The more different organisms are chosen, the longer the comparison will take. The results are displayed in a similar format to the precomputed Proteome Analysis statistics. The reference proteome is always shown in the first column, and rows are ordered according to the ranking in this proteome. The corresponding rank in the other proteomes selected for comparison may therefore not be ordered. These proteome comparisons are very useful ways of comparing protein families and domains across different organisms without having to do extensive sequence similarity searches. The proteins from the same family (InterPro entry) in each proteome can be identified and retrieved for further analysis.

This chapter has hopefully demonstrated the usefulness of InterPro and Integr8 for protein sequence classification and comparison. The tools are simple to use once the complexities of each database are apparent, and each has documentation and user support for additional help.

References

1. Finn, R. D., Mistry, J., Schuster-Bockler, B., et al. (2006) Pfam: clans, web tools and services. *Nucleic Acids Res.* **34,** D247–D251.
2. Hulo, N., Bairoch, A., Bulliard, V., et al. (2006) The PROSITE database. *Nucleic Acids Res.* **34,** D227–D230.
3. Attwood, T. K., Bradley, P., Flower, D. R., et al. (2003) PRINTS and its automatic supplement pre-PRINTS. *Nucleic Acids Res.* **31,** 400–402.

4. Letunic, I., Copley, R. R., Pils, B., Pinkert, S., Schultz, J., and Bork, P. (2006) SMART 5: domains in the context of genomes and networks. *Nucleic Acids Res.* **34,** D257–D260.
5. Haft, D. H., Selengut, J. D., and White, O. (2003) The TIGRFAMs database of protein families. *Nucleic Acids Res.* **31,** 371–373.
6. Bru, C., Courcelle, E., Carrere, S., Beausse, Y., Dalmar, S., and Kahn, D. (2005). The ProDom database of protein domain families: more emphasis on 3D. *Nucleic Acids Res.* **33,** D212–D215.
7. Wu, C. H., Nikolskaya, A., Huang, H., et al. (2004) PIRSF: family classification system at the Protein Information Resource. *Nucleic Acids Res.* **32,** D112–D114.
8. Madera, M., Vogel, C., Kummerfeld, S. K., Chothia, C., and Gough, J. (2004) The SUPERFAMILY database in 2004: additions and improvements. *Nucleic Acids Res.* **32,** D235–D239.
9. Pearl, F., Todd, A., Sillitoe, I., et al. (2005) The CATH Domain Structure Database and related resources Gene3D and DHS provide comprehensive domain family information for genome analysis. *Nucleic Acids Res.* **33,** D247–D251.
10. Mi, H., Lazareva-Ulitsky, B., Loo, R., et al. (2005) The PANTHER database of protein families, subfamilies, functions and pathways. *Nucleic Acids Res.* **32,** D284–D288.
11. Mulder, N. J., Apweiler, R., Attwood, T. K., et al. (2005) InterPro, progress and status in 2005. *Nucleic Acids Res.* **33,** D201–D205.
12. Harris, M. A., Clark, J., Ireland, A., et al. (2004) The Gene Ontology (GO) database and informatics resource. *Nucleic Acids Res.* **32,** 258–261.
13. Quevillon, E., Silventoinen, V., Pillai, S., et al. (2005) InterProScan: protein domains identifier. *Nucleic Acids Res.* **33,** W116–W120.
14. Wu, C. H., Apweiler, R., Bairoch, A., et al. (2006) The Universal Protein Resource (UniProt): an expanding universe of protein information. *Nucleic Acids Res.* **34,** D187–D191.
15. Pruess, M., Kersey, P., and Apweiler, R. (2005) The Integr8 project: a resource for genomic and proteomic data. *In Silico Biol.* **5,** 179–185.

6

Gene Annotation and Pathway Mapping in KEGG

Kiyoko F. Aoki-Kinoshita and Minoru Kanehisa

Summary

KEGG is a database resource (http://www.genome.jp/kegg/) that provides all knowledge about genomes and their relationships to biological systems such as cells and whole organisms as well as their interactions with the environment. KEGG is categorized in terms of building blocks in the genomic space, known as KEGG GENES, the chemical space, KEGG LIGAND, as well as wiring diagrams of interaction and reaction networks, known as KEGG PATHWAY. A fourth database called KEGG BRITE was also recently incorporated to provide computerized annotations and pathway reconstruction based on the current KEGG knowledgebase. KEGG BRITE contains KEGG Orthology (KO), a classification of ortholog and paralog groups based on highly confident sequence similarity scores, and the reaction classification system for biochemical reaction classification, along with other classifications for compounds and drugs. BRITE is also the basis for the KEGG Automatic Annotation Server (KAAS), which automatically annotates a given set of genes and correspondingly generates pathway maps. This chapter introduces KEGG and its various tools for genomic analyses, focusing on the usage of the KEGG GENES, PATHWAY, and BRITE resources and the KAAS tool (*see* **Note 1**).

Key Words: KEGG; KO; KEGG Orthology; gene annotation; pathway mapping; KAAS.

1. Introduction

1.1. The KEGG Resources

The *Kyoto Encyclopedia of Genes and Genomes* is most well known as KEGG, a publicly available (http://www.genome.jp/kegg/) knowledgebase of genomic and molecular information that is constantly being updated with the latest information from the published literature *(1,2)*. It consists of four

From: *Methods in Molecular Biology, vol. 396: Comparative Genomics, Volume 2*
Edited by: N. H. Bergman © Humana Press Inc., Totowa, NJ

main databases: GENES for genomic information, PATHWAY for pathways of biological processes, LIGAND for chemical compounds and reactions, and BRITE for ortholog/paralog information as well as other classification systems. The GENES database contains all complete genomes and some partially sequenced genomes, based on publicly available resources such as National Center for Biotechnology Information (NCBI) RefSeq *(3)* and GenBank *(4)*. Draft genomes and expressed sequence tag (EST) consensus contigs are stored in the auxiliary DGENES and EGENES databases, respectively. The PATHWAY database is a collection of manually drawn pathway diagrams for metabolism, genetic information processing, environmental information processing such as signal transduction, and various other cellular processes including human diseases. The LIGAND database consists of six components:

1. COMPOUND: metabolites and other chemical compound structures.
2. DRUG: chemical compound structures of drugs.
3. GLYCAN: carbohydrate structures.
4. REACTION: reaction formulas from ENZYME.
5. RPAIR: reactant pair transformation patterns in REACTION.
6. ENZYME: enzyme nomenclature.

Tools and interfaces for chemical compound structures are also available in LIGAND for querying the database.

The BRITE database consists of hierarchical classifications of various aspects of biological systems. Currently, it contains KEGG Orthology (KO), KEGG's own classification of orthologs and paralogs, the reaction classification system for biochemical reactions, and other classifications for compounds and drugs.

1.2. KEGG Orthology

The KO system is an attempt to organize the knowledge in the genomic space. The sequence similarity between all pairs of genes in KEGG GENES is calculated using the Smith-Waterman algorithm *(5)*. The resulting scores are then stored in the auxiliary KEGG SSDB database *(6)*. Based on these scores and the best-hit relations between genome pairs, orthologous genes (including orthologous relationships of paralogous genes) can be classified into various pathways. The KO identifier, or K number, is manually assigned to genes to link them with pathways, which enables the automatic generation of pathways from a given list of genes.

1.3. KEGG Automatic Annotation Server

KEGG Automatic Annotation Server (KAAS) is a web-based server (http://www.genome.jp/keg/kaas/) that can rapidly and automatically annotate genes with functions and reconstruct KEGG pathway maps *(7)*. This method has been effectively used to automatically annotate draft genomes and EST clusters in KEGG GENES. It is now available to users who wish to annotate their own list of genes or EST sequences with KO identifiers, based on which pathway maps can be generated.

2. Methods

The KEGG resources are accessible on the Internet from http://www.genome.jp/kegg/. From there, links are available to the KEGG Table of Contents as well as documentation and related database links. The Table of Contents link on this page provides the main entry point from which the KEGG resources may be customized and/or queried. For example, as shown near the bottom of **Fig. 1**, KEGG can be customized to display information for specified organisms by their KEGG organism code. The full organism list along with their codes can be seen by clicking on the "KEGG Organisms" link:

1. Click the "Organism" button to display a new popup window. This can be used to find a specific organism code by typing in keywords.
2. Type "salmonella" in the textfield. A list of all available salmonella genomes will be dynamically listed.
3. Select "sec" for *Salmonella enterica serovar Choleraesuis* and click the "Select" button.
4. "sec" is now entered in the textfield. Click the Go button to display all the pathways in KEGG for this organism.

2.1. Customizing KEGG

The pink Select button (which can also be found at the top of the individual pathway maps), will create a customized list of organisms, which will be used as the organism list for each pathway map. Let us select all the draft animal genomes, for example:

1. Click the pink "Select" button. A new dialog box will open (**Fig. 2**).
2. Mark the checkbox for "DGENES" and uncheck "GENES" and "EGENES."
3. Select "Animals."
4. Click the "Select" button. The Table of Contents will now display the current selection: "Animals in DGENES."
5. Click on the "Show currently selected organisms" link. The list of organisms that will be available in each pathway's organism list will be displayed along with their

KEGG - Table of Contents

KEGG2 PATHWAY GENES LIGAND BRITE XML API DBGET

KEGG Databases

Database	Content	Search & Compute	DBGET Search
KEGG PATHWAY	Wiring diagrams of biological systems (see statistics)	Search objects in KEGG pathways Color objects in KEGG pathways KEGG pathways in XML	PATHWAY
KEGG GENES	Building blocks in the genomic space (see statistics)	BLAST search against GENES/GENOME FASTA search against GENES/GENOME KEGG EXPRESSION	GENES DGENES / EGENES GENOME
KEGG LIGAND	Building blocks in the chemical space (see statistics)	Search similar compound structures Search similar glycan structures Predict reactions and assign EC numbers Generate possible reaction paths	COMPOUND DRUG GLYCAN REACTION RPAIR ENZYME LIGAND
KEGG BRITE	Functional hierarchies of biological systems	KEGG Orthology (KO) Automatic annotation (KO assignment) Therapeutic category of drugs	KO

See illustrations of KEGG databases and KEGG services.

Customized KEGG

KEGG for specific organisms

KEGG Organisms - the list of currently available organisms
Choose Organism Go Clear (examples) hsa mmu sce eco bsu syn

Customize the organism menu in pathway maps with selected organisms Select
Show currently selected organisms (All organisms in GENES)

KEGG for selected research areas

KEGG DRUG for drug information
KEGG GLYCAN for glycome informatics
KEGG REACTION for integration of genomics and chemistry
KEGG EXPRESSION for microarray data analysis
KEGG Auto Annotation for genome/EST annotation by KAAS

KEGG for software development

KGML - XML representation of KEGG pathways

Fig. 1. Screen shot of the KEGG Table of Contents page, from which all of KEGG's resources may be accessed.

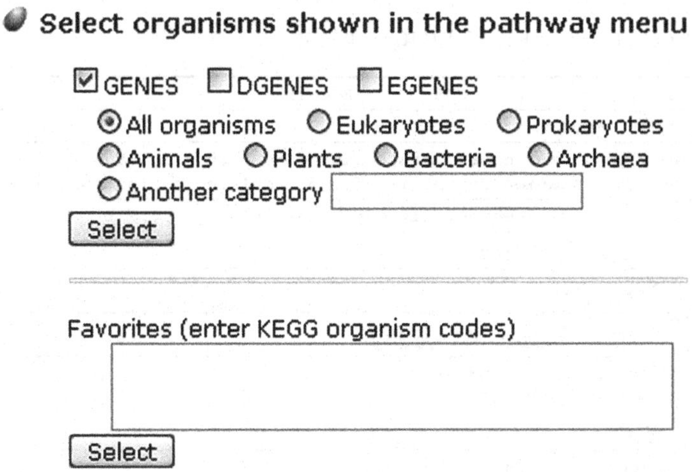

Select organisms shown in the pathway menu

GENES DGENES EGENES
All organisms Eukaryotes Prokaryotes
Animals Plants Bacteria Archaea
Another category
Select

Favorites (enter KEGG organism codes)
Select

Fig. 2. Dialog box for customizing the organism list that will be displayed for each pathway map.

hierarchical taxonomic categories and their KEGG organism code. The selection may be modified again by clicking on the pink "Select" button at the top right of this list.

2.2. Searching KEGG GENES

KEGG GENES currently contains over a million genes from almost 80 eukaryotic organisms (including draft genomes and EST contigs), more than 270 bacteria and 25 archaea genomes. The draft genomes are stored in the DGENES database, whereas EST contigs are stored in EGENES. The entry page at http://www.genome.jp/kegg/genes.html allows for direct keyword querying of these three databases, along with organism-specific keyword search and querying of other related databases such as NCBI using Entrez GeneIDs, GenBank IDs, and others. For example, let us look for all glycosyltransferase genes in KEGG GENES:

1. In the textfield under "Gene Catalogs," enter "glycosyltransferase" in the textbox for Search GENES. The pulldown menu gives the option to search draft genomes under DGENES and ESTs under EGENES.
2. Click the "Go" button. A list of all GENES entries will be returned. Clicking on any entry will display its sequence information as well as links to other databases such as NCBI.

An example of the GENES entry for the *Escherichia coli* glycosyltransferase gene mdoH is given in **Fig. 3**. From this entry listing, we see that this gene

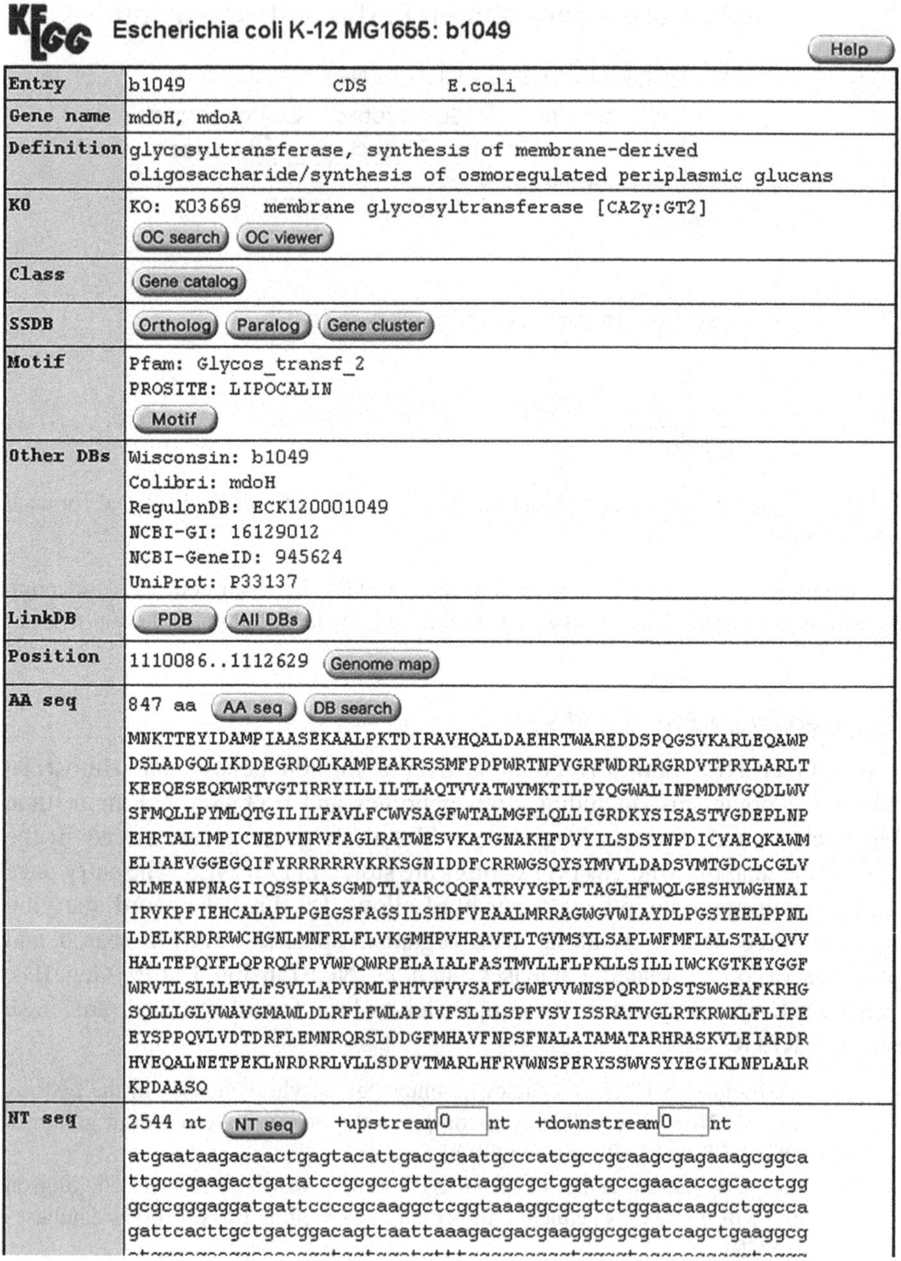

KEGG Escherichia coli K-12 MG1655: b1049

Help

Entry	b1049 CDS E.coli
Gene name	mdoH, mdoA
Definition	glycosyltransferase, synthesis of membrane-derived oligosaccharide/synthesis of osmoregulated periplasmic glucans
KO	KO: K03669 membrane glycosyltransferase [CAZy:GT2] OC search OC viewer
Class	Gene catalog
SSDB	Ortholog Paralog Gene cluster
Motif	Pfam: Glycos_transf_2 PROSITE: LIPOCALIN Motif
Other DBs	Wisconsin: b1049 Colibri: mdoH RegulonDB: ECK120001049 NCBI-GI: 16129012 NCBI-GeneID: 945624 UniProt: P33137
LinkDB	PDB All DBs
Position	1110086..1112629 Genome map
AA seq	847 aa AA seq DB search MNKTTEYIDAMPIAASEKAALPKTDIRAVHQALDAEHRTWAREDDSPQGSVKARLEQAWP DSLADGQLIKDDEGRDQLKAMPEAKRSSMFPDPWRTNPVGRFWDRLRGRDVTPRYLARLT KEEQESEQKWRTVGTIRRYILLILTLAQTVVATWYMKTILPYQGWALINPMDMVGQDLWV SFMQLLPYMLQTGILILFAVLFCWVSAGFWTALMGFLQLLIGRDKYSISASTVGDEPLNP EHRTALIMPICNEDVNRVFAGLRATWESVKATGNAKHFDVYILSDSYNPDICVAEQKAWM ELIAEVGGEGQIFYRRRRRRVKRKSGNIDDFCRRWGSQYSYMVVLDADSVMTGDCLCGLV RLMEANPNAGIIQSSPKASGMDTLYARCQQFATRVYGPLFTAGLHFWQLGESHYWGHNAI IRVKPFIEHCALAPLPGEGSFAGSILSHDFVEAALMRRAGWGVWIAYDLPGSYEELPPNL LDELKRDRRWCHGNLMNFRLFLVKGMHPVHRAVFLTGVMSYLSAPLWFMFLALSTALQVV HALTEPQYFLQPRQLFPVWPQWRPELAIALFASTMVLLFLPKLLSILLIWCKGTKEYGGF WRVTLSLLLEVLFSVLLAPVRMLFHTVFVVSAFLGWEVVWNSPQRDDDSTSWGEAFKRHG SQLLLGLVWAVGMAWLDLRFLFWLAPIVFSLILSPFVSVISSRATVGLRTKRWKLFLIPE EYSPPQVLVDTDRFLEMNRQRSLDDGFMHAVFNPSFNALATAMATARHRASKVLEIARDR HVEQALNETPEKLNRDRRLVLLSDPVTMARLHFRVWNSPERYSSWVSYYEGIKLNPLALR KPDAASQ
NT seq	2544 nt NT seq +upstream 0 nt +downstream 0 nt atgaataagacaactgagtacattgacgcaatgcccatcgccgcaagcgagaaagcggca ttgccgaagactgatatccgcgccgttcatcaggcgctggatgccgaacaccgcacctgg gcgcgggaggatgattccccgcaaggctcggtaaaggcgcgtctggaacaagcctggcca gattcacttgctgatggacagttaattaaagacgacgaagggcgcgatcagctgaaggcg

is a part of the K03669 ortholog group of membrane glycosyltransferases, as defined by the CAZy database. It also contains a motif defined under both Pfam and PROSITE. Other databases also store this gene's information.

2.3. Using KEGG SSDB

The sequences of all protein coding genes in KEGG GENES are compared with one another in a pairwise fashion using the Smith-Waterman algorithm. Because of the computational expense of executing this algorithm on over a million genes, these similarity scores are precomputed and stored under the KEGG SSDB database. The similarity scores and best-hit relations in this database are used for computing the ortholog groups that are entered into KEGG Orthology (*see* **Subheading 2.4.**). As can be seen in **Fig. 3** for the GENES entry of mdoH, the "SSDB" section provides links to list these orthologous and paralogous genes along with their similarity scores and other precalculated information. For example, to view the orthologous genes to this mdoH gene in *E. coli*:

1. Click the "Ortholog" button in the entry.
2. The list of orthologous genes sorted by similarity score will be displayed (**Fig. 4**).
3. Click the "Clear" button to uncheck all entries.
4. Mark the boxes for the entries with 90–99 % identity with the query gene (Entries ecc:c1314, ece:Z1684, spt:SPA1700, stm:STM1151, stt:t1769, sty:STY1188, and sec:SC1098 in **Fig. 4**).
5. Select "Search common motifs" from the pull-down menu next to the "Clear" button.
6. Click the "Select" button. A new window will appear listing the known motifs and their locations within the selected sequences, as in **Fig. 5**.

2.4. KEGG Orthology

KO is KEGG's own manually defined ortholog groups based on pathway information, specified by KO identifiers that start with "K." Orthologous genes can be linked to the pathway entries to which they correspond such that an

Fig. 3. Example of a KEGG GENES entry for gene mdoH of *Escherichia coli*. Many links are available in addition to the basic gene name, definition, and sequence information, such as ortholog group information linked to its KEGG Orthology group, motifs found in this gene with links to corresponding motif database entries as well as other database links such as to National Center for Biotechnology Information and UniProt.

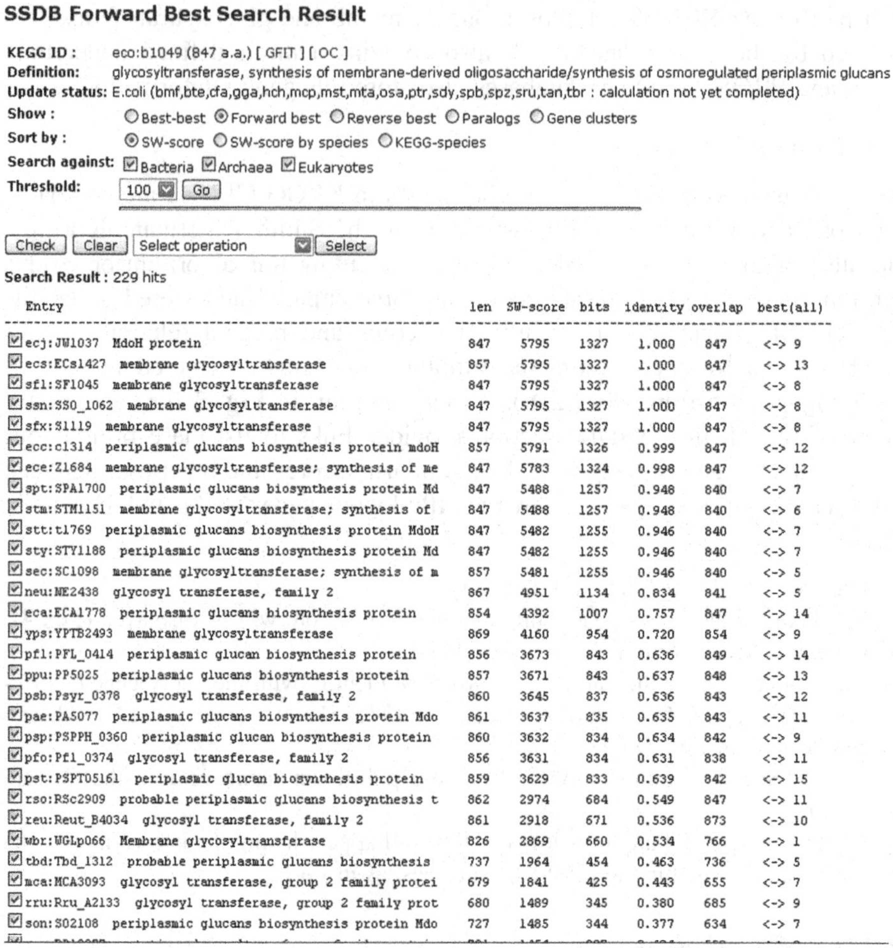

SSDB Forward Best Search Result

KEGG ID : eco:b1049 (847 a.a.) [GFIT] [OC]
Definition: glycosyltransferase, synthesis of membrane-derived oligosaccharide/synthesis of osmoregulated periplasmic glucans
Update status: E.coli (bmf,bte,cfa,gga,hch,mcp,mst,mta,osa,ptr,sdy,spb,spz,sru,tan,tbr : calculation not yet completed)
Show : Best-best ◉Forward best Reverse best Paralogs Gene clusters
Sort by : ◉SW-score SW-score by species KEGG-species
Search against: ☑Bacteria ☑Archaea ☑Eukaryotes
Threshold: 100 Go

Check Clear Select operation Select

Search Result : 229 hits

Entry	len	SW-score	bits	identity	overlap	best(all)
☑ecj:JW1037 MdoH protein	847	5795	1327	1.000	847	<-> 9
☑ecs:ECs1427 membrane glycosyltransferase	857	5795	1327	1.000	847	<-> 13
☑sfl:SF1045 membrane glycosyltransferase	847	5795	1327	1.000	847	<-> 8
☑ssn:SSO_1062 membrane glycosyltransferase	847	5795	1327	1.000	847	<-> 8
☑sfx:S1119 membrane glycosyltransferase	847	5795	1327	1.000	847	<-> 8
☑ecc:c1314 periplasmic glucans biosynthesis protein mdoH	857	5791	1326	0.999	847	<-> 12
☑ece:Z1684 membrane glycosyltransferase; synthesis of me	847	5783	1324	0.998	847	<-> 12
☑spt:SPA1700 periplasmic glucans biosynthesis protein Md	847	5488	1257	0.948	840	<-> 7
☑stm:STM1151 membrane glycosyltransferase; synthesis of	847	5488	1257	0.948	840	<-> 6
☑stt:t1769 periplasmic glucans biosynthesis protein MdoH	847	5482	1255	0.946	840	<-> 7
☑sty:STY1188 periplasmic glucans biosynthesis protein Md	847	5482	1255	0.946	840	<-> 7
☑sec:SC1098 membrane glycosyltransferase; synthesis of m	857	5481	1255	0.946	840	<-> 5
☑neu:NE2438 glycosyl transferase, family 2	867	4951	1134	0.834	841	<-> 5
☑eca:ECA1778 periplasmic glucans biosynthesis protein	854	4392	1007	0.757	847	<-> 14
☑yps:YPTB2493 membrane glycosyltransferase	869	4160	954	0.720	854	<-> 9
☑pfl:PFL_0414 periplasmic glucan biosynthesis protein	856	3673	843	0.636	849	<-> 14
☑ppu:PP5025 periplasmic glucans biosynthesis protein	857	3671	843	0.637	848	<-> 13
☑psb:Psyr_0378 glycosyl transferase, family 2	860	3645	837	0.636	843	<-> 12
☑pae:PA5077 periplasmic glucans biosynthesis protein Mdo	861	3637	835	0.635	843	<-> 11
☑psp:PSPPH_0360 periplasmic glucan biosynthesis protein	860	3632	834	0.634	842	<-> 9
☑pfo:Pfl_0374 glycosyl transferase, family 2	856	3631	834	0.631	838	<-> 11
☑pst:PSPTO5161 periplasmic glucan biosynthesis protein	859	3629	833	0.639	842	<-> 15
☑rso:RSc2909 probable periplasmic glucans biosynthesis t	862	2974	684	0.549	847	<-> 11
☑reu:Reut_B4034 glycosyl transferase, family 2	861	2918	671	0.536	873	<-> 10
☑wbr:WGLp066 Membrane glycosyltransferase	826	2869	660	0.534	766	<-> 1
☑tbd:Tbd_1312 probable periplasmic glucans biosynthesis	737	1964	454	0.463	736	<-> 5
☑mca:MCA3093 glycosyl transferase, group 2 family protei	679	1841	425	0.443	655	<-> 7
☑rru:Rru_A2133 glycosyl transferase, group 2 family prot	680	1489	345	0.380	685	<-> 9
☑son:SO2108 periplasmic glucans biosynthesis protein Mdo	727	1485	344	0.377	634	<-> 7

Fig. 4. Screen shot of a search result of the KEGG SSDB database for orthologs of a glycosyltransferase gene in *Escherichia coli.*

organism-specific pathway can be generated automatically once its genome has been annotated with KO identifiers.

KO groups are defined based on highly confident pairwise-similarity scores and best-hit relations of genes from the KEGG SSDB database (*see* **Subheading 2.3.**). These groups are then linked to the pathways via several methodologies: (1) Genomic information linked with pathways, (2) Cluster of Orthologous Groups information *(8)*, and (3) experts' classifications of protein

Search common motifs in selected sequences

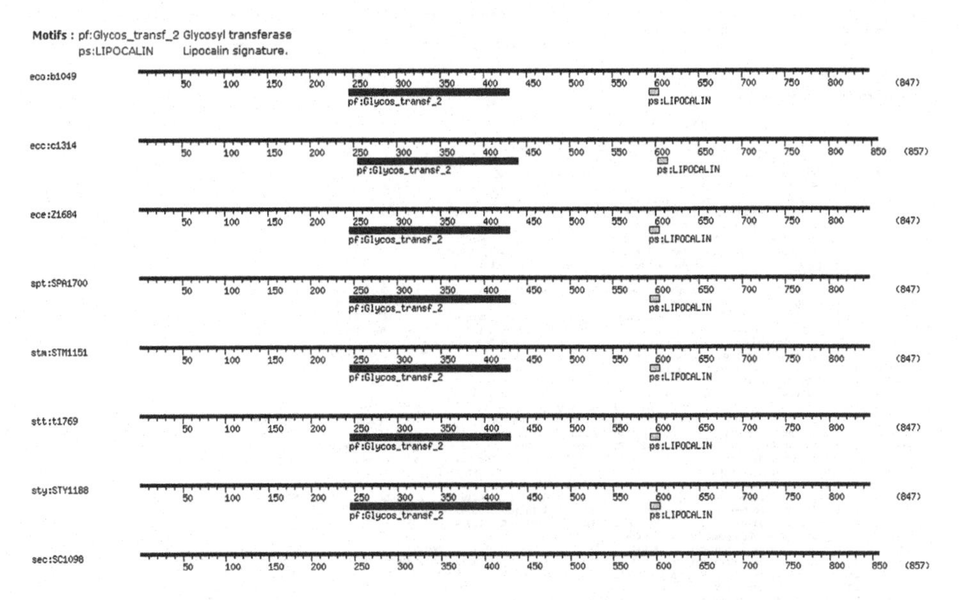

Fig. 5. Screen shot of a search result for common motifs to the query gene, a glycosyltransferase gene in *Escherichia coli* with motifs in both Pfam and Prosite.

families. The current list of KOs can be found in the KEGG BRITE database at http://www.genome.jp/kegg/brite.html:

1. Go to the BRITE database on the web.
2. Click on the KO link to display the list of KOs organized by pathway hierarchy, as in **Fig. 6**.
3. Click on "4th Level" to see the full details of the KO hierarchy.
4. The details of any KO group can be viewed by clicking on its entry ID. For example, click on K00845. The entry for each KO group contains the following information:
   a. The pathways in which this KO group may be found.
   b. Links to other related databases in KEGG such as REACTION and ENZYME, as well as links to its corresponding Cluster of Orthologous Groups ID and Gene Ontology ID *(9)*.
   c. The list of specific genes within each organism that belongs to this group.

2.5. KEGG PATHWAY

The KEGG PATHWAY database provides manually generated wiring diagrams of genes and compounds for various biological processes ranging

KEGG KEGG Orthology (KO)

[1st Level | 2nd Level | 3rd Level | 4th Level | Text Search]

01100 Metabolism

01110 Carbohydrate Metabolism

```
00010 Glycolysis / Gluconeogenesis [PATH:ko00010] [GO:0006096 0006094]

K00845  E2.7.1.2, glk; glucokinase [EC:2.7.1.2] [COG:COG0837] [GO:0004340]
K00844  E2.7.1.1; hexokinase [EC:2.7.1.1] [GO:0004396]
K01084  E3.1.3.9, G6PC; glucose-6-phosphatase [EC:3.1.3.9] [GO:0004346]
K01810  E5.3.1.9, pgi; glucose-6-phosphate isomerase [EC:5.3.1.9] [COG:COG0166] [GO:0004347]
K06859  E5.3.1.9A; glucose-6-phosphate isomerase [EC:5.3.1.9] [COG:COG2140] [GO:0004347]
K00850  E2.7.1.11, pfk; 6-phosphofructokinase [EC:2.7.1.11] [COG:COG0205 COG1105] [GO:0003872]
K01086  E3.1.3.11; fructose-1,6-bisphosphatase [EC:3.1.3.11] [GO:0042132]
K03841  FBP1, FBP2, fbp; fructose-1,6-bisphosphatase I [EC:3.1.3.11] [COG:COG0158] [GO:0042132]
K02446  GLPX; fructose-1,6-bisphosphatase II [EC:3.1.3.11] [COG:COG1494] [GO:0042132]
K04041  FBP3, fbp; fructose-1,6-bisphosphatase III [EC:3.1.3.11] [GO:0042132]
K01622  E4.1.2.13; fructose-bisphosphate aldolase [EC:4.1.2.13] [GO:0004332]
K01624  E4.1.2.13B, fbaA; fructose-bisphosphate aldolase, class II [EC:4.1.2.13] [COG:COG0191] [GO:0004332]
K01623  E4.1.2.13A, fbaB; fructose-bisphosphate aldolase, class I [EC:4.1.2.13] [COG:COG1830 COG3588] [GO:0004332]
K01803  E5.3.1.1, tpiA; triosephosphate isomerase (TIM) [EC:5.3.1.1] [COG:COG0149] [GO:0004807]
K00134  E1.2.1.12, GAPD, gapA; glyceraldehyde 3-phosphate dehydrogenase [EC:1.2.1.12] [COG:COG0057] [GO:0004365]
K00927  E2.7.2.3, pgk; phosphoglycerate kinase [EC:2.7.2.3] [COG:COG0126] [GO:0004618]
K01834  E5.4.2.1, gpm; phosphoglycerate mutase [EC:5.4.2.1] [COG:COG0406 COG0588 COG0696] [GO:0004619]
K01689  E4.2.1.11, eno; enolase [EC:4.2.1.11] [COG:COG0148] [GO:0004634]
K00873  E2.7.1.40, pyk; pyruvate kinase [EC:2.7.1.40] [COG:COG0469] [GO:0004743]
K00160  E1.2.4.1, pdh; pyruvate dehydrogenase [EC:1.2.4.1] [GO:0004739]
K00163  E1.2.4.1C, aceE; pyruvate dehydrogenase E1 component [EC:1.2.4.1] [COG:COG2609] [GO:0004739]
K00161  E1.2.4.1A, pdhA; pyruvate dehydrogenase E1 component, alpha subunit [EC:1.2.4.1] [COG:COG1071] [GO:0004739]
K00162  E1.2.4.1B, pdhB; pyruvate dehydrogenase E1 component, beta subunit [EC:1.2.4.1] [COG:COG0022] [GO:0004739]
K00627  E2.3.1.12, pdhC; pyruvate dehydrogenase E2 component (dihydrolipoamide acetyltransferase) [EC:2.3.1.12] [COG:COG0508] [GO:0004742]
K00382  E1.8.1.4, pdhD; dihydrolipoamide dehydrogenase [EC:1.8.1.4] [COG:COG1249] [GO:0004148]
K00016  E1.1.1.27, ldh; L-lactate dehydrogenase [EC:1.1.1.27] [COG:COG0039] [GO:0004459]
K01568  E4.1.1.1, pdc; pyruvate decarboxylase [EC:4.1.1.1] [COG:COG0028] [GO:0004737]
K00001  E1.1.1.1, adh; alcohol dehydrogenase [EC:1.1.1.1] [COG:COG1012 COG1062 COG1064 COG1454] [GO:0004022 0004023 0004024 0004025]
K00002  E1.1.1.2, adh; alcohol dehydrogenase (NADP+) [EC:1.1.1.2] [COG:COG0656] [GO:0008106]
K00114  E1.1.99.8; alcohol dehydrogenase (acceptor) [EC:1.1.99.8] [GO:0018468]
K04022  EUTG; alcohol dehydrogenase [COG:COG1454]
K00128  E1.2.1.3; aldehyde dehydrogenase (NAD+) [EC:1.2.1.3] [COG:COG1012] [GO:0004029]
K00129  E1.2.1.5; aldehyde dehydrogenase (NAD(P)+) [EC:1.2.1.5] [COG:COG1012] [GO:0004030]
K01895  E6.2.1.1, acs; acetyl-CoA synthetase [EC:6.2.1.1] [COG:COG0365] [GO:0003987]
K01512  E3.6.1.7, acyP; acylphosphatase [EC:3.6.1.7] [COG:COG1254] [GO:0003998]
K01837  E5.4.2.4, BPGM; bisphosphoglycerate mutase [EC:5.4.2.4] [GO:0004082]
K01088  E3.1.3.13, BPGP; bisphosphoglycerate phosphatase [EC:3.1.3.13] [GO:0004083]
K05715  2PGK; 2-phosphoglycerate kinase [EC:2.7.2.-] [COG:COG2074]
K05716  CPGS; cyclic 2,3-diphosphoglycerate synthetase [EC:4.6.1.-] [COG:COG2403]
K01785  E5.1.3.3, galM; aldose 1-epimerase [EC:5.1.3.3] [COG:COG2017] [GO:0004034]
K01085  E3.1.3.10, agp; glucose-1-phosphatase [EC:3.1.3.10] [GO:0008877]
```

Fig. 6. Hierarchical list of the KEGG Orthology (KO). The current screen shot lists the KO groups in the glycolysis/gluconeogenesis pathway of KEGG.

from metabolic pathways such as carbohydrate and energy metabolism and the biosynthesis of secondary metabolites, through regulatory pathways and human diseases. The entry page to this database can be found at http://www.genome.jp/kegg/pathway.html, where the entire list of available pathways can be found. Keyword search for specific pathways can be executed using the textfield in the top section of this page. A pathway may have complementary information listed to the right of its name, such as literary references.

Each metabolic pathway map is a diagram of boxes containing EC numbers (for genes) and circles (for chemical compounds) connected by directed edges. Different types of arrows and lines indicate different relationships, as listed in the Help, which can be found at http://www.genome.jp/kegg/document/help_pathway.html or by clicking the Help button at the top right of each pathway map. Each box is linked to its KEGG GENES entry (unless the Reaction reference pathway is selected, in which it is linked to its KEGG REACTION entry in KEGG LIGAND), and each circle is linked to its KEGG COMPOUND entry in the KEGG LIGAND database. Ovals are linked to other pathway maps. The top left of each pathway map has links back to the main Pathway list as well as the corresponding Ortholog table, below which is the customizable organism list (*see* **Subheading 2.1.**). In addition to the list of organisms, the organism list also contains options to view the basic reference pathway, the KO reference pathway, the Reaction reference pathway, and the "all organisms" pathway:

1. The basic reference pathway provides a global picture of all known interactions based on the literature.
2. The KO reference pathway colors the genes in the basic reference pathway that are assigned to ortholog groups, or KO numbers.
3. The Reaction reference pathway links the genes in the pathway to the KEGG REACTION entries in which each gene participates as opposed to its KEGG ENZYME entry.
4. The species-specific (e.g., *Homo sapiens*) pathways, which can be selected using the organism list, are colored versions of the basic reference pathway, with the colored genes corresponding to those that can be found in the chosen species, so the pathway can be found via the colorings.
5. The "all organisms" pathways color the genes corresponding to all organisms in KEGG in the basic reference pathway and links them to the KO groups to which they belong.

As an example, let us look at the Galactose metabolism pathway, which can be found under the "Carbohydrate Metabolism" section (**Fig. 7**).

1. Select "Homo sapiens" under the organism list pull-down menu to the top-left of the pathway. Then click the "Go" button. The pathway diagram will now display the pathway with the genes expressed in human colored in green.
2. Click on "Nucleotide sugars metabolism" near the top-left of the pathway, linked to UDPgalactose. The "Nucleotide sugars metabolism" pathway map will be displayed for human.
3. In contrast, now select the "Reference pathway (KO)" option from the organism list pull-down menu. Then click the "Go" button. The genes that have KO annotations will be colored in lavender.
4. Click the box labeled "2.7.7.24" for example, to display the KO entry, which lists all ortholog genes belonging to this group. The pathways in which this group can be found will also be listed, in addition to related links to other databases of ortholog information.

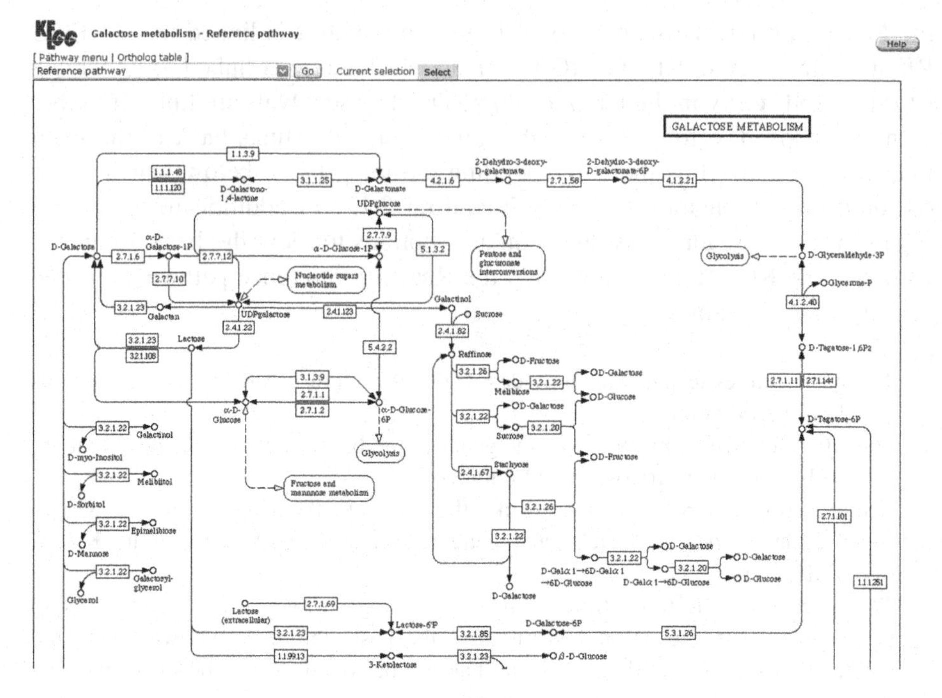

Fig. 7. The galactose metabolism reference pathway.

2.6. KEGG Automatic Annotation Server

KAAS was released in 2005 to aid researchers who have sequences that they wish to annotate in bulk *(7)*. The input genes are annotated with KO numbers (*see* **Subheading 2.4.**), based on its sequence information. Specifically, given a genome to annotate, it is compared against each specified target genome in the GENES database using BLAST ***(10,11)*** in both forward and reverse directions. The ortholog candidates are then classified into groups by their K numbers. Ortholog candidates without K numbers are put into one group. Finally, a score is determined for each ortholog group in order to assign the best fitting K numbers to the query gene.

To use KAAS, first go to the KAAS web server at http://www.genome.jp/kegg/kaas/ (**Fig. 8**). There are three options to use this server: (1) using a complete set of genes from a single genome, (2) using a subset of genes, (3) using a set of ESTs. Depending on the content of the input set, either the BBH (bi-directional best hit) or SBH (single-directional best hit) assignment method should be used. BBH attempts to find homologous genes that are BBHs to each other. That is, two

KEGG Automatic Annotation Server

for ortholog assignment and pathway mapping

About KAAS

KAAS (KEGG Automatic Annotation Server) provides functional annotation of genes by BLAST comparisons against the manually curated KEGG GENES database. The result contains KO (KEGG Orthology) assignments and automatically generated KEGG pathways.

- KAAS Help

Complete or Draft Genome

KAAS works best when a complete set of genes in a genome is known. Prepare query amino acid sequences and use the BBH (bi-directional best hit) method to assign orthologs.

- KAAS job request (BBH method)

Partial Genome

KAAS can also be used for a limited number of genes. Prepare query amino acid sequences and use the SBH (single-directional best hit) method to assign orthologs.

- KAAS job request (SBH method)
- KAAS interactive

ESTs

When ESTs are comprehensive enough, a set of consensus contigs can be generated by the Eassembler server and used as a gene set for KAAS with the BBH method. Otherwise, use ESTs as they are with the SBH method.

- Eassembler job request (available soon)
- KAAS job request (BBH method)
- KAAS job request (SBH method)

Example of Results

KO assignment

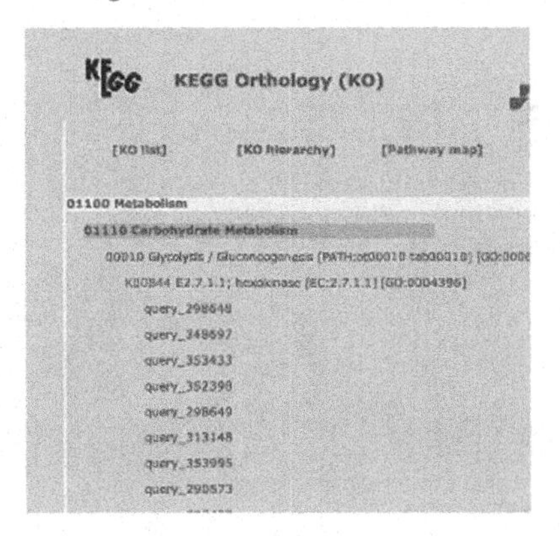

KEGG pathway mapping

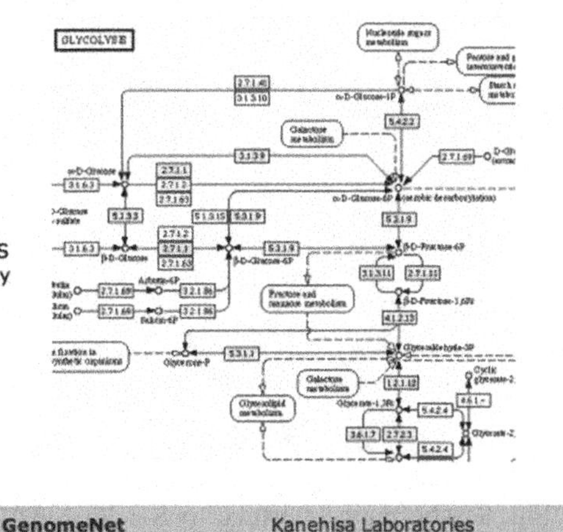

Last updated: July 27, 2005

Feedback **BRITE** **KEGG** **GenomeNet** Kanehisa Laboratories

Fig. 8. The initial interface to KAAS, the KEGG Automatic Annotation Server, from which different annotation jobs may be executed based on the type of input data (gene lists) available.

BBH genes are the top-matching genes for one another. On the other hand, SBH is less restrictive in that, given two SBH genes *a* and *b*, even if the top-matching gene for *b* is *a*, the top matching gene for *a* may not necessarily be *b*. Thus, BBH can provide more highly confident scores given that the original input gene list is that of a complete genome, whereas SBH should suffice for partial gene lists (*see also* **Note 2**). Each of these options can be modified from the main KAAS server input screen (shown in **Fig. 9**).

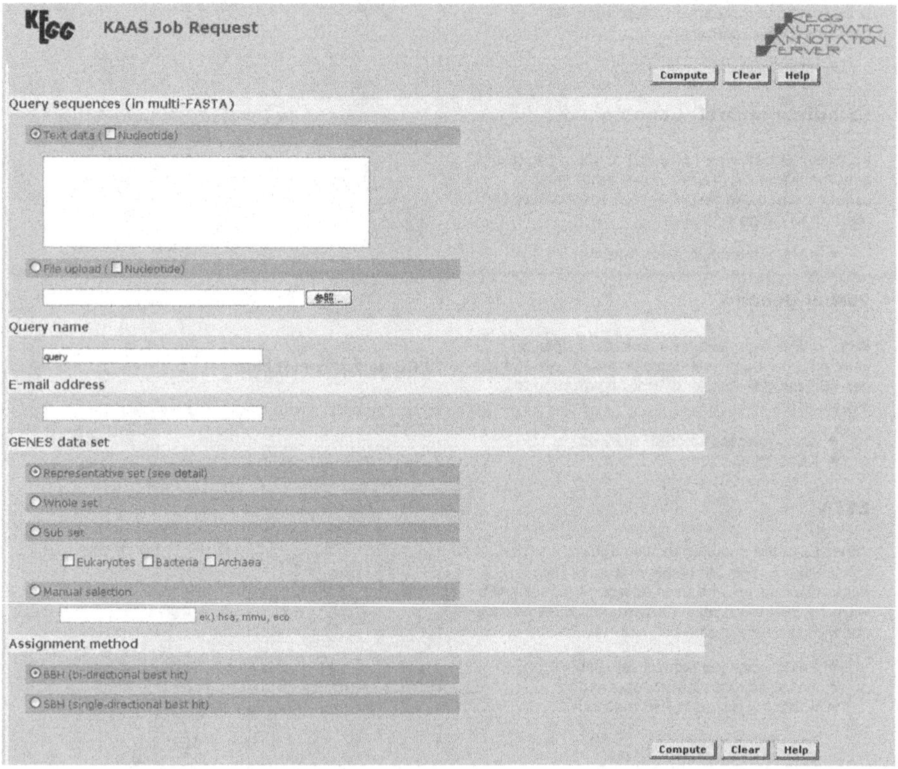

Fig. 9. The main interface to KEGG KAAS. A file of sequences in multi-FASTA format can either be copy-and-pasted or specified using the "Text data" or "File upload" options, respectively. The name for this query can be specified in the "Query name" textfield. An email address to which status information should be sent is required. The set of genomes with which to compare the given input sequences can be selected. Either the BBH or SBH assignment methods should be selected, and when all options are specified, the "Compute" button can be clicked to execute the query.

1. A file of sequences can either be copy-and-pasted or specified using the "Text data" or "File upload" options, respectively. Note that the sequences should be in multi-FASTA format, and the "Nucleotide" checkbox should be checked accordingly.
2. The name for this query may be modified from the default by specifying it in the "Query name" textfield.
3. An email address to which status information should be sent is required.
4. The set of genomes with which to compare the given input sequences can be selected from one of the following:
   a. Representative set: consists of a select set of organisms that are considered representative of their kingdoms.
   b. Whole set: consists of all organisms in KEGG.
   c. Sub set: one or more of the following taxonomic groups may be selected: Eukaryotes, Bacteria, or Archaea. All genomes in the selected groups will be used for comparison.
   d. Manual selection: a comma-delimited list of KEGG organism codes of genomes with which to compare may be specified manually in the given textfield.
5. Either the BBH or SBH assignment methods should be selected. It is recommended that BBH be used if the input sequences represent a whole genome. Otherwise, SBH should be used as it is less restrictive and returns results more quickly (*see* **Note 2**).
6. When all options are specified, click the "Compute" button to execute the query.

Once the query job is complete, an email containing a link to the results page will be sent to the specified email address. The results list will look like

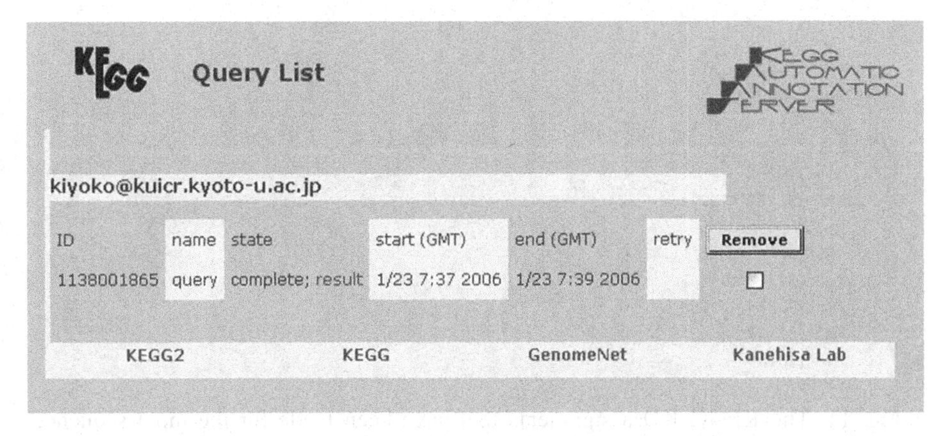

Fig. 10. Results of a KAAS query. A listing of all queries that have been sent to KAAS using this email address will be shown.

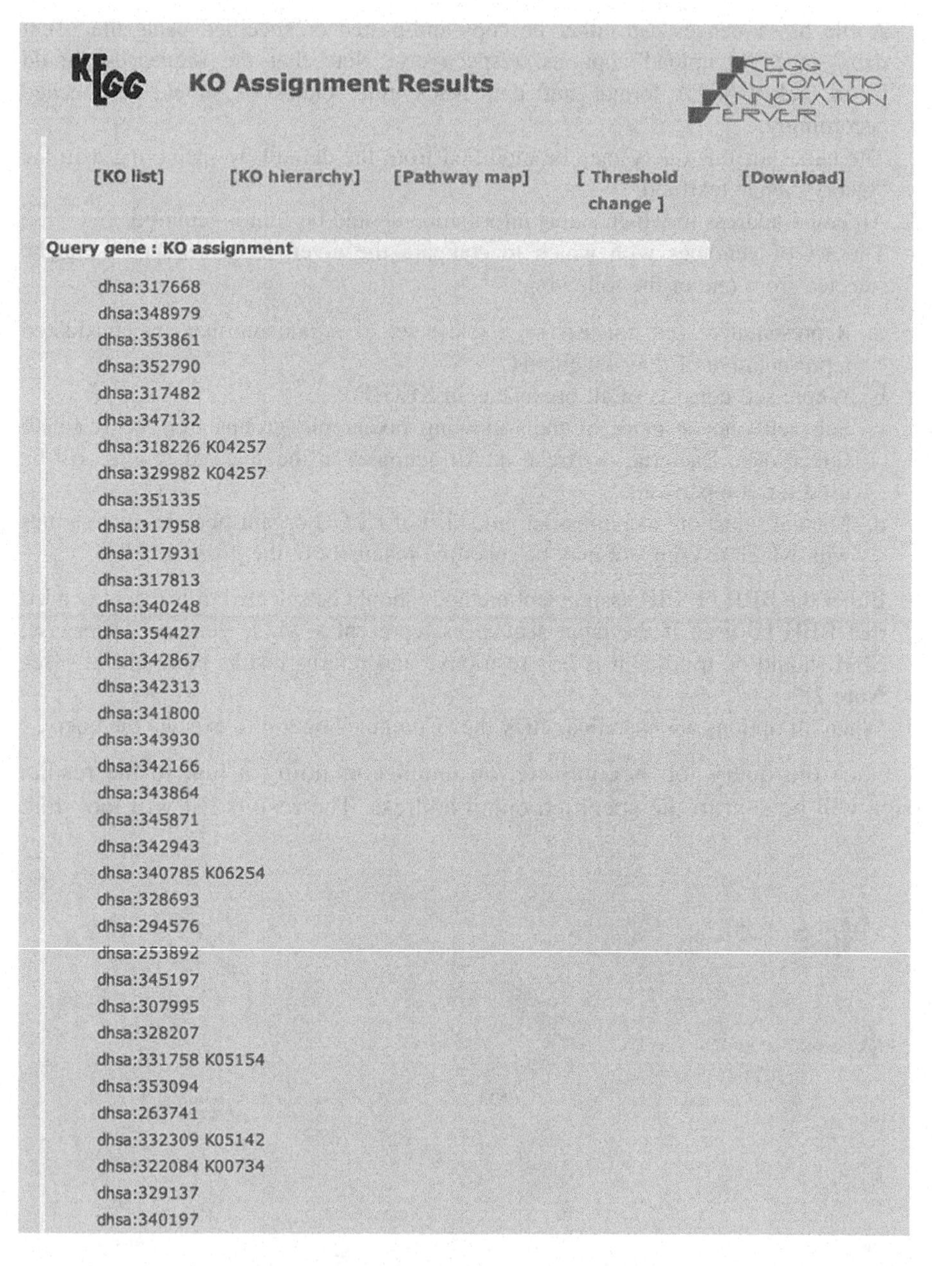

Fig. 11. The KEGG KO assignments that have been made for the input sequence list is listed.

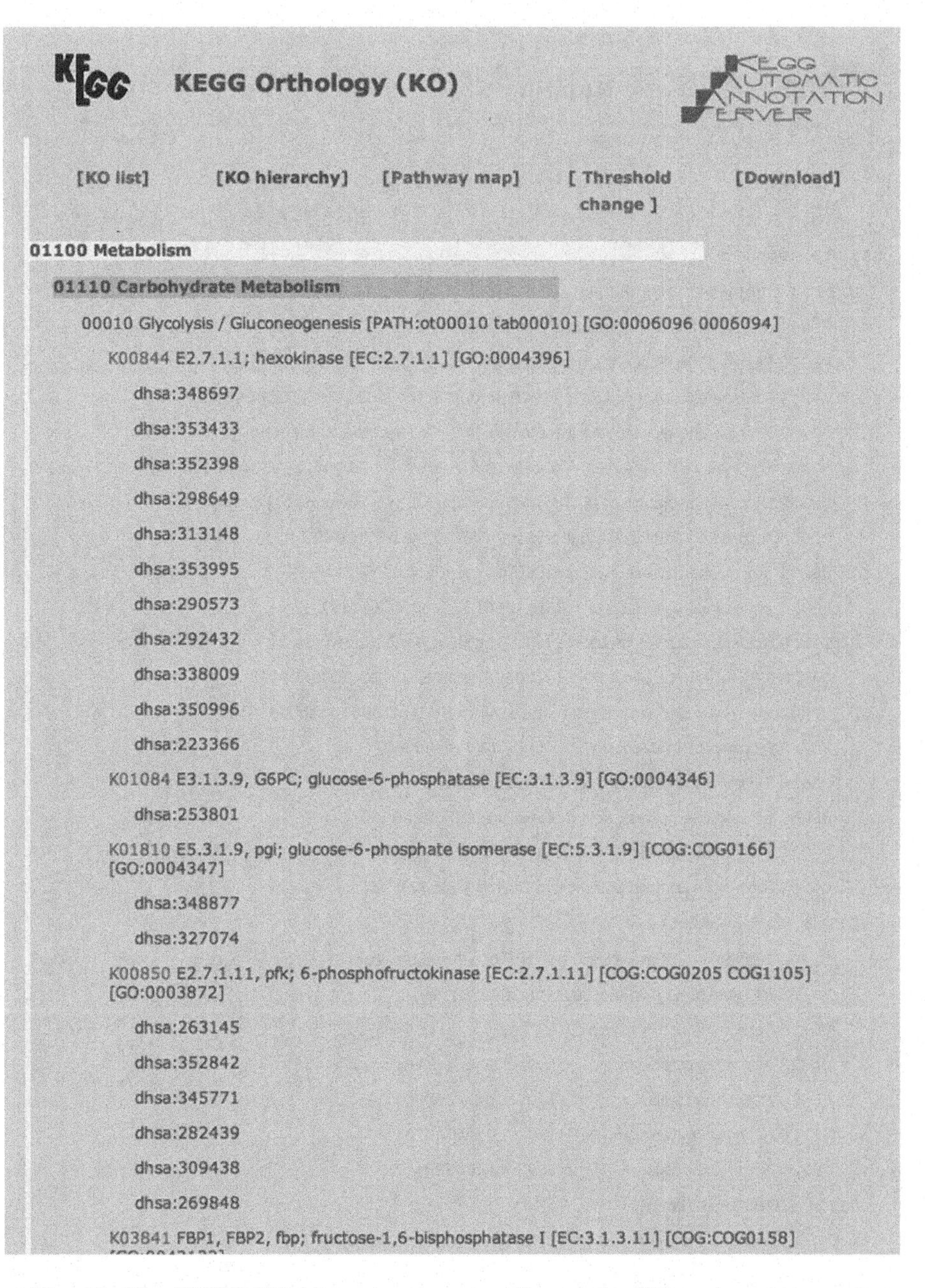

Fig. 12. The KEGG KO hierarchy corresponding to the KO assignments made to the input is displayed.

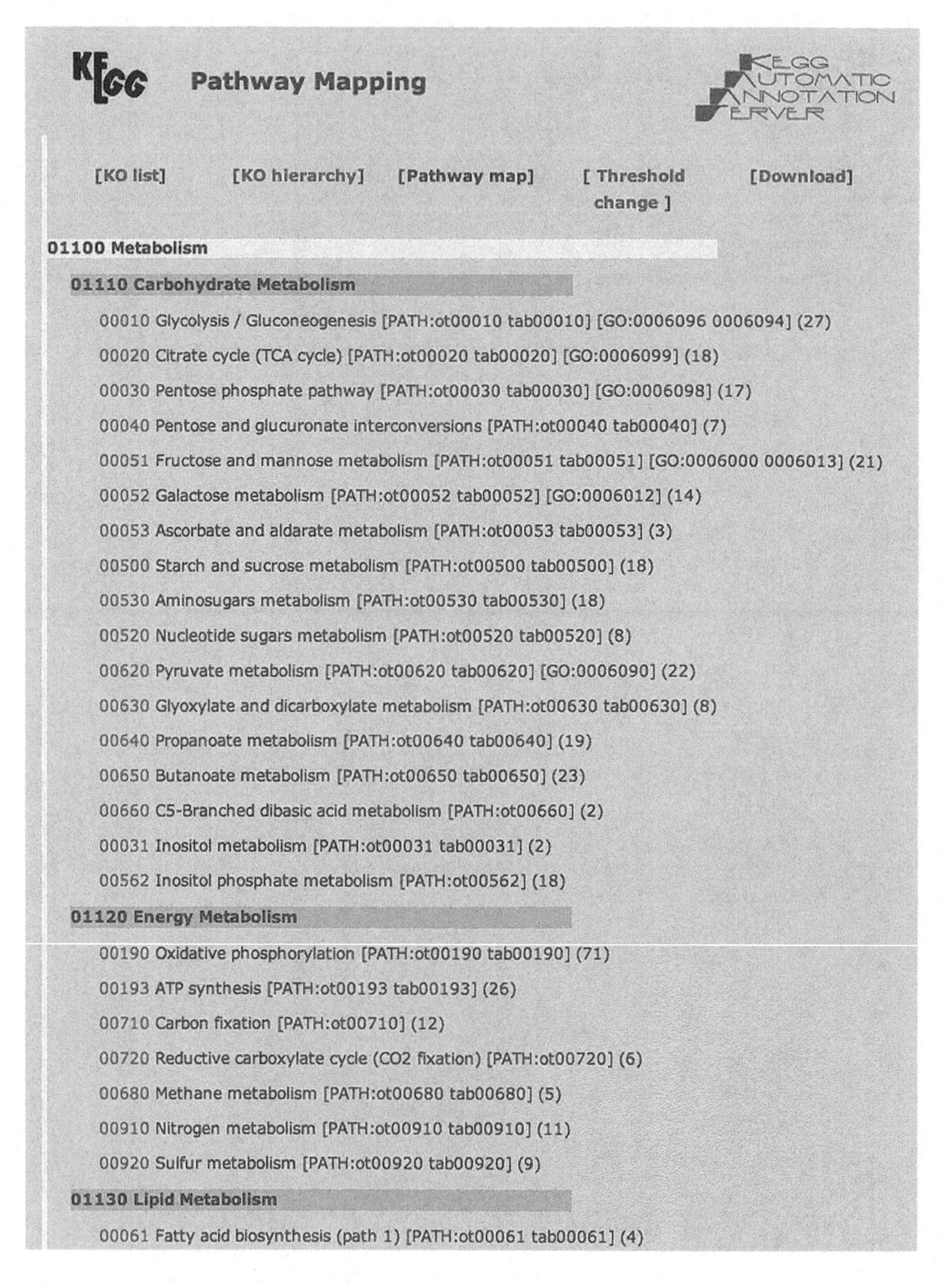

Fig. 13. The list of pathways in which KEGG KO assignments were found is listed. Pathways can then be seen with the input sequences colored accordingly.

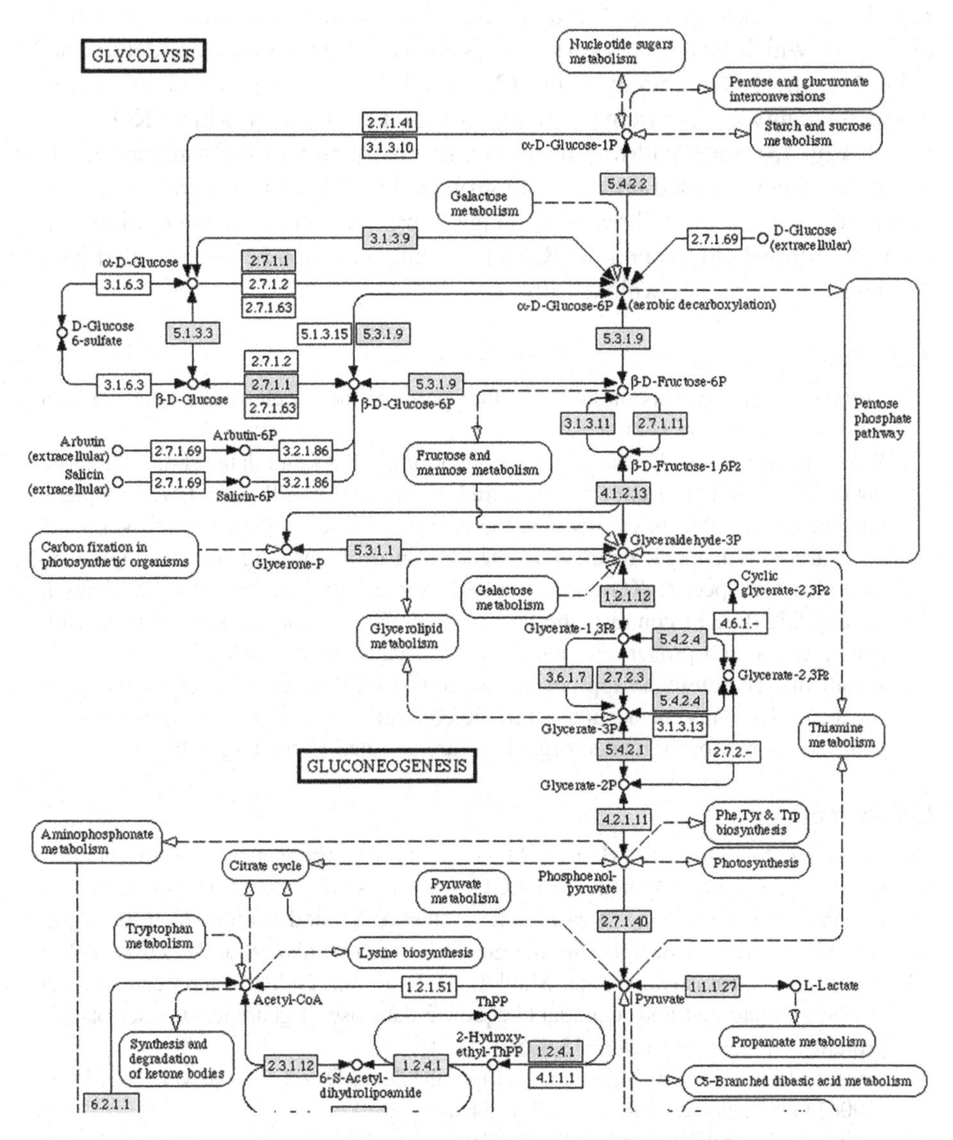

Fig. 14. An example of a pathway colored according to KEGG KO assignment.

Fig. 10. By clicking on the "results" link, the assignment results page will be displayed, which has three views: the list of KO groups that were determined (**Fig. 11**), the KO hierarchy (**Fig. 12**), which lists the KO's and its content based on pathway hierarchy, and the list of pathways in which KOs were found (**Fig. 13**). Each pathway result is a reconstruction of the pathway for the given set of genes based on their KO groups (**Fig. 14**). The assignment can be modified by using the "Threshold change" option to rerun the annotation with a different threshold to send to BLAST. Results can also be downloaded from this page using the "Download" link.

3. Notes

1. Further detailed methodologies on other parts of the KEGG resource have been published elsewhere *(12)*.
2. When using KAAS, depending on the kingdom of the input data set and the target data set with which it is to be compared, performance will vary. For example, it has been found that in eukaryotes, the accuracy of reassignment improves when a eukaryote-rich target data set is used because of fewer incorrect assignments to prokaryote-specific KO entries caused by the large number of prokaryotes in KEGG GENES. In contrast, the accuracy for *E. coli* and *Synechocystis* slightly decreases using a prokaryote-rich data set, because of the lack of closely related organisms. Therefore, an appropriate target data set should be chosen for the query organism. Not only the accuracy, but the computation time, should improve if the genes of more closely related organisms are selected as the target data set.

References

1. Kanehisa, M., Goto, S., Hattori, M., et al. (2006) From genomics to chemical genomics: new developments in KEGG. *Nucleic Acids Res.* **34,** D354–D357.
2. Kanehisa, M., Goto, S., Kawashima, S., Okuno, Y., and Hattori, M. (2004) The KEGG resource for deciphering the genome. *Nucleic Acids Res.* **32,** D277–D280.
3. Pruitt, K. D., Tatusova, T., and Maglott, D. R. (2005) NCBI Reference Sequence (RefSeq): a curated non-redundant sequence database of genomes, transcripts and proteins. *Nucleic Acids Res.* **33,** D501–D504.
4. Benson, D. A., Karsch-Mizrachi, I., Lipman, D. J., Ostell, J., and Wheeler, D. L. (2006) GenBank. *Nucleic Acids Res.* **34,** D16–D20.
5. Smith, T. F. and Waterman, M. S. (1981) Identification of common molecular subsequences. *J. Mol. Biol.* **147,** 195–197.
6. Kanehisa, M., Goto, S., Kawashima, S., and Nakaya, A. (2002) The KEGG databases at GenomeNet. *Nucleic Acids Res.* **30,** 42–46.
7. Moriya, Y., Itoh, M., Okuda, S., Yoshizawa, A., and Kanehisa, M. (2007) KAAS: an automatic genome annotation and pathway reconstruction server. *Nucleic Acids Res.* in press.

8. Tatusov, R. L., Fedorova, N. D., Jackson, J. D., et al. (2003) The COG database: an updated version includes eukaryotes. *BMC Bioinformatics* **4,** 41.
9. (2006) The Gene Ontology (GO) project in 2006. *Nucleic Acids Res.* **34,** D322–D326.
10. Altschul, S. F., Gish, W., Miller, W., Myers, E. W., and Lipman, D. J. (1990) Basic local alignment search tool. *J. Mol. Biol.* **215,** 403–410.
11. Altschul, S. F., Madden, T. L., Schaffer, A. A., et al. (1997) Gapped BLAST and PSI-BLAST: a new generation of protein database search programs. *Nucleic Acids Res.* **25,** 3389–3402.
12. Aoki, K. F. and Kanehisa, M. (2005) in *Current Protocols in Bioinformatics,* (Andreas D. and Baxevanis, L. D. S., ed.), John Wiley and Sons, Ltd.

II

Orthologs, Synteny, and Genome Evolution

7

Ortholog Detection Using the Reciprocal Smallest Distance Algorithm

Dennis P. Wall and Todd DeLuca

Summary

All protein coding genes have a phylogenetic history that when understood can lead to deep insights into the diversification or conservation of function, the evolution of developmental complexity, and the molecular basis of disease. One important part to reconstructing the relationships among genes in different organisms is an accurate method to find orthologs as well as an accurate measure of evolutionary diversification. The present chapter details such a method, called the reciprocal smallest distance algorithm (RSD). This approach improves upon the common procedure of taking reciprocal best Basic Local Alignment Search Tool hits (RBH) in the identification of orthologs by using global sequence alignment and maximum likelihood estimation of evolutionary distances to detect orthologs between two genomes. RSD finds many putative orthologs missed by RBH because it is less likely to be misled by the presence of close paralogs in genomes. The package offers a tremendous amount of flexibility in investigating parameter settings allowing the user to search for increasingly distant orthologs between highly divergent species, among other advantages. The flexibility of this tool makes it a unique and powerful addition to other available approaches for ortholog detection.

Key Words: Orthologs; orthology; reciprocal smallest distance; reciprocal BLAST hit; maximum likelihood; molecular phylogenetics; phylogeny.

1. Introduction

The number of fully sequenced genomes is growing at an unprecedented rate. Presently there are 339 completed genomes and 1867 in various stages of construction (588 of these are Eukaryotic genomes) *(1)*. The sheer number of

From: *Methods in Molecular Biology, vol. 396: Comparative Genomics, Volume 2*
Edited by: N. H. Bergman © Humana Press Inc., Totowa, NJ

newly sequenced genomes brings exciting opportunities and challenges to the field of comparative genomics. Paramount to this field is the ability to accurately detect orthologs across genomes. Orthologs are used to answer, partially or completely, many important questions in a vast majority of biological fields *(2)*. Aside from the important task of providing functional annotation to protein coding genes *(2–4)*, orthologs have been used in understanding the evolution of Eukaryotes *(5–8)*, predicting novel protein interactions *(9–12)*, finding disease genes in "model" organisms *(13)*, determining the composition of gene families *(14)*, understanding coevolution of interacting proteins *(15)*, comprehending the degree of compensation for deleterious mutations *(16,17)*, predicting deleterious alleles in humans *(18,19)*, as well as in measuring the effects of functional genomic variables on protein evolutionary rate to converge on a general model of protein evolution *(20–29)*, and other areas of evolutionary genomics *(30)*.

In all cases, being able to differentiate between orthologs (sequences that diverged from each other at the same time as the species divergence) and paralogs (sequences that diverged at another time) is critical. For example, when comparing the evolutionary rates of proteins in the absence of a normalizing molecular clock, such as the rate of synonymous substitutions, estimates of evolutionary rate must be based upon comparisons between sequences that are strictly orthologous. Only if all sequence comparisons share the same time of divergence are protein evolutionary distances expected to be proportional to relative evolutionary rates.

The importance of this distinction between orthologs and paralogs is illustrated in **Fig. 1**. The gene duplication event denoted by a black circle significantly predates the speciation event and results in two paralogous gene lineages. Treating as orthologs genes that are the result of this split, namely comparing either D or E with A, or comparing C with B, would yield inflated measures of evolutionary divergence between the two species and cause misinterpretation of functional similarity. Only the orthologous comparisons (A compared with C and B compared with D or E) would yield evolutionary distances indicative of the relative rates of protein evolution and an appropriate interpretation of functional equivalency.

A common procedure for identifying proteins that are putatively orthologous involves the identification of reciprocal best Basic Local Alignment Search Tool (BLAST) hits (RBH) *(23,25)*. Protein i in genome I is a reciprocal best hit of protein j in genome J if a forward search of genome J with protein i yields as the top hit protein j, and a reciprocal query of genome I with protein j yields as the top hit protein i. However, BLAST searches often return as the

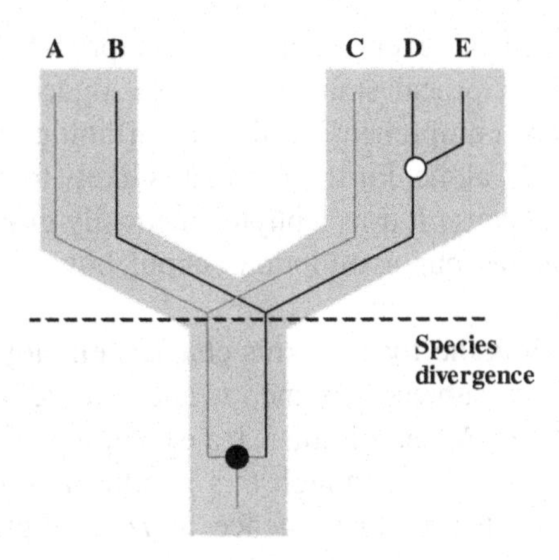

Fig. 1. Conflating paralogs and orthologs. An ortholog is defined as a gene in two species that has evolved directly from a single gene in their last common ancestor ***(52)***, that is, that arose at the evolutionary split between two species, and that conveys the same function in both species. Conversely paralogs are sequences that diverged from each other at some other time, and that either evolve new function or lose function entirely via effects of genetic drift. Here, **D** and **E** are orthologs to **B**, but paralogs to **A** and **C**. Distinguishing between paralogs and orthologs is important to many areas of study, including functional annotation of genes in newly sequenced genomes and to accurately estimating the rates of protein diversification in genomes. Thick lines represent the organism tree, thin lines represent the gene tree.

highest scoring hit a protein that is not the nearest phylogenetic neighbor of the query sequence ***(31)***. If the forward BLAST yields a paralogous best hit, regardless of whether the reciprocal BLAST corrects the error by recovering an actual ortholog, both pairs will be excluded. Thus, although RBH will rightfully prevent admission of the paralogous pair to the set of proteins for which relative evolutionary rates are estimated, it will also wrongly exclude an authentically orthologous pair from consideration. Despite this potential pitfall, perhaps the most commonly used orthology resources—Clusters of Orthologous Genes (C/KOGs) ***(14,32)***, INPARANOID ***(33,34)***, and the National Center For Biotechnology Information's (NCBI) resource, HomoloGene ***(35)***—are based on reciprocal best BLAST hits, and are thus likely to be incomplete or possibly erroneous.

In an effort to correct for this source of error, Wall et al. ***(36)*** developed a novel algorithm that preserves the safeguard of reciprocal genome queries, but is less susceptible to excluding an ortholog if a paralog is returned as the top hit in either the forward or reverse steps of a reciprocal BLAST query. This approach, termed the reciprocal smallest distance algorithm (RSD), has been shown to provide more comprehensive lists of orthologs than other methods that are based on BLAST alone. Furthermore, it is likely to be more accurate for identifying orthologs because it uses a phylogenetically grounded measurement of similarity that matches our assumptions about how orthologs in different species have evolved.

RSD (**Fig. 2**), which is the focus of this chapter, employs BLAST *(37)* as a first step, starting with a subject genome, J, and a protein query sequence, i, belonging to genome I. A set of hits, H, exceeding a predefined significance threshold (e.g., $e < 10^{-20}$, though this is adjustable) is obtained. Then, using clustalW *(38)*, each protein sequence in H is aligned separately with the original query sequence i. If the alignable region of the two sequences exceeds a threshold fraction of the alignment's total length (0.8 was the working cutoff in the original paper *[36]*), the program PAML *(39)* (specifically, the package codeml) is used to obtain a maximum likelihood estimate of the number of amino acid substitutions separating the two protein sequences, given an empirical amino acid substitution rate matrix *(40)*. The model under which a maximum likelihood estimate is obtained in RSD may include variation in evolutionary rate among protein sites, by assuming a γ-distribution of rate across sites and by setting the shape parameter of this distribution, α, to a level appropriate for the phylogenetic distance of the species being compared *(41)*.

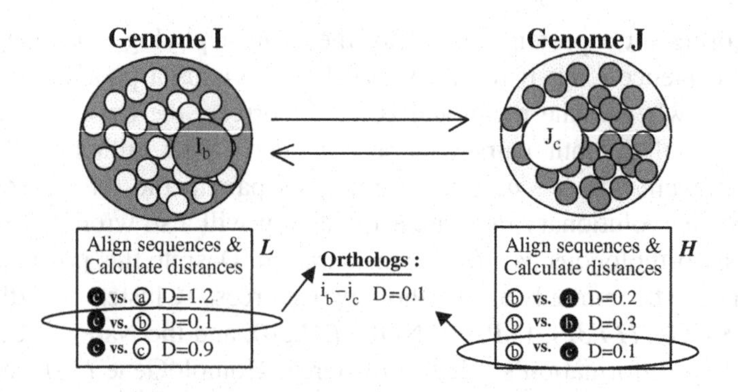

Fig. 2. The reciprocal smallest distance algorithm (RSD). Arrows denote bidirectional Basic Local Alignment Search Tool runs. After each run, hits are paired with the query to calculate evolutionary distances. If the same pair produces the smallest distance in both search directions, it is assumed to be orthologous. The specifics of the algorithm are provided in **Subheading 1.**

(This parameter, α, may be altered to accommodate different degrees of phylogenetic distance.) Of all sequences in *H* for which an evolutionary distance is estimated, only *j*, the sequence yielding the shortest distance, is retained. This sequence *j* is then used for a reciprocal BLAST against genome *I*, retrieving a set of high scoring hits, *L*. If any hit from *L* is the original query sequence, *i*, the distance between *i* and *j* is retrieved from the set of smallest distances calculated previously. The remaining hits from *L* are then separately aligned with *j* and maximum likelihood distance estimates are calculated for these pairs as described above. If the protein sequence from *L* producing the shortest distance to *j* is the original query sequence, *i*, it is assumed that a true orthologous pair has been found and their evolutionary distance is retained (**Fig. 2**).

RSD has been shown to find a substantially larger list of orthologs than reciprocal best BLAST *(36)*. Thus, relative to BLAST alone, BLAST followed by evolutionary distance estimation based on pairwise global alignment is more likely to identify the same pair of sequences in both directions of a reciprocal query. This is presumably because BLAST can often return a paralog in at least one direction of a reciprocal query. When this occurs, the orthologous sequence is likely to be among the sequences that are high-scoring, but not best, BLAST hits, an issue expected to be even more common in genomes that have high rates of gene duplication or gene conversion. In such cases, the global alignment and evolutionary distance estimation in RSD can recover the putative ortholog, revealing that it is in fact the nearest evolutionary neighbor of the query, even if not the best BLAST hit *(36)*.

The original RSD algorithm has already been used in a variety of contexts to increase our understanding of both the nature of genome organization and to identify phylogenetic trends among different species *(42–46)*. To enable its efficient usage, we have packaged the RSD algorithm into a user-friendly software suite, the operation of which is described in the next sections. RSD is available as a local standalone package, written primarily in Python and suitable for any Unix-based operating system. This program provides a tremendous amount of flexibility to navigate through parameter space, allowing alteration to the number of hits that are returned from the forward and reverse BLAST steps of the algorithm, changes to the BLAST *e*-value threshold for the hits returned, the divergence parameter, and the shape parameter of the γ-distribution, α. Together these parameters can be used to fine-tune the accuracy of the distance calculations and ortholog assignments, an aspect that differentiates it from other ortholog detection applications. For these and other reasons, we expect RSD to make a solid addition to the arsenal of tools available for ortholog identification.

2. Materials

This section outlines the software download and installation steps required to run the RSD package on the local computer, as well as the software dependencies of RSD. RSD may be used only on Unix-based operating systems (e.g., Linux, Mac OS X), thus its installation and execution assumes a rudimentary understanding of Unix on the part of the user.

1. Download whole genome sequences in amino acid format from reliable public repositories. Specifically, download the complete set of peptides from at least two species in fastA-formatted files, one file per species (*see* **Notes 1** and **2**).
2. Download and install the Washington University BLAST 2.0 executables (*see* **Note 3**). The programs blastp (the protein-based BLAST tool), xdget, and xdformat are required to run RSD.
3. Download and install the multiple sequence alignment program clustalW v1.83 for Unix *(47)* (*see* **Notes 4** and **5**). The clustalW program is required to run RSD.
4. Download and install the phylogenetic analysis package PAML v3.14 *(39)* (*see* **Note 6**). The codeml program is required to run RSD.
5. Download and install Python 2.4, including the bsddb module (*see* **Note 7**).
6. Download RSD from http://cbi.med.harvard.edu/RSD.tar.gz (*see* **Note 8**).
7. Unpack the RSD archive to the directory where it should be installed (*see* **Note 8**).
8. After unpacking RSD, there will be a folder called "RSD_standalone."
9. Change directory to RSD_standalone and list its contents. The following files and two directories should be present (the README file contains similar instructions).

BioUtilities.py	ReadFasta.py	Examples
Blast_compute.py	Utility.py	execute.py
README	clustal2phylip	***formatter.py***
RSD.py	codeml.ctl_cp	jones.dat
RSD_common.py		

10. Three files previously listed—RSD.py, Blast_compute.py, and formatter.py shown in ***bold*** and ***italics***—are executed by the user. The remaining files are subsidiary components of the programs that should not be altered. The directory called examples contains example data described later in this tutorial.
11. Blast_compute.py precomputes all-against-all reciprocal BLASTs building a forward BLAST hits file (FBH) and a reverse BLAST hits file (RevBH) (*see* **Note 9**). At the Unix prompt type "python Blast_compute.py- -help" to see a list of options (*see also* **Table 1** for a description of these options).
12. RSD.py is the central program of the RSD package. At the Unix prompt type "python RSD.py- -help" to see a list of those options (*see* **Table 1** for descriptions).
13. Formatter.py can be used to format a fastA-formatted proteome for use with RSD.py and Blast_compute.py (*see* **Note 10**).

Table 1
The Two User-Operated RSD Programs and Their Associated Options

Program	Options	Description
Blast_compute.py	-h, --help	Shows the parameter options for the program
	-q QUERYDB, --querydb=QUERYDB	Path to the query database (fastA format)
	-s SUBJECTDB, --subjectdb=SUBJECTDB	Path to the subject db (fastA format)
	-o OUTFILE,--outfile=OUTFILE	Path to the BLAST db output file (any name)
RSD.py	-h, --help	Shows the parameter options for the program
	--thresh=THRESH	Required. *e*-value threshold for BLAST subcomponent (e.g., 0.01, 1e-10)
	--div=DIV	Required. Sequence divergence for alignment subcomponent. (e.g., 0.2, 0.5, 0.8)
	-u USERSEQS, --userseqs=USERSEQS	Optional. Path to file containing FastA formatted sequences from the query database. Useful if the user is interested in only a small subset of genes as opposed to the entire list of orthologs between two genomes.
	-q QUERYDB, --querydb=QUERYDB	Required. Path to the query database (fastaA format) and, if forward BLAST hits option is not given, wublast formatted index files with same base name
	-s SUBJECTDB, --subjectdb=SUBJECTDB	Required. Path to the subject database (fastA format) and, if reverse BLAST hits option not given, wublast formatted index files with same base name
	-o OUTFILE, --outfile=OUTFILE	Required. Path to RSD output file (will be overwritten)

(Continued)

Table 1
(Continued)

Program	Options	Description
	--fbh=FBH	Optional. Path to forward BLAST hits file, a file (generated by Blast_compute) that caches the results from the BLASTing every sequence in the query db against the subject db and vice versa. Used to avoid repeated BLASTing when RSD run multiple times with the same query db and subject but different threshold and divergence parameters.
	--revbh=RevBH	Optional. Path to reverse BLAST hits file, otherwise instructions the same as previously listed.
	-n NHITS, --nhits=NHITS	Optional. Default = 3. The number of top BLAST hits to be investigated as possible orthologs.
	--fa=FIXALPHA	Optional. Default = 1 (α fixed at specified or default value). If set to 0, α will be estimated by the program codeml.
	-a ALPHA, --alpha=ALPHA	Optional. Default=1.53. α = value for shape parameter of the gamma distribution.

3. Methods

This section details the remaining formatting steps and Unix commands required to run the RSD package. We have streamlined RSD to require as few steps as possible for proper execution, distilling it down to two straightforward pieces—properly formatting the genomic databases and manipulating RSD's command-line options.

3.1. Preparing to Run RSD From a Local Computer

1. Confirm that python, blastp, xdget, xdformat, clustalW, and codeml are functional and are in the path (*see* **Note 11**).

2. Preformat the genomes by removing everything except for alphanumeric characters from the name lines of each sequence. Characters such as "|,?,$,%.#,^,(,)" can cause the program to quit unexpectedly. Name lines must also be truncated to no more than 22 characters. See the examples/Blast/blast_dbs subdirectory for examples of properly formatted databases (*see also* **Note 10**).
3. Format the genomes by using the xdformat program from the Washington University BLAST 2.0 executables. The command is "xdformat –p –I name_of_your_fastA_formatted_database." This will create the indices used by the program blastp (*see also* **Note 10**).

3.2. Running RSD

RSD has four chief modes of operation: (1) running BLAST within the algorithm (hereafter referred to as "on the fly" computation mode) for an all-against-all search between two genomes. (2) Running an all-against-all query using precomputed BLAST databases (thereby removing BLAST steps from the algorithm and speeding up the computation of reciprocal smallest distance significantly). (3) On the fly using user input sequences (*see* **Note 13**) to detect orthologs. (4) Using user input sequences together with precomputed BLAST databases. Precomputed BLAST databases are generated via Blast_compute.py.

1. The RSD package comes with example files to test the installation and to give an idea of how to set program options. The following steps use these example files to demonstrate the setup of common argument strings for running RSD.
2. Test computing BLAST hits between two genomes by typing: "python Blast_compute.py -q examples/Blast/blast_dbs/AGRO1_format.faa -s examples/Blast/blast_dbs/SINO1_format.faa -o example_AGRO1_vs_SINO1_blast_hits" (*see* **Note 14**).
3. Test operation mode 1 by typing: "python RSD.py --thresh=1e-10--div=0.5-q examples/Blast/blast_dbs/SINO1_format.faa -s examples/Blast/blast_dbs/AGRO1_format.faa -o examples/Orthologs/example_output_complete" (*see* **Note 14**).
4. Test operation mode 2 by typing: "python RSD.py --thresh=1e-10--div=0.5-q examples/Blast/blast_dbs/SINO1_format.faa-s examples/Blast/blast_dbs/AGRO1_format. faa-o examples/Orthologs/example_output_complete-fbh=examples/Blast/blast_results/example_blast_resultsF--revbh=examples/Blast/blast_results/example_blast_resultsR" (*see* **Note 14**).
5. Test operation mode 3 by typing: "python RSD.py --thresh=1e-10--div=0.5 -u examples/userseqs_example -q examples/Blast/blast_dbs/SINO1_format.faa -s examples/Blast/blast_dbs/AGRO1_format.faa -o examples/Orthologs/example_output_userseqs"(*see* **Note 14**).
6. Test operation mode 4 by typing: "python RSD.py --thresh=1e-10--div = 0.5-u examples/userseqs_example -q examples/Blast/blast_dbs/SINO1_format.faa -s examples/Blast/blast_dbs/AGRO1_format.faa -o examples/Orthologs/

example_output- -fbh=examples/Blast/blast_results/example_blast_resultsF- -revbh =examples/Blast/blast_results/example_blast_resultsR"(*see* **Note 14**).

7. In all of the test runs previously listed, the adjustable parameters, threshold (thresh), and divergence (div) are left unchanged. Note also that the other adjustable parameters nhits, fixalpha, and alpha are not included in the command-line arguments at all, and are thus left at their default values (**Table 1**). Nevertheless, these parameters can have a large effect on the size and content of the list of orthologs identified between two genomes and on the estimated evolutionary distance between proteins. Try altering these parameters variously to see how the lists of orthologs change (*see* **Notes 15** and **16**).

4. Notes

1. A sequence in fastA format begins with a single-line description, followed by lines of sequence data. The description line is distinguished from the sequence data by a greater-than (">") symbol in the first column. An example of an amino acid sequence in fastA format is: >gi|4502175|ref|NP_001640.1| apical protein of Xenopus-like; APX homolog of Xenopus; apical protein, *Xenopus laevis*-like [*Homo sapiens*]
MEGAEPRARPERLAEAETRAADGGRLVEVQLSGGAPWGFTLKGGREHGE
PLVITKIEEGSKAAAVDKLLAGDEIVGINDIGLSGFRQEAICLVKGSHKTLK
LVVKRRSELGWRPHSWHATKFSDSHPELAASPFTSTSGCPSWSGRHHASS
SSHDLSSSWEQTNLQRTLDHFSSLGSVDSLDHPSSRLSVAKSNSSIDHLGSH
SKRDSAYGSFSTSSSTPDHTLSKADTSSAENILYTVGLWEAPRQGGRQAQ
AAGDPQGSEEKLSCFPPRVPGDSGKGPRPEYNAEPKLAAPGRSNFGPVWY
VPDKKKAPSSPPPPPPPLRSDSFAATKSHEKAQGPVFSEAAAAQHFTALAQ
AQPRGDRRPELTDRPWRSAHPGSLGKGSGGPGCPQEAHADGSWPPSKDG
ASSRLQASLSSSDVRFPQSPHSGRHPPLYSDHSPLCADSLGQEPGAASFQN
DSPPQVRGLSSCDQKLGSGWQG
Note that blank lines are not allowed in the middle of fastA input.
2. We use the word genome in this chapter to refer to an organism's complete set of protein coding genes encoded into amino acids. Complete genomes are available from many online resources including:
http://www.ensembl.org/info/data/download.html ***(48)***
ftp://ftp.ncbi.nih.gov/genomes/
http://www.genomesonline.org/ ***(49)***
3. The Washington University BLAST 2.0 executables require that a license agreement be established at the following website http://blast.wustl.edu/licensing before downloading.
4. clustalW v1.83 is available for download at ftp://ftp.ebi.ac.uk/pub/software/unix/clustalw (the relevant file is "clustalw1.83.UNIX.tar.gz"). After downloading, gunzip will be needed to extract and compile the program. See the README included with the download for details (*see also* **ref. *38***).

5. In principle, other alignment programs could be used so long as they can produce output in clustalW format.
6. PAML v3.15 is available for download at http://abacus.gene.ucl.ac.uk/software/paml.html. After downloading the user will need to compile the source code. The binary file used by RSD is codeml. Once compiled this binary executable may be moved to a directory of choice (e.g., /usr/bin or /usr/local/bin).
7. Python v2.4 can be downloaded from http://www.python.org/download/, except for Mac OS X users. See http://www.python.org/download/download_mac.html for details on downloading and installing python, including a working bsddb module.
8. After downloading RSD.tar.gz, use tar and gunzip to decompress and extract the archive. Note that wget, gunzip, and tar are common Unix commands that might be needed when downloading genomes as well as the other software packages.
9. The Blast_compute.py will conduct all-against-all queries between the two genomes building a forward blats hits file and a reverse BLAST hits file. These results contain all hits irrespective of the *e*-value or score for each hit. Thus, the user is free to explore the effects of different *e*-value thresholds when using these precomputed BLAST databases in RSD processes. Using RSD in a precompute mode significantly speeds the search time, especially when testing the effects of multiple, different parameter combinations on the identification of orthologs between two genomes.
10. formatter.py is a simple python script that can format whole genomic databases downloaded from NCBI. To run this program simply type "python formatter.py- -infile name_of_database_to_format –outdir dir_to_place_formatted_db_and_indices." The script will eliminate any offending characters and create a new, properly formatted database for use with Washington University BLAST 2.0. If there are databases in formats other than NCBI's standard format that also have offending characters, the formatter.py script may need to be altered.
11. Test the functionality of the external programs blastp, xdget, xdformat, clustalW, and codeml by typing the names of the packages at the Unix command prompt. If the commands are recognized, the packages were installed correctly and are in the path.
12. The jones matrix is one of many possible matrices that can be used by the package codeml. PAML ships with several including dayhoff.dat, wag.dat, and mtmam.dat. The user may also want to create an amino acid substitution matrix ***(50,51)***.
13. A user may create a fastA formatted file that contains select sequences from a query genome for which they would like to find orthologs in a particular subject genome. For example, a user may be interested in orthologous relationships for a set of proteins that belong to a particular biological pathway. The computation time for this calculation is significantly less than an all-against-all search.
14. Please note that the text in this command is wrapped (i.e., no hard returns).
15. The orthologous pair found by RSD is frequently not the top BLAST hit; 16% of the time it is the second hit on the list and 7% of the time it is the third hit

(Wall and DeLuca, unpublished). This highlights the differences between BLAST-based approaches and the RSD method for discovery of orthologs and stresses the importance of considering hits within a BLAST report deeper than the first or even the third best hit.

16. α, the shape parameter of the γ-distribution can make a significant impact on the estimates of divergence between two sequences, and in some cases can alter the composition of the list of orthologous pairs. For example, consider the following lists of orthologs between two species of fungus, *Schizosaccharomyces pombe* (*S. pombe*) and *Saccharomyces cerevisiae* (*S. cerevisiae*):

S.pombe	S.cerev.	D (α=0.2)	S.pombe	S.cerevisiae	D (α=1.5)
19075528	6320000	2.668			
19075302	6320003	0.9366			
19112374	6320006	19.0			
19113007	6320008	19.0			
19075193	6320010	3.0685			
19114542	6320011	19.0			
19112732	6320013	0.6942			
19114337	6320016	0.9877			
19075528	6320000	0.7918			
19075302	6320003	0.4386			
19112374	6320006	2.6379			
19113007	6320008	2.1115			
19075193	6320010	0.8305			
19114542	6320011	2.1817			
19112732	6320013	0.3818			
19114337	6320016	0.4624			

Because these species last shared a common ancestor more that 330 million years ago and are quite distantly related, choosing a value of 1.5 for the shape parameter of the γ-distribution will yield more reliable results ***(41)*** than the value of 0.2, which produces spurious distance estimates (the distance value of 19 is an upper limit of the codeml package and should not be trusted).

Acknowledgments

Many thanks to past and present members of the Computational Biology Initiative who provided advice and expertise, I-Hsien Wu, Tom Monaghan,

Jian Pu, Saurav Singh, and Leon Peshkin. This material is based upon work supported by the National Science Foundation under Grant No. DBI 0543480.

References

1. Bernal, A., Ear, U., and Kyrpides N. (2001) Genomes OnLine Database (GOLD): a monitor of genome projects world-wide. *Nucleic Acids Res.* **29,** 126–127.
2. Turchin, A. and Kohane, I. S. (2002) Gene homology resources on the World Wide Web. *Physiol Genomics* **11,** 165–177.
3. Pellegrini, M., Marcotte, E. M., Thompson, M. J., Eisenberg, D., and Yeates, T. O. (1999) Assigning protein functions by comparative genome analysis: protein phylogenetic profiles. *Proc. Natl. Acad. Sci. USA* **96,** 4285–4288.
4. Marcotte, E. M., Pellegrini, M., Ng, H. L., Rice, D. W., Yeates, T. O., and Eisenberg, D. (1999) Detecting protein function and protein-protein interactions from genome sequences. *Science* **285,** 751–753.
5. Dacks, J. B. and Doolittle, W. F. (2001) Reconstructing/deconstructing the earliest eukaryotes: how comparative genomics can help. *Cell* **107,** 419–425.
6. Lang, B. F., Seif, E., Gray, M. W., O'Kelly, C. J., and Burger, G. (1999) A comparative genomics approach to the evolution of eukaryotes and their mitochondria. *J. Eukaryot. Microbiol.* **46,** 320–326.
7. Rubin, G. M., Yandell, M. D., Wortman, J. R., et al. (2000) Comparative genomics of the eukaryotes. *Science* **287,** 2204–2215.
8. Ureta-Vidal, A., Ettwiller, L., and Birney, E. (2003) Comparative genomics: genome-wide analysis in metazoan eukaryotes. *Nat. Rev. Genet.* **4,** 251–262.
9. Espadaler, J., Aragues, R., Eswar, N., et al. (2005) Detecting remotely related proteins by their interactions and sequence similarity. *Proc. Natl. Acad. Sci. USA* **102,** 7151–7156.
10. Espadaler, J., Romero-Isart, O., Jackson, R. M., and Oliva, B. (2005) Prediction of protein-protein interactions using distant conservation of sequence patterns and structure relationships. *Bioinformatics* **21,** 3360–3368.
11. Matthews, L. R., Vaglio, P., Reboul, J., et al. (2001) Identification of potential interaction networks using sequence-based searches for conserved protein-protein interactions or "interologs". *Genome Res.* **11,** 2120–2126.
12. Kim, W. K., Bolser, D. M., and Park, J. H. (2004) Large-scale co-evolution analysis of protein structural interlogues using the global protein structural interactome map (PSIMAP). *Bioinformatics* **20,** 1138–1150.
13. O'Brien, K. P., Westerlund, I., and Sonnhammer, E. L. (2004) OrthoDisease: a database of human disease orthologs. *Hum. Mutat.* **24,** 112–119.
14. Tatusov, R. L., Koonin, E. V., and Lipman, D. J. (1997) A genomic perspective on protein families. *Science* **278,** 631–637.
15. Fraser, H. B., Hirsh, A. E., Wall, D. P., and Eisen, M. B. (2004) Coevolution of gene expression among interacting proteins. *Proc. Natl. Acad. Sci. USA* **101,** 9033–9038.

16. Kulathinal, R. J., Bettencourt, B. R., and Hartl, D. L. (2004) Compensated deleterious mutations in insect genomes. *Science* **306,** 1553–1554.
17. Kondrashov, A. S., Sunyaev, S., and Kondrashov, F. A. (2002) Dobzhansky-Muller incompatibilities in protein evolution. *Proc. Natl. Acad. Sci. USA* **99,** 14,878–14,883.
18. Sunyaev, S., Kondrashov, F. A., Bork, P., and Ramensky, V. (2003) Impact of selection, mutation rate and genetic drift on human genetic variation. *Hum. Mol. Genet.* **12,** 3325–3330.
19. Sunyaev, S., Ramensky, V., Koch, I., Lathe, W., 3rd, Kondrashov, A. S., and Bork, P. (2001) Prediction of deleterious human alleles. *Hum. Mol. Genet.* **10,** 591–597.
20. Fraser, H. B., Wall, D. P., and Hirsh, A. E. (2003) A simple dependence between protein evolution rate and the number of protein-protein interactions. *BMC Evol. Biol.* **3,** 11.
21. Herbeck, J. T. and Wall, D. P. (2005) Converging on a general model of protein evolution. *Trends Biotechnol.* **23,** 485–487.
22. Wall, D. P., Hirsh, A. E., Fraser, H. B., et al. (2005) Functional genomic analysis of the rates of protein evolution. *Proc. Natl. Acad. Sci. USA* **102,** 5483–5488.
23. Hirsh, A. E. and Fraser, H. B. (2001) Protein dispensability and rate of evolution. *Nature* **411,** 1046–1049.
24. Hurst, L. D. and Smith, N. G. (1999) Do essential genes evolve slowly? *Curr. Biol.* **9,** 747–750.
25. Jordan, I. K., Rogozin, I. B., Wolf, Y. I., and Koonin, E. V. (2002) Essential genes are more evolutionarily conserved than are nonessential genes in bacteria. *Genome Res.* **12,** 962–968.
26. Krylov, D. M., Wolf, Y. I., Rogozin, I. B., and Koonin, E. V. (2003) Gene loss, protein sequence divergence, gene dispensability, expression level, and interactivity are correlated in eukaryotic evolution. *Genome Res.* **13,** 2229–2235.
27. Yang, J., Gu, Z., and Li, W. H. (2003) Rate of protein evolution versus fitness effect of gene deletion. *Mol. Bol. Evol.* **20,** 772–774.
28. Zhang, J. and He, X. (2005) Significant impact of protein dispensability on the instantaneous rate of protein evolution. *Mol. Biol. Evol.* **22,** 1147–1155.
29. Rocha, E. P. and Danchin, A. (2004) An analysis of determinants of amino acids substitution rates in bacterial proteins. *Mol. Biol. Evol.* **21,** 108–116.
30. Koonin, E. V., Aravind, L., and Kondrashov, A. S. (2000) The impact of comparative genomics on our understanding of evolution. *Cell* **101,** 573–576.
31. Koski, L. B. and Golding, G. B. (2001) The closest BLAST hit is often not the nearest neighbor. *J. Mol. Evol.* **52,** 540–542.
32. Tatusov, R. L., Galperin, M. Y., Natale, D. A., and Koonin, E. V. (2000) The COG database: a tool for genome-scale analysis of protein functions and evolution. *Nucleic Acids Res.* **28,** 33–36.

33. O'Brien, K. P., Remm, M., and Sonnhammer, E. L. (2005) Inparanoid: a comprehensive database of eukaryotic orthologs. *Nucleic Acids Res.* **33,** D476–D480.
34. Remm, M., Storm, C. E., and Sonnhammer, E. L. (2001) Automatic clustering of orthologs and in-paralogs from pairwise species comparisons. *J. Mol. Biol.* **314,** 1041–1052.
35. Wheeler, D. L., Barrett, T., Benson, D. A., et al. (2005) Database resources of the National Center for Biotechnology Information. *Nucleic Acids Res.* **33,** D39–D45.
36. Wall, D. P., Fraser, H. B., and Hirsh, A. E. (2003) Detecting putative orthologs. *Bioinformatics* **19,** 1710–1711.
37. Altschul, S. F., Gish, W., Miller, W., Myers, E. W., and Lipman, D. J. (1990) Basic local alignment search tool. *J. Mol. Biol.* **215,** 403–410.
38. Chenna, R., Sugawara, H., Koike, T., et al. (2003) Multiple sequence alignment with the Clustal series of programs. *Nucleic Acids Res.* **31,** 3497–3500.
39. Yang, Z. (1997) PAML: a program package for phylogenetic analysis by maximum likelihood. *Comput. Appl. Biosci.* **13,** 555–556.
40. Jones, D. T., Taylor, W. R., and Thornton, J. M. (1992) The rapid generation of mutation data matrices from protein sequences. *Comput. Appl. Biosci.* **8,** 275–282.
41. Nei, M., Xu, P., and Glazko G. (2001) Estimation of divergence times from multiprotein sequences for a few mammalian species and several distantly related organisms. *Proc. Natl. Acad. Sci. USA* **98,** 2497–2502.
42. Degnan, P. H., Lazarus, A. B., and Wernegreen, J. J. (2005) Genome sequence of Blochmannia pennsylvanicus indicates parallel evolutionary trends among bacterial mutualists of insects. *Genome Res.* **15,** 1023–1033.
43. Gasch, A. P., Moses, A. M., Chiang, D. Y., Fraser, H. B., Berardini, M., and Eisen, M. B. (2004) Conservation and evolution of cis-regulatory systems in ascomycete fungi. *PLoS Biol.* **2,** e398.
44. Nayak, S., Goree, J., and Schedl T. (2005) fog-2 and the evolution of self-fertile hermaphroditism in *Caenorhabditis. PLoS Biol.* **3,** e6.
45. Wu, H., Su, Z., Mao, F., Olman, V., and Xu, Y. (2005) Prediction of functional modules based on comparative genome analysis and Gene Ontology application. *Nucleic Acids Res.* **33,** 2822–2837.
46. Wuchty, S. (2004) Evolution and topology in the yeast protein interaction network. *Genome Res.* **14,** 1310–1314.
47. Thompson, J. D., Higgins, D. G., and Gibson, T. J. (1994) CLUSTAL W: improving the sensitivity of progressive multiple sequence alignment through sequence weighting, position-specific gap penalties and weight matrix choice. *Nucleic Acids Res.* **22,** 4673–4680.
48. Birney, E., Andrews, D., and Caccamo, M. (2006) Ensembl 2006. *Nucleic Acids Res.* **34,** D556–D5561.
49. Liolios, K., Tavernarakis, N., Hugenholtz, P., and Kyrpides, N. C. (2006) The Genomes On Line Database (GOLD) v.2: a monitor of genome projects worldwide. *Nucleic Acids Res.* **34,** D332–D334.

50. Bastien, O., Roy, S., and Marechal, E. (2005) Construction of non-symmetric substitution matrices derived from proteomes with biased amino acid distributions. *C R Biol.* **328,** 445–453.
51. Olsen, R. and Loomis, W. F. (2005) A collection of amino acid replacement matrices derived from clusters of orthologs. *J. Mol. Evol.* **61,** 659–665.
52. Fitch, W. M. (1970) Distinguishing homologous from analogous proteins. *Syst. Zool.* **19,** 99–113.

8

Finding Conserved Gene Order Across Multiple Genomes

Giulio Pavesi and Graziano Pesole

Summary

The ever increasing amount of annotated genomic sequences permits now to shed light on the molecular dynamics at the basis of evolution. Genome evolution involves both changes at the sequence level, but also rearrangements on the gene organization along the genome. Comparative analysis of gene order in multiple genomes may be thus of help in the investigation of general or lineage-specific genome plasticity, as well as in the inference of phylogenetic relationships. In particular, conserved gene contiguity in a chromosome, even if interrupted by intervening genes, may suggest potential functional couplings.

This unit explains the usage of GeneSyn, a software tool for the automatic identification of conserved gene order across multiple annotated genomes.

Key Words: Conserved gene order; genome annotation; synteny analysis; gene rearrangements; orthologous genes; paralogous genes.

1. Introduction

The increasing amount of fully annotated genomic sequences permits now to gain essential information on the molecular dynamics at the basis of evolution. Genome evolution involves variation at the sequence level (base substitutions or insertion/deletions), but also changes on a larger scale, like rearrangements on the gene organization along the genome. Comparative analysis of gene order in multiple genomes containing homologous genes may be thus of significant help in the investigation of general or lineage-specific genome plasticity, as well as in the inference of phylogenetic relationships ***(1,2)***. In particular, conserved gene contiguity within a chromosome (sometimes also referred to as conserved gene

From: *Methods in Molecular Biology, vol. 396: Comparative Genomics, Volume 2*
Edited by: N. H. Bergman © Humana Press Inc., Totowa, NJ

synteny), even if interrupted by intervening genes, suggest potential functional couplings of the genes involved *(3)*.

GeneSyn *(4)* is a software tool for the automatic identification of conserved gene order across multiple annotated genomes, that detects sets of genes appearing in the same order (measure that can be defined in different ways) in any subset of the genomes.

2. Materials

2.1. Hardware and Software

1. Any Pentium-class (or equivalent) computer with at least 256 Mb of RAM (for typical executions of the program), running a standard distribution of any of these operating systems: Unix (any version), Linux (any), MAC OS (any). The computer has to be connected to the Internet to download the software package.
2. A C compiler must be installed on the computer to compile the program. The program can be compiled also under Windows with any Unix-Linux emulator (like Cygwin) installed.

2.2. Input File Format

GeneSyn needs as input at least two sequences of gene identifiers (that from now on we will call improperly "genomes"). Each genome is defined in FASTA-like format: that is, it must start with a line with ">" followed by the sequence name. Then, on a new line, there must be the sequence of gene identifiers annotated in the genome. An example is the following:

```
>cin
cox2 cob pro nad4l his arg gln
>csa
cox2 cob nad1 pro nad4l his arg gln
```

Gene identifiers can contain letters, numbers or any other symbol, and can be separated by any number of white spaces or newline characters. It is essential that the same genes (i.e., orthologous genes) are denoted in the same way in their respective genomes. The program is case-insensitive, that is, it reads for example `Gln` and `gln` as if they were the same gene. In the previous example, all the genes are defined as appearing on the same strand (*see also* **Note 1g**). Double stranded output can be defined by putting a minus sign before the identifier of genes appearing on the complementary DNA strand, without spaces:

```
>cin
cox2 cob pro nad4l his arg gln
>csa
cox2 cob -nad1 pro nad4l his arg gln
```

Thus, in the example, gene `nad1` of the second genome appears on the reverse strand. The assessment of orthologous genes, to which the same name is assigned, can be made through retrieval of specialized databases, like COG *(5)*, or by using specific algorithms *(6)*. Genes do not have to be present in all the input genomes, and, vice versa, the same gene name can appear more than once in a single genome (e.g., for denoting paralogous genes). Notice however that some advanced parameters are available only on input without duplicated genes in a single genome (*see* **Subheading 3.3.**).

3. Methods

In the following, we first introduce (1) how to download and install the software on a local computer, then explain (2) how to run the program with standard parameter settings, (3) how to set additional parameters, and (4) how to read the output of the program.

3.1. Downloading and Installing GeneSyn

The GeneSyn package is available free of charge for noncommercial use. Because the software is written in the C language, a C compiler (e.g., gcc) must be installed on the computer to compile the program. No other special library is needed. The program can be compiled also under Windows with any Unix-Linux emulator (like Cygwin) installed.

1. Connect to http://www.pesolelab.it/, and click on the "Tools" link appearing in the page. Another page with a list of several tools will appear, including GeneSyn.
2. Follow the corresponding link, and you will reach a page that permits to download the program. Click on the appropriate link, and select the "Save" option.
3. The file downloaded will be named `GenesynX.X.tar.gz`, where X.X are the numbers denoting the version of the software available. Current version is 1.1, thus, for example, the package is called `Genesyn1.1.tar.gz`. In the following, replace 1.1 with the digits of the version number you downloaded. The commands described in the following have to be entered from a command-line prompt. Note that most of the Linux/Unix systems are case-sensitive, and thus commands entered should respect the upper/lowercase characters appearing in file names.
4. To unpack the file, enter gunzip Genesyn1.1.tar.gz. File Genesyn1.1.tar.gz now has become Genesyn1.1.tar. Enter tar xvf Genesyn1.1.tar. At this point, a directory named Genesyn1.1 should have appeared.
5. Enter `cd Genesyn1.1` to enter the newly created directory. It contains a file named "`compile`," and a directory named "`src`" (containing the source code).

6. Edit file `compile`, and put after `CC=` the name of the C compiler available on your computer (the compiler is set to `gcc`, if it is available on your computer you do not have to edit the file). To check whether `gcc` is available on your computer, enter `gcc -v`. If a message showing the version number of `gcc` appears, then it is available. Otherwise, consult your system andministrator to find out whether any C compiler is installed on your system.
7. Enter`./compile`.
8. If everything went well, two executables named `genesyn` and `grunner` will have appeared. Make sure that, wherever you place the executables, `genesyn` and `grunner` are located in the same directory.
9. To test if the program works, enter `./test`. It will launch the program on a sample input file contained in the GeneSyn package (example.tfa), so you can also get familiar with the input and output format.

3.2. Running GeneSyn With Default Parameters

The following commands have to be entered from a command-line prompt.

1. To run the program with default settings, enter

   ```
   ./genesyn input-file <percent>
   ```

   where

   a. `input-file` is the name of the input file,
   b. `<percent>` must be an integer number between 1 and 100 denoting the minimum percentage of input genomes in which genes in the same order have to appear. In other words, the program will output subsequences of genes common to the given percentage of the input genomes. A subsequence of a sequence is a sequence that can be obtained by deleting any number of items (in this case, genes) from the original sequence. Therefore, the subsequences will consist of genes appearing in the same order in the genomes where the subsequences appear.

2. The two previously described parameters are required, exactly in that order. With default parameters, the program will:

   a. Assume that the input genomes are linear.
   b. Look for genes (subsequences) appearing in the same order on the same strand.
   c. Look for genes (subsequences) appearing in the same order and separated by any number of interleaving genes.
   d. Output solutions at least three genes long.

3.3. Setting Additional Parameters

After the basic command-line elements:

```
./genesyn input-file percent
```

a number of additional parameters can be specified. More than one parameter of the ones described in the following can be added after the required ones, in any order.

3.3.1. Running the Program on Circular Genomes

In some cases, it is useful to assume that input genomes are circular, e.g., in the case of bacterial or mitochondrial genomes. To set this option, add `-c` to the command-line:

```
./genesyn input-file percent -c
```

Thus, given input file

```
>Sequence 1
gene1 -gene4 gene3 -gene2 gene5
>Sequence 2
-gene4 gene3 -gene2 gene5 gene1
```

with percent = 100, the program will report `gene1 -gene4 gene3 -gene2 gene5` as a common subsequence. Without the `-c` option, it reports only `-gene4 gene3 -gene2 gene5`. Notice that this option can be set only if the genomes do not contain duplicated genes.

3.3.2. Running the Program on Both Strands of the Input

It might happen that the same genes appear in the same order in two or more genomes, but on different strands (as explained before the reverse strand can be specified by putting a minus sign ahead of gene names). With default parameters the program assumes that common subsequences must appear on the same strand. To set this option, add `-r` to the command-line:

```
./genesyn input-file percent -r
```

In this way, for example, given as input file

```
>Sequence 1 gene1 gene4 gene3 gene2 gene5
>Sequence 2
-gene5 gene2 -gene3 gene4 -gene1
```

with percent = 100, the program will report `gene1 gene3 gene5` as appearing in both genomes. In Sequence 2, it appears on the reverse strand.

3.3.3. Finding Strand-Independent Subsequences

Even if input genes are annotated on both strands of the genomes, the program can be run to find conserved gene order regardless of the strand on which single genes are annotated. To set this option, add `-n` to the command-line:

```
./genesyn input-file percent -n
```

In this way, for example, given as input

```
>Sequence 1
gene1 gene4 gene3 gene2 gene5
>Sequence 2
-gene1 gene2 -gene5 gene4 -gene3
```

with percent = 100 the program will report subsequence `gene1 gene4 gene3` as appearing in both genomes.

3.3.4. Setting a Maximum Gap Value for Subsequences

With default parameters, the program reports subsequences common to a number of the input genomes, as long as the genes appear in the same order, regardless of the number of interleaving genes separating them in their genomes. A maximum gap value (expressed as a number of genes) can be set on the command-line:

```
./genesyn input-file percent -g <value>
```

where `<value>` must be a non-negative integer number. For example, by setting "`-g 0`" the program will report only subsequences where the genes are adjacent in their respective genomes (maximum gap is zero). Note that this option can be set only if the genomes do not contain duplicated genes.

3.3.5. Setting a Maximum Gap Value in Basepairs

A maximum value for gaps separating genes can be defined according to the distance of the genes in the genomes, defined as the number of basepairs separating them. In this case, however, also the input file has to be formatted in a different way. An example is the following:

>Sequence 1
gene1 s_{11} gene4 s_{12} gene3 s_{13} gene2
>Sequence 2
-gene5 s_{21} gene2 s_{22} -gene3 s_{23} gene4

that is, gene names must be separated by an integer number (s_{11}, s_{12}, and so on in the example) denoting the number of basepairs between them (e.g., it can be the distance separating their respective TSSs). Clearly, the distance between nonadjacent genes will be the sum of the distances associated with the genes separating them. In the example, the distance between gene1 and gene3 in Sequence 1 will be given by $s_{11} + s_{12}$.

3.4. Reading the Output

The output of the program appears on the screen and it is simultaneously written into a file whose name is displayed on the screen at the end of the execution. The output file name has the form `gs<number>.res`, where `<number>` is an integer that changes from execution to execution.

The output file starts first of all with the command-line that had been used to produce the output, followed by a brief description of the parameter settings that were used, and finally the list of the subsequences that satisfied the input constraints. An example is presented in **Fig. 1**, showing the output of the program obtained on the example.tfa file provided with the package and with the parameters set in the test script.

As explained in the preamble, the program looked for subsequences appearing in 17 genomes out of 34 (the percentage was set to 50%), at least 9 genes long (-m 9 in the command-line). The `-r` option was also set,

```
./genesyn example.tfa 50 -r -m 9
Looking for subsequences appearing in at least 17 genomes out of 34
Subsequences have to be at least 9 elements long
Processing both strands of input genomes

COX1 COX2 K ATP8 ATP6 COX3 NAD3 COB M
17 occurrences
           123456789 0123456789 0123456789 01234
Forward  : 110111111 1010111111 1000000000 00000|
Backward : 000000000 0000000000 0000000000 00000

V RRNL L1 NAD1 NAD4L F -N -NAD3 -COX3
17 occurrences
           123456789 0123456789 0123456789 01234
Forward  : 000000000 0000000000 0001111100 00000|
Backward : 000011111 1010100111 1000000000 00000
```

Fig. 1. A sample output of GeneSyn.

so the program looked for subsequences appearing on either strand of the input genomes. After the heading, the output lists the subsequences satisfying the input constraints, followed by the number of genomes containing them (17 occurrences), and a schematic representation of where the subsequence was found (one if it was found, zero otherwise) and on which strand. In the example, the first subsequence was found in input genomes 1, 2, from 4 to 10, and from 14 to 20 on their "forward" strand. The second subsequence, instead, was found in genomes from 4 to 10, 12, 14, and from 17 to 20 on the reverse strand; and in genomes from 23 to 27 on the forward strand. Genomes are numbered according to the order they appear in the input file, starting from 1.

4. Notes

1. The algorithm GeneSyn is based on finds subsequences of the input genomes by building an indexing structure called directed acyclic word graph (DASG). More in detail, a DASG for a set of sequences is a directed acyclic graph such that:
   a. On vertex (called source) has only outgoing edges.
   b. Each edge is labeled with an element of the input sequences (in this case, the name of a gene).
   c. On each path from the source node, the concatenation of the labels of the edges encountered corresponds to a subsequence of at least one of the input sequences.
   d. Because the graph is acyclic, for any vertex V of the graph, no path starting at V can return back to V itself.

      A q-DASG can be defined in a similar way, with the only difference that subsequences corresponding to paths in the graph must appear in at least q of the input sequences. Thus, as a first step, the algorithm builds a q-DASG according to the percentage parameter provided as input. Further details on how the graph is constructed can be found in **ref. 7**. Once the graph is built, finding subsequences common to at least q of the input sequences is straightforward, because it involves performing a traversal of the graph. Although, in general, finding the longest common subsequence of a set of sequences is a NP-Hard problem (that is, no solution can be obtained in reasonable time), the problem becomes tractable if an element appears at most once in a sequence. Anyway, the program accepts as input genomes in which a gene identifier can appear any number of times, but in this case only default parameters can be used.
   e. In case an error occurs during the execution of the program, usually a self-explanatory message appears. The most frequent cause of error is because of the misplacement of the executable files genesyn and grunner. If the program crashes for any other reason, users are welcome to contact the authors via email: giulio.pavesi@unimi.it. If the program reports that no subsequences were found, and instead a result was expected, the most common reason is that different names were provided for the same gene in different genomes (*see also* **Note 1g**).

f. The input file must be a plain text file. A common source of trouble, often ignored, is the fact that text files have a different format in different operating systems (e.g., a text MS-DOS/Windows file is different from a Unix/Linux one). Although most of Unix/Linux systems usually recognize MS-DOS files and process them correctly, it is often a good idea to process a MS-DOS file transferred to a Unix/Linux computer with the "dos2unix" utility. Also, do not run the program on a MS-Word document (that contains all the Word formatting characters, and it is not a text file) or do not prepare the text input file by cutting and pasting the text from a Word document: most often than not, paste invisible Word formatting characters that will cause the input to be misread by the program. Rather, save Word file as a text file.

g. The duration of a run of the program strongly depends on the input parameters that were provided, especially the percentage and gap values. In any case, the time seldom exceeds a few minutes on tens of genomes with a few hundred genes with no gap threshold and percentage set to around half of the sequences. If input genomes contain duplicated genes, instead, the time and memory usage might grow exponentially (on the number of duplicated genes).

References

1. Boore, J. L. and Brown, W. M. (1998) Big trees from little genomes: mitochondrial gene order as a phylogenetic tool. *Curr. Opin. Genet. Dev.* **8,** 668–674.
2. Blanchette, M., Kunisawa, T., and Sankoff, D. (1999) Gene order breakpoint evidence in animal mitochondrial phylogeny. *J. Mol. Evol.* **49,** 193–203.
3. Overbeek, R., Fonstein, M., D'Souza, M., Pusch, G. D., and Maltsev, N. (1999) The use of gene clusters to infer functional coupling. *Proc. Natl. Acad. Sci. USA* **96,** 2896–2901.
4. Pavesi, G., Mauri, G., Iannelli, F., Gissi, C., and Pesole, G. (2004) GeneSyn: a tool for detecting conserved gene order across genomes. *Bioinformatics* **20,** 1472–1474.
5. Tatusov, R. L., Fedorova, N. D., Jackson, J. D., et al. (2003) The COG database: an updated version includes eukaryotes. *BMC Bioinformatics* **4,** 41.
6. Li, L., Stoeckert, C. J., Jr., and Roos, D. S. (2003) OrthoMCL: identification of ortholog groups for eukaryotic genomes. *Genome Res.* **13,** 2178–2189.
7. Crochemore, M. and Tronicek, Z. (2002) On the size of DASG for multiple texts. *Lecture Notes in Computer Science* **2476,** 58–64.

9

Analysis of Genome Rearrangement by Block-Interchanges

Chin Lung Lu, Ying Chih Lin, Yen Lin Huang, and Chuan Yi Tang

Summary

Block-interchanges are a new kind of genome rearrangements that affect the gene order in a chromosome by swapping two nonintersecting blocks of genes of any length. More recently, the study of such rearrangements is becoming increasingly important because of its applications in molecular evolution. Usually, this kind of study requires to solve a combinatorial problem, called the block-interchange distance problem, which is to find a minimum number of block-interchanges between two given gene orders of linear/circular chromosomes to transform one gene order into another. In this chapter, we shall introduce the basics of block-interchange rearrangements and permutation groups in algebra that are useful in analyses of genome rearrangements. In addition, we shall present a simple algorithm on the basis of permutation groups to efficiently solve the block-interchange distance problem, as well as ROBIN, a web server for the online analyses of block-interchange rearrangements.

Key Words: Bioinformatics; genome rearrangement; transpositions; block-interchanges; permutations.

1. Introduction

With an increasing number of genomic data (DNA, RNA, and protein sequences) being available, the study of genome rearrangement has received a lot of attention in computational biology and bioinformatics because of its applications in the measurement of evolutionary difference between two species. One of the most promising ways to do this research currently is to compare the orders of the identical genes in two different genomes. Different from the

From: *Methods in Molecular Biology, vol. 396: Comparative Genomics, Volume 2*
Edited by: N. H. Bergman © Humana Press Inc., Totowa, NJ

traditional point mutations (e.g., insertions, deletions, and substitutions) acting on individual base, various large-scale mutations acting on genes within or among chromosomes have been proposed for measuring the rearrangement distance between two related genomes, such as the so-called intrachromosomal rearrangements (e.g., reversals/inversions ***(1–4)***, transpositions ***(5)***, and block-interchanges ***(6–8)***) and the so-called interchromosomal rearrangements (e.g., translocations ***(9,10)***, fusions, and fissions ***(11)***).

Block-interchanges, introduced by Christie, are a new kind of rearrangements that affect a chromosome by swapping two nonintersecting blocks of genes of any length ***(6)***. Examples of block-interchange rearrangements are shown in **Fig. 1**. More recently, the study of block-interchange rearrangements is becoming increasingly important because of its applications in molecular evolution ***(7,8)***. From the biological perspective, block-interchanges can be considered as a generalization of transpositions (one of the most frequent rearrangements) because the blocks exchanged by transpositions must be contiguous (or adjacent) on the chromosome, whereas those by block-interchanges need not be (*see* **Fig. 1** for the difference).

The study of block-interchange rearrangements usually requires to solve a combinatorial problem, called the block-interchange distance problem, which is to find a minimum number of block-interchanges between two given gene orders of linear/circular chromosomes such that these block-interchanges can transform one gene order into another. Christie first proposed an $O(n^2)$ time algorithm to solve this problem on linear chromosomes using breakpoint diagram approach ***(6)***, where n is the number of genes on chromosomes. Lin et al. ***(7)*** later made use of permutation groups in algebra to design a simpler algorithm of $O(\delta n)$ time for solving the block-interchange distance problem

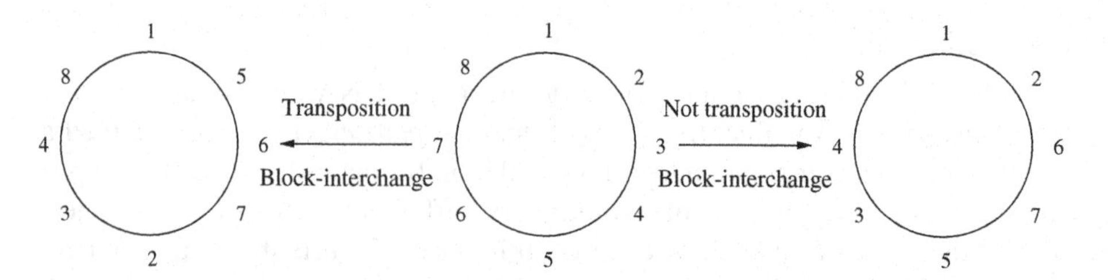

Fig. 1. Rearrangements acting on a circular chromosome. The left rearrangement, swapping two adjacent blocks [2, 3, 4] and [5, 6, 7], is a transposition, but the right one is not because the exchanged [3, 4] and [6, 7] are not adjacent. However, both of them are block-interchanges.

on linear chromosomes as well as on circular chromosomes *(7)*, where δ is the block-interchange distance that can be calculated in $O(n)$ time in advance. They also demonstrated that block-interchange rearrangements seem to play a significant role in the genetic evolution of bacterial species (e.g., vibrio pathogens). Recently, Lu et al. *(8)* have further implemented the algorithms designed by Lin et al. *(7)* into a web server, called ROBIN (short for rearrangement of block-interchanges), for allowing the public to conduct the online analyses of block-interchange rearrangements between two linear/circular chromosomes *(8)*. Particularly, not only gene orders but also bacterial-size sequences are allowed to be the input of ROBIN. If the input is sequence data, ROBIN can automatically search for the identical landmarks, which are the homologous/conserved regions shared by all the input sequences.

In this chapter, we shall first introduce the basics of permutations and their relationships with genome rearrangements, then describe a simple algorithm on the basis of permutations for calculating the block-interchange distance between two given gene orders of linear/circular chromosomes, and finally introduce ROBIN and describe its detailed steps to conduct the analyses of block-interchange rearrangements for chromosomal sequences or gene orders.

2. Permutations vs Genome Rearrangements

In group theory, a *permutation* is defined to be a one-to-one mapping from a set $E = \{1, 2, \ldots, n\}$ into itself, where n is a positive integer. For example, we may define a permutation α of the set {1, 2, 3, 4, 5, 6, 7} by specifying $\alpha(1) = 4, \alpha(2) = 3, \alpha(3) = 1, \alpha(4) = 2, \alpha(5) = 7, \alpha(7) = 6$, and $\alpha(6) = 5$. The above mapping can be expressed using a *cycle notation* as illustrated in **Fig. 2** and simply denoted by $\alpha = (1, 4, 2, 3)(5, 7, 6)$. A cycle of length k, say $(a_1, a_2, \ldots, a_k)$, is simply called *k-cycle* and can be rewritten as $(a_i, a_{i+1}, \ldots, a_k, a_1, \ldots, a_{i-1})$, where $2 \leq i < k$, or $(a_k, a_1, a_2, \ldots, a_{k-1})$. Any two cycles are said to be *disjoint* if they have no element in common. In fact, any permutation, say α, can be written in a unique way as the product of disjoint cycles, which is called the cycle decomposition of α, if we ignore the order of the cycles in the product *(12)*. Usually, a cycle of length one in α is not explicitly written and its element, say x, is said to be *fixed* by α because $\alpha(x) = x$. Especially, the permutation whose elements are all fixed is called an identity permutation and is denoted by **1** (i.e., $\mathbf{1} = (1)\ (2) \ldots (n)$).

Given two permutations α and β of E, the *composition* (or *product*) of α and β, denoted by $\alpha\beta$, is defined to be a permutation of E with $\alpha\beta(x) = \alpha(\beta[x])$ for all $x \cdot E$. For instance, if we let $E = \{1, 2, 3, 4, 5, 6, 7\}, \alpha = (1, 5)$ and $\beta = (1, 4, 2, 3, 5, 7, 6)$, then $\alpha\beta = (1, 4, 2, 3)(5, 7, 6)$. If α and β are disjoint cycles,

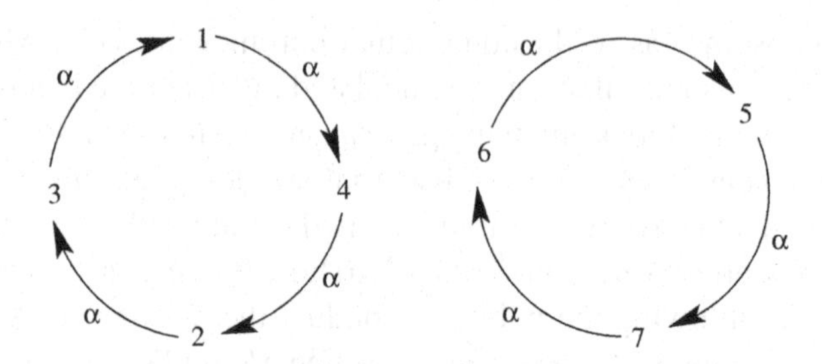

Fig. 2. The illustration of a permutation $\alpha = (1, 4, 2, 3)(5, 7, 6)$ meaning that $\alpha(1) = 4$, $\alpha(2) = 3$, $\alpha(3) = 1$, $\alpha(4) = 2$, $\alpha(5) = 7$, $\alpha(7) = 6$, and $\alpha(6) = 5$.

then $\alpha\beta = \beta\alpha$. The *inverse* of α is defined to be a permutation, denoted by α^{-1}, such that $\alpha\alpha^{-1} = \alpha^{-1}\alpha = \mathbf{1}$. If a permutation is expressed by the product of disjoint cycles, then its inverse can be obtained by just reversing the order of the elements in each cycle. For example, if $\alpha = (1, 4, 2, 3)(5, 7, 6)$, then $\alpha^{-1} = (3, 2, 4, 1)(6, 7, 5)$. Clearly, $\alpha^{-1} = \alpha$ if α is a two-cycle.

Meidanis and Dias first noted that each cycle of a permutation may represent a circular chromosome of a genome with each element of the cycle corresponding to a gene, and the element order of the cycle corresponding to the gene order of the chromosome ***(13)***. **Figure 2**, for example, denotes a genome with two circular chromosomes, one represented by (1, 4, 2, 3) and the other by (5, 7, 6). Moreover, they observed that rearrangement events, such as fusions and fissions (respectively, transpositions), correspond to the composition of a two-cycle (respectively, three-cycle) and the permutation representing a circular genome. For instance, let α be any permutation with cycle decomposition of $\alpha_1\alpha_2 \ldots \alpha_r$.

1. If $\rho = (x, y)$ is a two-cycle and x and y are in the different cycles of α, say $\alpha_p = (a_1 \equiv x, a_2, \ldots, a_i)$ and $\alpha_q = (b_1 \equiv y, b_2, \ldots, b_j)$ where $1 \leq p, q \leq r$, then in the composition $\rho\alpha$, α_p, and α_q are joined into a cycle $(a_1, a_2, \ldots, a_i, b_1, b_2, \ldots, b_j)$. For example, if $\rho = (1, 5)$ and $\alpha = (1, 4, 2, 3)(5, 7, 6)$, then $\rho\alpha = (1, 4, 2, 3, 5, 7, 6)$, meaning that two circular chromosomes (1, 4, 2, 3) and (5, 7, 6) in α are merged into one larger circular chromosome (1, 4, 2, 3, 5, 7, 6) by a fusion event ρ (i.e., ρ acts on α as a join operation).
2. If $\rho = (x, y)$ is a two-cycle and x and y are in the same cycle of α, say $\alpha_p = (a_1 \equiv x, a_2, \ldots, a_i \equiv y, a_{i+1}, \ldots, a_j)$ where $1 \leq p \leq r$, then in the composition $\rho\alpha$, this cycle α_p is broken into two disjoint cycles $(a_1, a_2, \ldots, a_{i-1})$ and $(a_i, a_{i+1}, \ldots, a_j)$. For example, if $\rho = (1, 5)$ and $\alpha = (1, 4, 2, 3, 5, 7, 6)$, then $\rho\alpha = (1, 4, 2, 3)(5, 7, 6)$, meaning that a circular chromosome (1, 4, 2, 3, 5, 7, 6) in α is broken into two

smaller circular chromosomes (1, 4, 2, 3) and (5, 7, 6) by a fission event ρ (i.e., ρ acts on α as a split operation).

3. If $\rho = (x, y, z)$ is a three-cycle and x, y and z are in the same cycle of α, say $\alpha_p = (a_1 \equiv x, a_2, \ldots, a_i, b_1 \equiv y, b_2, \ldots, b_j, c_1 \equiv z, c_2, \ldots, c_k)$ where $1 \le p \le r$, then in the composition $\rho\alpha$, the cycle α_p becomes $(a_1, a_2, \ldots, a_i, c_1, c_2, \ldots, c_k, b_1, b_2, \ldots, b_j)$. For example, if $\rho = (1, 3, 7)$ and $\alpha = (1, 4, 2, 3, 5, 7, 6)$, then $\rho\alpha = (1, 4, 2, 7, 6, 3, 5)$, meaning that two adjacent blocks [3, 5] and [7, 6] in α are exchanged by a transposition event ρ.

In fact, we further observed that a block-interchange event affecting α corresponds to the composition of two two-cycles, say ρ_1 and ρ_2, and α under the condition that ρ_1 acts on α as a split operation and ρ_2 acts on $\rho_1\alpha$ as a join operation *(7)*. More clearly, let $\alpha_p = (a_1, a_2, \ldots, a_k)$ be a cycle of $\alpha, \rho_1 = (a_1, a_i)$ and $\rho_2 = (a_h, a_j)$, where $1 < i \le k, 1 \le h \le i-1$ and $i \le j \le k$. Then $\rho_2\rho_1\alpha$ is a permutation obtained from α by exchanging the blocks $[a_h, a_{i-1}]$ and $[a_j, a_k]$ in α_p. For instance, if $\rho_1 = (1, 3), \rho_2 = (4, 7)$ and $\alpha = (1, 4, 2, 3, 5, 7, 6)$, then $\rho_2\rho_1\alpha = (1, 7, 6, 3, 5, 4, 2)$, a resulting permutation by exchanging blocks [4, 2] and [7, 6] in (1, 4, 2, 3, 5, 7, 6).

3. Sorting a Permutation by Block-Interchanges

Given two permutations $\alpha = (a_1, a_2, \ldots, a_n)$ and $\beta = (b_1, b_2, \ldots, b_n)$ of $E = \{1, 2, \ldots, n\}$, the block-interchange distance problem is to find a minimum series of block-interchanges $\sigma_1, \sigma_2, \ldots, \sigma_t$ such that $\sigma_t\sigma_{t-1}\ldots\sigma_1\alpha = \beta$ (i.e., transforming α into β), and the number t is then called the block-interchange distance between α and β. Usually, β is replaced with $I = (1, 2, \ldots, n)$. In this case, the block-interchange distance problem can be regarded as a problem of sorting α using the minimum number of block-interchanges and hence this problem is also called as sorting (a permutation) by block-interchanges.

In fact, the problem of sorting by block-interchanges on circular chromosomes is equivalent to that on linear chromosomes, because the algorithm of solving the former can be used to solve the latter, and vice versa *(7)*. In the following, we shall introduce an algorithm on the basis of permutation groups to solve the sorting circular chromosomes by block-interchanges. In the next section, we then describe how to use it to solve the same problem on linear chromosomes, and vice versa.

Now, we let α and I be permutations representing two given circular genomes. Then it is clear that $I\alpha^{-1}\alpha = I$, intuitively implying that $I\alpha^{-1}$ contains all information needed to transform α into I. Actually, we are indeed able to derive a minimum number of block-interchanges from $I\alpha^{-1}$ required to transform α into I. Moreover, the block-interchange distance between α and

I equals to $\delta = (n - f(I\alpha^{-1}))/2$, where $f(I\alpha^{-1})$ denotes the number of the disjoint cycles in the cycle decomposition of $I\alpha^{-1}$. It should be noted that $f(I\alpha^{-1})$ counts also the nonexpressed cycles of length one. For example, if $I\alpha^{-1} = (1, 4, 2, 3)(5, 7)$ and $E = \{1, 2, \ldots, 7\}$, then $f(I\alpha^{-1}) = 3$, instead of $f(I\alpha^{-1}) = 2$, because $I\alpha^{-1} = (1, 4, 2, 3)(5, 7)(6)$, a product of three disjoint cycles.

The algorithm we designed to compute a minimum series of block-interchanges for the transformation between α and I is very simple *(7)*. It just calculates $I\alpha^{-1}$ in the beginning and then iteratively repeats the following steps for the purpose of finding the ith block-interchange that is represented by a product of two two-cycles, say $(u, v)(x, y)$, where $1 \leq i \leq \delta$.

1. Arbitrarily choose two adjacent elements x and y in a cycle of $I\alpha^{-1}$. Then the composition of (x, y) and α leads α to be broken into two disjoint cycles.
2. Find two adjacent elements u and v in $I\alpha^{-1}(x, y)$ such that the product of (u, v) and $(x, y)\alpha$ becomes a single cycle.
3. Replace α with $(u, v)(x, y)\alpha$ and recalculate $I\alpha^{-1}$, meaning that the original α affected by a block-interchange $(u, v)(x, y)$ has become a new chromosome $(u, v)(x, y)\alpha$.

Let us take $\alpha = (5, 2, 7, 4, 1, 6, 3)$ and $I = (1, 2, \ldots, 7)$ for an instance. We first calculate $I\alpha^{-1}$, which equals to $(1, 5, 4)(2, 6)(3, 7)$. Clearly, $f(I\alpha^{-1}) = 3$ and hence we understand immediately that the block-interchange distance between α and I is $\delta = (7 - 3)/2 = 2$, which means that α can be transformed into I using only two block-interchanges. Next, we show how to find these two block-interchanges, denoted by $\sigma_1 = \rho_2^1\rho_1^1$ and $\sigma_2 = \rho_2^2\rho_1^2$. Initially, we arbitrarily choose two adjacent elements in $I\alpha^{-1}$, say 2 and 6, and let $\rho_1^1 = (2, 6)$. Then ρ_1^1 affects α as a split operation because $\rho_1^1\alpha = (2, 7, 4, 1)(6, 3, 5)$. Then we continue to choose two adjacent elements 3 and 7 in $I\alpha^{-1}\rho_1^1 = (1, 5, 4)(3, 7)$ and let $\rho_2^1 = (3, 7)$ because $\rho_2^1\rho_1^1\alpha$ becomes a single cycle (5, 6, 7, 4, 1, 2, 3). As a result, we obtain a new chromosome $\rho_2^1\rho_1^1\alpha = (5, 6, 7, 4, 1, 2, 3)$ from $\alpha = (5, 2, 7, 4, 1, 6, 3)$ by exchanging blocks (2) and (6) using the first block-interchange $\rho_2^1\rho_1^1 = (3, 7)(2, 6)$. Next, we proceed with the second iteration by first replacing α with $\rho_2^1\rho_1^1\alpha$ and recalculating $I\alpha^{-1}$, which is now (1, 5, 4) = (1, 4)(1, 5). Similarly, we arbitrarily choose $\rho_1^2 = (1, 5)$ and then let $\rho_2^2 = (1, 4)$. Consequently, $\rho_2^2\rho_1^2\alpha = (1, 2, 3, 4, 5, 6, 7)$ is equal to I, the target chromosome obtained from $\alpha = (5, 6, 7, 4, 1, 2, 3)$ by exchanging blocks (4) and (1, 2, 3) using the second block-interchange $\rho_2^2\rho_1^2 = (1, 4)(1, 5)$.

In the previously described algorithm, there are totally δ iterations to be proceeded with each iteration costing $O(n)$ time. Hence, the time-complexity of this algorithm is $O(\delta n)$, where the calculation of $\delta = (n - f(I\alpha^{-1}))/2$ can be done in $O(n)$ time in advance.

4. Circular vs Linear Chromosomes

In the previous section, we have introduced an algorithm to solve the block-interchange distance problem on circular chromosomes by representing permutations as circular chromosomes. Of course, permutations can also be used to represent linear chromosomes. In this situation, a permutation $(a_1, a_2, \ldots, a_n)$ is not equal to $(a_i, \ldots, a_n, a_1, \ldots, a_{i-1})$, where $2 \le i \le n$. In the following, we shall describe how to slightly modify this algorithm to deal with the case of representing $(a_1, a_2, \ldots, a_n)$ and $(1, 2, \ldots, n)$ as linear chromosomes α and I, respectively.

First of all, we add a new element 0 into the beginning of α and denote this new permutation by $\alpha' = (0 \equiv a_0, a_1, a_2, \ldots, a_n)$. Then we regard α' as a circular chromosome and apply the modified algorithm (we will introduce later) to α' such that the obtained block-interchanges, say $\sigma_1', \sigma_2', \ldots, \sigma_\delta'$, for optimally transforming α' into $I' = (0, 1, \ldots, n)$ satisfy the property that none of the two blocks interchanged by each σ_i' contains a_0, where $1 \le i \le \delta$. The purpose of this property is to ensure that for each σ_i' affecting circular chromosome $\sigma_{i-1}' \cdots \sigma_1'\alpha'$, we can find a corresponding block-interchange σ_i to affect linear chromosome $\sigma_{i-1} \cdots \sigma_1\alpha$ such that the blocks interchanged by σ_i are the same as those interchanged by σ_i'. Because of this property, $\sigma_1, \sigma_2, \ldots, \sigma_\delta$ can be used to transform linear chromosome α into I and moreover, their number is minimum, because any series of block-interchanges of transforming α into I must also be a series of block-interchanges of transforming α' into I'.

Now, we describe how to modify the algorithm in the previous section such that none of the two blocks interchanged by each σ_i' contains a_0, where $1 \le i \le \delta$. Notice that when applying the original algorithm to α', the chosen (x, y) and (u, v) in the ith iteration (i.e., $\sigma_i' = (u, v)(x, y)$) may lead one of the exchanged blocks, say B_1 and B_2, to contain a_0. If this situation occurs, then we consider the following two cases to modify the algorithm.

Case 1: B_1 and B_2 are not adjacent. Then we just interchange the roles of (x, y) and (u, v) by letting (u, v) as the split operation and (x, y) as the join operation, and then apply $(x, y)(u, v)$, instead of $(u, v)(x, y)$, to $\sigma_{i-1}' \cdots \sigma_1'\alpha'$. We illustrate its reason by just considering the first iteration of producing σ_1' as follows. Let $(x, y) = (a_i, a_j)$ and $(u, v) = (a_k, a_l)$, where $1 \le i < j \le n, i < k < j$ and $j < l$. Then $(u, v)(x, y)\alpha' = (a_i, \ldots, a_{k-1}, \boxed{a_l, \ldots, a_n, a_0, \ldots, a_{i-1}}, a_j, a_{j+1}, \ldots, a_{l-1}, \boxed{a_k, \ldots, a_{j-1}})$, and $(x, y)\ (u, v)\alpha' = (a_k, \ldots, a_{j-1},$

$\boxed{a_i, \ldots, a_{k-1}}, a_l, \ldots, a_n, a_0, \ldots, a_{i-1}, \boxed{a_j, \ldots, a_{l-1}}$). From the viewpoint of circular chromosome, they are equivalent (i.e., $(u, v)(x, y)\alpha' = (x, y)(u, v)\alpha'$). However, both the blocks interchanged using $(x, y)(u, v)$ contains no a_0.

Case 2: B_1 and B_2 are adjacent. In this case, we let $B_1 = [a_l, \ldots, a_n, a_0, a_1, \ldots a_{i-1}]$ denote the block of containing a_0 and B_3 be the remaining block. It is not hard to see that either $x = u$ or $y = v$. If $x = u$, then $(u, v)(x, y) = (x, v)(x, y)$, which also equals to $(x, y)(v, y)$. Then we modify the algorithm by applying $(x, y)(v, y)$, instead of $(u, v)(x, y)$, to $\sigma'_{i-1} \ldots \sigma'_1\alpha'$. Because it leads to the exchange of B_2 and B_3, which is equivalent to the exchange of B_1 and B_2 from the circular viewpoint, and neither B_2 nor B_3 contains a_0. If $y = v$, then $(u, v)(x, y) = (u, y)(x, y)$ that equals to $(u, x)(u, y)$. Then we modify the algorithm by applying $(u, x)(u, y)$ to $\sigma'_{i-1} \ldots \sigma'_1\alpha'$, because it leads B_2 and B_3 to be exchanged.

In fact, we can use the algorithm of sorting linear chromosomes by block-interchanges to solve the same problem on circular chromosomes as follows. Now, let $\alpha = (a_1, a_2, \ldots, a_n)$ be represented as a circular chromosome, instead of a linear chromosome. Then we can linearize it by removing an arbitrary element, say a_1, from α so that the resulting permutation $\alpha' = (a_2, a_3, \ldots, a_n)$ becomes a linear chromosome. The removed element a_1 actually plays the same role as the a_0 we have introduced in the previous discussion. With the same arguments, the optimal solution for linear α' corresponds to a minimum series of block-interchanges required to transform circular α into I.

5. The Web Server of ROBIN

Based on the algorithms we described in the previous sections, we have developed a web server, called ROBIN, for analyzing genome rearrangements of block-interchanges between two linear/circular chromosomes *(7)*. ROBIN takes two or more linear/circular chromosomes as its input and outputs the block-interchange distance between any pair of the input chromosomes, as well as an optimal scenario of the needed block-interchanges. The input can be either bacterial-size sequence data or gene-/landmark-order data. If the input is a set of sequences, ROBIN will automatically search for the identical landmarks that are the homologous/conserved regions shared by all the input sequences.

5.1. Locally Collinear Blocks

In analyses of genome rearrangements, the commonly used landmarks can be exact matches such as the maximal unique matches (MUMs) as in MUMmer *(14)*, approximate matches without gaps such as the yielding fragments as in DIALIGN *(15)* and LAGAN *(16)*, approximate matches with gaps such as the hit fragments as in BLASTZ ***(17)***, or regions of local collinearity such as the locally collinear blocks (LCBs) as in Mauve ***(18)***.

A LCB is a collinear set of the multi-MUMs, where *multi-MUMs* are exactly matching subsequences shared by all the considered genomes that occur only once in each of genomes and that are bounded on either side by mismatched nucleotides. In ROBIN, we adopt LCBs as landmarks because each LCB may correspond to a homologous region of sequence shared by all the input chromosomes and does not contain any genome rearrangements.

Usually, each LCB is associated with a weight that can serve as a measure of confidence that it is a true homologous region rather than a random match. The weight of an LCB is defined as the sum of the lengths of multi-MUMs in this LCB. By selecting a high minimum weight, users can identify the larger LCBs that are truly involved in genome rearrangement, whereas by selecting a low minimum weight, users can trade some specificity for sensitivity to identify the smaller LCBs that are possibly involved in genome rearrangement.

5.2. Usage of ROBIN

ROBIN can be easily accessed via a simple web interface (*see* **Fig. 3**) and it also provides an online help to help users to run it. The input of ROBIN can be two or more linear/circular chromosomes that are either genomic sequences with bacterial size or unsigned integer sequences with each integer representing an identical gene/landmark on all the chromosomes being considered. If the input is a set of genomic sequences, ROBIN will identify all the LCBs meeting the user-specified minimum weight. The minimum LCB weight is a user-definable parameter and its default is set to be three times the minimum multi-MUM length. The output of ROBIN is the block-interchange distance between any two input chromosomes and an optimal scenario of block-interchanges for the transformation.

5.2.1. ROBIN Input

Users can choose to input sequence data or gene-/landmark-order data to run ROBIN. If the input is the sequence data, then users can run ROBIN by following the steps described next.

1. Input or paste two or more sequences in FASTA format in the top field of web interface, or simply upload a plain text file of sequences in FASTA format.
2. Specify the minimum multi-MUM length, whose default is set to $\log_2 n$, where n is the average sequence length, and the minimum LCB weight, whose default is set to 3 * (minimum multi-MUM length).

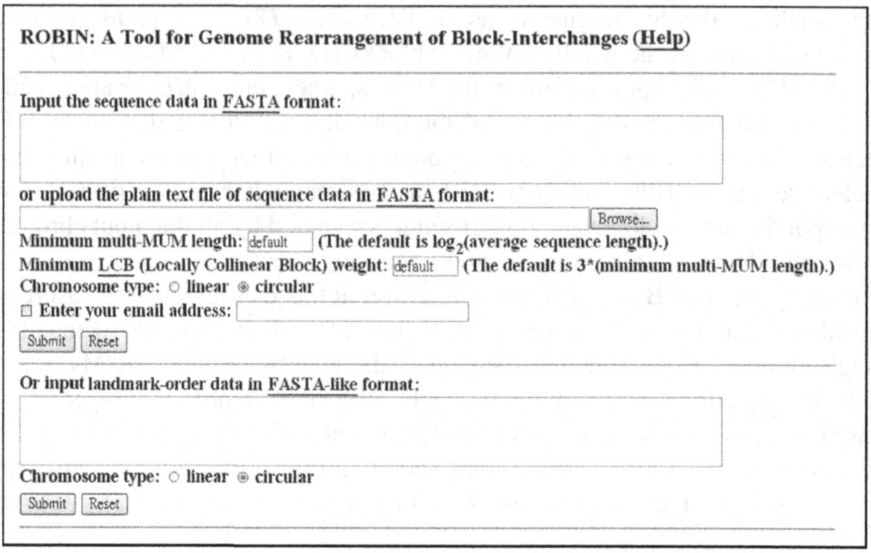

ROBIN: A Tool for Genome Rearrangement of Block-Interchanges (Help)

Input the sequence data in FASTA format:

or upload the plain text file of sequence data in FASTA format:
Browse...
Minimum multi-MUM length: default **(The default is $\log_2$(average sequence length).)**
Minimum LCB (Locally Collinear Block) weight: default **(The default is 3*(minimum multi-MUM length).)**
Chromosome type: ○ **linear** ◉ **circular**
☐ **Enter your email address:**
Submit Reset

Or input landmark-order data in FASTA-like format:

Chromosome type: ○ **linear** ◉ **circular**
Submit Reset

Fig. 3. The web interface of ROBIN. A description of the options is given in the text.

3. Select the type of chromosomes being considered, which can be either linear or circular.
4. Check email box and enter an email address simultaneously, if users would like to run ROBIN in a batch way. In the batch way, users will be notified of the output via email when the submitted job is finished. This email check is optional but recommended if the sequences users enter/upload are large-scale.
5. Click "Submit" button to run ROBIN.

   If users choose gene-/landmark-order data as the input of ROBIN, then they just follow the steps next.
6. Input or paste two or more unsigned gene/landmark orders in FASTA-like format in the bottom field of web interface. The FASTA-like format (*see* **Fig. 4** for

```
>NC_005140|Vibrio vulnificus chromosome II
1 2 3 4 5 6 7 8 9 10 11 12 13 14 15 16 17 18 19 20
>NC_004605|Vibrio parahaemolyticus chromosome II
1 2 9 10 8 20 11 12 13 14 15 16 18 17 19 4 7 6 5 3
>NC_002506|Vibro cholerae chromosome II
1 3 6 18 13 15 12 16 9 17 19 14 11 20 5 8 7 10 2 4
```

Fig. 4. An example of three landmark-order data in FASTA-like format.

an example) contains a single-line description followed by lines of unsigned integers, which are separated by space(s), with each integer representing an identical gene/landmark on the chromosomes being considered.

7. Select the chromosome type.
8. Click "Submit" button to run ROBIN.

5.2.2. ROBIN Output

If the input is sequence data, ROBIN will first show the values of the parameters specified by users, then output the order of the identified common LCBs that shared by all the input sequences, and finally display the calculated distance matrix with each entry indicating the block-interchange distance between any two input sequences (*see* **Fig. 5**). In addition, users can further view the detailed information of each identified LCB order just by clicking the associated link, such as the position, length, and weight of each LCB and the overall coverage of all LCBs on the chromosome (*see* **Fig. 6**). The position of each identified LCB is denoted by its left and right end coordinates on the chromosome. If both the left and right coordinates are negative values, it implies that the identified LCB is the inverted region on the opposite strand of the given chromosome. It should be noted that users can also view and check the optimal scenario of

```
Minimum multi-MUM size: 21
LCB Weight: default (Which equals to 63)
Chromosome type: circular

Computed common LCB orders:
S1: NC_005140|Vibrio vulnificus chromosome II
LCB order: 1 2 3 4 5 6 7 8 9 10 11 12 13 14 15 16 17 18 19 20
S2: NC_004605|Vibrio parahaemolyticus chromosome II
LCB order: 1 2 9 10 8 20 11 12 13 14 15 16 18 17 19 4 7 6 5 3
S3: NC_002506|Vibrio cholerae chromosome II
LCB order: 1 3 6 18 13 15 12 16 9 17 19 14 11 20 5 8 7 10 2 4

Computed block-interchange distance matrix:
     S1   S2   S3
S1   -    5    9
S2   5    -    8
S3   9    8    -
```

Fig. 5. The identified locally collinear block orders of the second circular chromosomes of three *Vibrio* species and their calculated block-interchange distances by running ROBIN with default parameters.

```
LCB number, left and right end coordinates of LCB (LCB's length, LCB's weight)
 1. (-64511) : 1933 (66445, 2244)   11. 69457 : 703008 (8436, 386)
 2. 26295 : 27064 (770, 141)        12. 765208 : 766062 (855, 143)
 3. 82621 : 89126 (6506, 629)       13. 782406 : 786122 (3717, 703)
 4. 298416 : 298509 (94,69)         14. 807648 : 819746 (12099, 522)
 5. 35313 : 346503 (11191, 438)     15. 844213 : 845713 (1501, 123)
 6. 413258 : 437461 (24204, 1235)   16. 1193459 : 1193541 (83, 83)
 7. 466467 : 471912 (5446, 839)     17. 1218701 : 1230415 (11715, 913)
 8. 476046 : 491903 (15858, 183)    18. 1302985 : 1304148 (1164, 284)
 9. 540354 : 562937 (22584, 1086)   19. 1311460 : 1330114 (18655, 917)
10. 582930 : 585836 (2907, 375)     20. 1409999 : 1411253 (1255, 99)

The total LCBs cover 11.6% of the genome.
```

Fig. 6. The detailed information of the identified locally collinear blocks (LCBs) for the second chromosome of *Vibrio vulnificus* in **Fig. 5**. Notice that the left end coordinate of the first LCB is negative because this LCB is a region across the position 1 of *V. vulnificus* chromosome that is circular.

```
S2= ([1 2] 9 10 8 20 11 12 13 14 15 16 18 17 19 [4 7 6 5] 3)
->  (4 7 6 5 9 10 [8] 20 [11 12 13 14 15 16 18 17 19] 1 2 3)
->  (4 7 [6 5] 9 10 11 12 13 14 15 16 18 17 19 20 [8] 1 2 3)
->  (4 [7 8 9 10 11 12 13 14 15 16 18 17 19 20] 6 [5] 1 2 3)
->  (4 5 6 7 8 9 10 11 12 13 14 15 16 [18] [17] 19 20 1 2 3)
->  (4 5 6 7 8 9 10 11 12 13 14 15 16 17 18 19 20 1 2 3) =S1
```

Fig. 7. An optimal scenario of five block-interchanges between *Vibrio vulnificus* and *Vibrio parahaemolyticus* in which the blocks of consecutive integers in square brackets are exchanged.

block-interchanges for any two input sequences by clicking the link associated with the entry of the distance matrix (*see* **Fig. 7**).

If the input is gene/landmark data, ROBIN will only show the values of the specified parameters and output the calculated block-interchange distance matrix along with the optimal scenarios of block-interchanges.

6. Notes

1. Block-interchanges can be considered as a generalization of transpositions from the biological perspective. From the computational perspective, the block-interchange distance problem is solvable in polynomial time, whereas the transposition distance problem is still open. As introduced in this chapter, permutation groups in algebra are a useful tool in the design of an efficient algorithm to the block-interchange distance problem. Hence, it is interesting to further study whether or not they can still be applied to the transposition distance problem.

2. Each LCB identified by ROBIN can be considered as a homologous/conserved region shared by all chromosomes without any genome rearrangements to occur. The weight of each identified LCB can be used to measure its confidence of being a true homologous/conserved region. The larger the weight, the greater the confidence. By selecting a high minimum weight, users can use ROBIN to identify the larger LCBs that are truly involved in genome rearrangement, whereas by selecting a low minimum weight, they can trade some specificity for sensitivity to identify the smaller LCBs that are possibly involved in genome rearrangement.
3. A phylogenetic reconstruction based on the block-interchange distances can be easily carried out using PHYLIP, a package of programs for inferring phylogenies. PHYLIP is distributed by Joe Felsenstein (University of Washington) and available for free from http://evolution.genetics.washington.edu/phylip.html.

References

1. Bafna, V. and Pevzner, P. A. (1996) Genome rearrangements and sorting by reversals. *SIAM Journal on Computing* **25,** 272–289.
2. Hannenhalli, S. and Pevzner, P. A. (1999) Transforming cabbage into turnip: polynomial algorithm for sorting signed permutations by reversals. *Journal of the ACM* **46,** 1–27.
3. Kaplan, H., Shamir, R., and Tarjan, R. E. (2000) Faster and simpler algorithm for sorting signed permutations by reversals. *SIAM Journal on Computing* **29,** 880–892.
4. Bader, D. A., Yan, M., and Moret, B. M. W. (2001) A linear-time algorithm for computing inversion distance between signed permutations with an experimental study. *Journal of Computatonal Biology* **8,** 483–491.
5. Bafna, V. and Pevzner, P. A. (1998) Sorting by transpositions. *SIAM J Appl Math* **11,** 221–240.
6. Christie, D. A. (1996) Sorting by block-interchanges. *Information Processing Letters* **60,** 165–169.
7. Lin, Y. C., Lu, C. L., Chang, H. -Y., and Tang, C. Y. (2005) An efficient algorithm for sorting by block-interchanges and its application to the evolution of vibrio species. *Journal of Computational Biology* **12,** 102–112.
8. Lu, C. L., Wang, T. C., Lin, Y. C., and Tang, C. Y. (2005) ROBIN: a tool for genome rearrangement of block-interchanges. *Bioinformatics* **21,** 2780–2782.
9. Kececioglu, J. D. and Ravi, R. (1995) Of mice and men: algorithms for evolutionary distances between genomes with translocation, in *Proceedings of the 6th ACM-SIAM Symposium on Discrete Algorithms (SODA1995),* ACM/SIAM, San Francisco, CA, pp. 604–613.
10. Hannenhalli, S. (1996) Polynomial algorithm for computing translocation distance between genomes. *Discrete Applied Mathematics* **71,** 137–151.
11. Hannenhalli, S. and Pevzner, P. A. (1995) Transforming men into mice (polynomial algorithm for genomic distance problem), in *Proceedings of the 36th IEEE*

Symposium on Foundations of Computer Science (FOCS1995), IEEE Computer Society, pp. 581–592.

12. Fraleigh, J. B. (2003) *A First Course in Abstract Algebra.* 7th edition, Addison-Wesley, Boston, MA.
13. Meidanis, J. and Dias, Z. (2000) An alternative algebraic formalism for genome rearrangements, in *Comparative Genomics: Empirical and Analytical Approaches to Gene Order Dynamics, Map Alignment and Evolution of Gene Families,* (Sankoff, D. and Nadeau, J. H., eds.), Kluwer Academic Publisher, pp. 213–223.
14. Delcher, A. L., Kasif, S., Fleischmann, R. D., Peterson, J., White, O., and Salzberg, S. L. (1999) Alignment of whole genomes. *Nucleic Acids Res.* **27,** 2369–2376.
15. Morgenstern, B., Frech, K., Dress, A., and Werner, T. (1998) DIALIGN: finding local similarities by multiple sequence alignment. *Bioinformatics* **14,** 290–294.
16. Brudno, M., Do, C. B., Cooper, G. M., et al. (2003) LAGAN and Multi-LAGAN: efficient tools for large-scale multiple alignment of genomic DNA. *Genome Res.* **13,** 721–731.
17. Schwartz, S., Kent, W. J., Smit, A., et al. (2003) Human-mouse alignments with BLASTZ. *Genome Res.* **13,** 103–107.
18. Darling, A. C. E., Mau, B., Blattner, F. R., and Perna, N. T. (2004) Mauve: multiple alignment of conserved genomic sequence with rearrangements. *Genome Res.* **14,** 1394–1403.

10

Analyzing Patterns of Microbial Evolution Using the Mauve Genome Alignment System

Aaron E. Darling, Todd J. Treangen, Xavier Messeguer, and Nicole T. Perna

Summary

During the course of evolution, genomes can undergo large-scale mutation events such as rearrangement and lateral transfer. Such mutations can result in significant variations in gene order and gene content among otherwise closely related organisms. The Mauve genome alignment system can successfully identify such rearrangement and lateral transfer events in comparisons of multiple microbial genomes even under high levels of recombination. This chapter outlines the main features of Mauve and provides examples that describe how to use Mauve to conduct a rigorous multiple genome comparison and study evolutionary patterns.

Key Words: Microbial evolution; sequence alignment; comparative genomics; genome alignment; genome rearrangement; lateral transfer; *Yersinia pestis*.

1. Introduction

As genomes evolve, mutational forces and selective pressures introduce rearrangements via inversion, transposition, and duplication/loss processes. Lateral transfer can introduce novel gene content, or in the case of homologous recombination, introduce more subtle changes such as allelic substitution. Through genome comparison we hope to first identify differences among organisms at the genome level and then infer the biological significance of differences and similarities among related organisms.

From: *Methods in Molecular Biology, vol. 396: Comparative Genomics, Volume 2*
Edited by: N. H. Bergman © Humana Press Inc., Totowa, NJ

Traditional sequence alignment methods were developed to accurately align individual gene sequences where rearrangement rarely occurs ***(1–6)***. When aligning two or more genomes, shuffled regions of orthologous sequence may be interspersed with paralogous and novel sequence regions, forcing a genome alignment method to map segmental homology among genomes ***(7–14)***. Once a segmental homology map exists, the alignment task can be approached using traditional alignment methods based on dynamic programming and pair-hidden Markov model heuristics ***(15)***.

In genome alignment, it is important to go beyond simple classification of sequence regions as either homologous or unrelated. Local alignment programs such as Basic Local Alignment Search Tool provide such functionality. We define a global genome alignment to be a complete catalog of orthologous, xenologous ***(16)***, paralogous, and unrelated sites among a group of genome sequences. All sites must be assigned to one of the given categories, and in the cases of orthology, paralogy, and xenology, the related sites must be identified. Furthermore, homologous sites should be grouped into regions of maximal collinearity such that the leftmost and rightmost site in any group defines a breakpoint of rearrangement. We refer to such maximal collinear sets of homologous sites as locally collinear blocks (LCBs) because they cover a "block" of sequence without any internal genome rearrangement. Such a categorization of sites implicitly defines breakpoints of rearrangement, recombination, duplication, and insertion and deletion processes. A genome alignment lends itself to downstream evolutionary inferences such as rearrangement history ***(17–21)***, phylogeny ***(22)***, ancestral state prediction, and detection of selective pressure in coding sequence ***(23)*** and in noncoding sequence ***(24)***.

Most current genome alignment systems, including Mauve ***(25)***, construct an incomplete form of genome alignment as defined above. Shuffle-LAGAN ***(26)*** aligns both single-copy and repetitive regions in pairs of genomes, but does not classify them as either orthologous or paralogous. Mauve aligns orthologous and xenologous regions, but does not distinguish between the two cases. Mauve also aligns orthologous repeats, but does not align paralogous repeats. Mulan ***(27)*** and M-GCAT ***(28)*** both construct alignments similar to Mauve. Recent versions of MUMmer ***(29)*** identify and align orthologous and paralogous sequence, but among pairs of genomes only. Some earlier genome alignment systems have treated segmental homology mapping as a separate step and thus assume sequence collinearity ***(30–33)***.

Mauve performs five basic steps when constructing a genome alignment:

1. Search for local multiple alignments (approximate multi-MUMs).
2. Calculate a phylogenetic guide tree using the local alignments.

3. Select a subset of the local alignments to use as anchors.
4. Conduct recursive anchoring to identify additional alignment anchors.
5. Perform a progressive alignment of each LCB.

The METHODS section describes each of these five steps in greater detail.

Before performing an analysis with Mauve, a researcher should first ask whether genome alignment is the right analysis. If the answer is "yes," the next question should be "Is Mauve the right tool for the job?" Mauve performs best when aligning a relatively small number of closely related genomes. It can align genomes that have undergone rearrangement and lateral transfer. However, as mentioned previously, the current version of Mauve constructs only an incomplete form of genome alignment. Mauve genome alignments alone do not provide a suitable basis for inferences on paralogous gene families. Mauve is also limited in its ability to align rearranged and large (e.g., > 10 Kbp) collinear segments that exist in only a subset of the genomes under study. And, in general, the level of nucleotide similarity among all taxa should be greater than 60%. **Subheading 3.** explains the reasons for these limitations in more detail and gives hints on choosing alignment parameters so as to mitigate any potential problems.

2. Materials

1. Windows, Mac OS X 10.3+, or Linux Operating System.
2. Mauve Multiple Genome Alignment software (http://gel.ahabs.wisc.edu/mauve/download.php).
3. Four *Yersinia* genomes in GenBank format:
   a. *Yersinia pestis* KIM, accession number: **AE009952** (http://www.ncbi.nlm.nih.gov/entrez/viewer.fcgi?db=nucleotide&val=AE009952).
   b. *Y. pestis* CO92, accession number: **AL590842** (http://www.ncbi.nlm.nih.gov/entrez/viewer.fcgi?db=nucleotide&val=AL590842).
   c. *Y. pestis* 91001, accession number: **AE017042** (http://www.ncbi.nlm.nih.gov/entrez/viewer.fcgi??db=nucleotide&val=AE017042).
   d. *Yersinia pseudotuberculosis* IP32953, accession number: **BX936398** (http://www.ncbi.nlm.nih.gov/entrez/viewer.fcgi?db=nucleotide&val=BX936398).
4. Two drosophila genomes:
   a. *Drosophila melanogaster* assembly 4.0 in GenBank format: (http://gel.ahabs.wisc.edu/mauve/chapter/dmel.gbk).
   b. *Drosophila yakuba* assembly 2.1 in FastA format: (http://gel.ahabs.wisc.edu/mauve/chapter/dyak.fas).
5. Example output files located at http://gel.ahabs.wisc.edu/mauve/chapter/.

3. Methods

The following sections provide a deeper look into how the five main steps in Mauve's alignment algorithm contribute to the global genome alignment process.

3.1. Search for Local Multiple Alignments (Approximate Multi-MUMs)

Mauve uses a seed-and-extend hashing method to simultaneously identify highly similar unique regions among all genomes. The method used in current Mauve releases remains similar to that described in ***(14,25)*** but has been extended to approximate matching using spaced seeds ***(34)***. In addition to finding matching regions that exist in all genomes, the algorithm identifies matches that exist only among a subset of the genomes being aligned. The local multiple alignment method is very efficient in practice and typically requires less than 1 min per bacterial-size genome to find local alignments, and around 3 h for each mammalian genome on a standard workstation computer.

3.2. Calculate a Phylogenetic Guide Tree Using Local Alignments

As part of the alignment process, Mauve calculates a guide tree via genome distance phylogeny ***(35)***. Rather than using pairwise BLAST hits to estimate distance, Mauve uses the local multiple alignments identified in the previous step. The average amount of novel sequence among each pair of genomes is calculated using the local multiple alignments. This value is used as a pairwise distance measure to construct a phylogenetic tree via Neighbor Joining ***(36)***.

3.3. Selecting a Set of Anchors

In addition to local multiple alignments that are part of truly homologous regions, the set of approximate multi-MUMs may contain spurious matches arising as a result of random sequence similarity. This step attempts to filter out such spurious matches while determining the boundaries of LCBs. Local alignments are clustered together into LCBs and each LCB is required to meet a minimum weight criteria, calculated as the sum of lengths of its constituent local alignments. Local alignments contained in LCBs that do not meet the weight criteria are deleted.

3.4. Recursive Anchoring and Gapped Alignment

The initial anchoring step may not be sensitive enough to detect the full region of homology within and surrounding the LCBs. Using the existing anchors as a guide, two types of recursive anchoring are performed repeatedly. First, regions outside of LCBs are searched to extend the boundaries of existing LCBs and identify new LCBs. Second, unanchored regions within LCBs are searched for additional alignment anchors.

3.5. Progressive Alignment

After gathering a complete set of alignment anchors among all genomes, Mauve then performs either a MUSCLE or a CLUSTAL W progressive alignment. CLUSTAL alignments use the previously calculated genome guide tree. The progressive alignment algorithm is executed once for each pair of adjacent anchors in every LCB, calculating a global alignment over each LCB. Tandem repeats <10 Kbp in total length are aligned during this phase. Regions >10 Kbp without an anchor are ignored. For additional details and a more in-depth algorithmic analysis refer to **ref.** *(25)*.

3.6. Understanding the Alignment Parameters

To accurately align a set of genomes using the aforementioned steps, it can be helpful to carefully tailor the provided Mauve alignment parameters to better suit the characteristics of the genomes being compared. The following sections provide a detailed explanation of each configurable alignment parameter.

3.6.1. Match Seed Size

The seed size parameter sets the minimum length of local multiple alignments used during the first pass of anchoring the alignment. When aligning divergent genomes or aligning more genomes simultaneously, lower seed sizes may provide better sensitivity. However, because Mauve also requires the matching seeds to be unique in each genome, setting this value too low will reduce sensitivity (small k-mers are less likely to be unique).

3.6.2. Default Seed Size

Setting this option will allow Mauve to automatically select an initial match seed size that is appropriate for the length of sequences being aligned. The default seed size for 1 MB genomes is typically around 11, around 15 for 5 MB genomes, and continues to grow with the size of the genomes being aligned up to 21 for mammalian genomes. The defaults may be conservative (too large), especially when aligning more divergent genomes (*see* **Note 1** for suggestions).

3.6.3. LCB Weight

The LCB weight sets the minimum number of matching nucleotides identified in a collinear region for that region to be considered true homology vs random similarity. Mauve uses a greedy breakpoint elimination algorithm to compute a set of LCBs that have the given minimum weight. By default an LCB weight of three times the seed size will be used. For many genome comparisons the default LCB weight is too low, and a higher value will be desired. The ideal LCB weight can be determined interactively in the alignment display using the LCB weight slider.

3.6.4. Determine LCBs

If this option is disabled Mauve will identify local multiple alignments, but will not cluster them into LCBs. See the description of match generation in the command-line interface chapter.

3.6.5. Assume Collinear Genomes

Select this option if it is certain that there are no rearrangements among the genomes to be aligned. Using this option when aligning collinear genomes can result in improved alignment accuracy.

3.6.6. Island and Backbone Sizes

An island is a region of the alignment where one genome has a sequence element that one or more others lack. This parameter sets the alignment gap sizes used to calculate islands and backbone segments.

3.6.7. Full Alignment

Selecting the "Full alignment" option causes Mauve to perform a recursive anchor search and a full gapped alignment of the genome sequences using either the MUSCLE or the ClustalW progressive alignment method. Disabling this option will allow Mauve to rapidly generate a homology map without the time consuming gapped-alignment process (*see* **Note 2** for performance tips for full alignments).

3.7. Examples

3.7.1. A Detailed Example Using Four Yersinia Genomes

The *Yersinia* genus is responsible for diseases as common as gastroenteritis and as infamous as the plague. Of the 11 known *Yersinia* species, *Y. pestis* became the most notorious when identified as the cause of the bubonic and pneumonic plague *(37)*. *Y. pestis* can be further classified into three main biovars according to three historically recognized pandemics: Antiqua (~1000 AD), Medievalis (1300–1800), and Orientalis (1900–Present). Additionally, it is believed that *Y. pestis* is a clone that has evolved from *Y. pseudotuberculosis* as recent as 1500 yr ago. The recent and rapid evolution of *Yersinia* species provides an excellent example for study with Mauve. In this example we will align and analyze the four currently finished genomes of *Yersinia*: *Y. pseudotuberculosis* IP32953, *Y. pestis* KIM, *Y. pestis* 910001, and *Y. pestis* CO92.

3.7.2. Running Mauve

Under Windows, Mauve can be launched directly from the Start Menu. On Mac OS X, Mauve is distributed as a stand-alone application and can be run from any location. On Linux and other Unix variants, simply run. `/Mauve` from within the Mauve directory to start the Mauve Java GUI (*see* **Note 3** for further tips on configuring Mauve to avoid Java heap space limitations).

3.7.3. Locating the Input Data

To align these four *Yersinia* genomes, we first need to download the four input genome sequence files (as listed in **Subheading 2.**) and load them into Mauve. Mauve accepts the following genome sequence file formats: FastA, Multi-FastA, GenBank flat file, and raw format. FastA and GenBank format files with the genome of your organism can usually be downloaded from NCBI at ftp://ftp.ncbi.nih.gov/genomes/. The .fna files are in FastA format and the .gbk files are in GenBank format; for this example we will use the .gbk files.

3.7.4. Loading the Input Data

Once Mauve has started up, simply select "Align…" from the "File" menu to access the Alignment dialog box, as shown in **Fig. 1**:

The top entry area lists the sequence file(s) containing the genomes that will be aligned. To add a sequence file, click the "Add sequence…" button and select the file to add. The Windows and Mac OS X versions of Mauve support drag-and-drop, allowing sequence files to be added by dragging them in from the Windows Explorer or Mac OS Finder.

3.7.5. Configuring the Output Data Location

The location where Mauve stores its alignment results can be set using the "Output file:" text entry field. By default Mauve creates output files in the system's

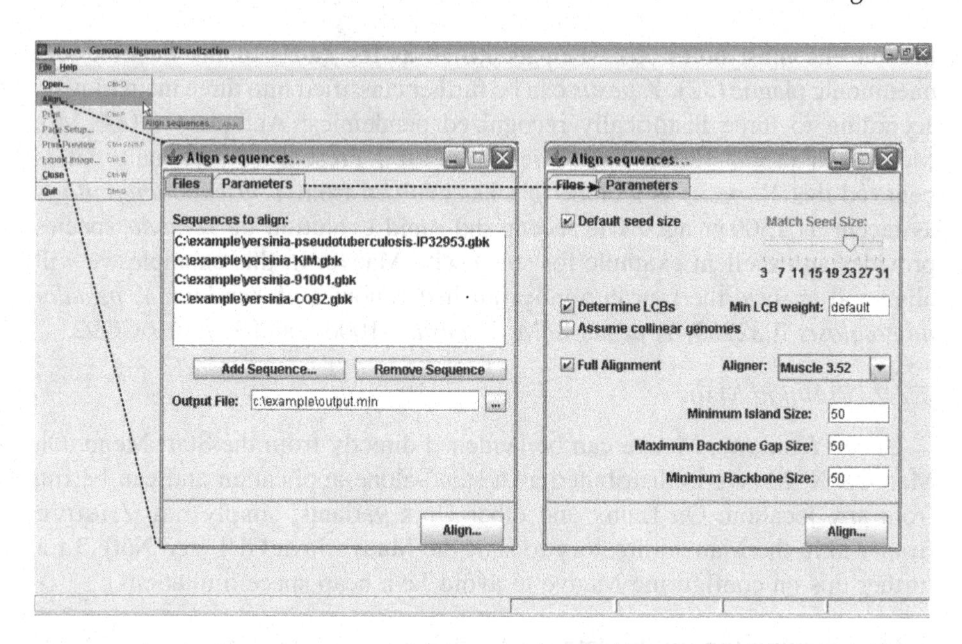

Fig. 1. The sequence alignment dialog. Selecting the "Align..." option from the "File..." menu will display the "Align sequences" dialog. The "Files" panel allows for selection of files containing genome sequences to be aligned, and when switching to the "Parameters" panel, configurable alignment parameters are displayed.

temporary storage directory. Under Windows this is usually `C:\Documents and Settings\ <username> \Local Settings\Temp\` or `/tmp` under Mac OS X and Unix. Because the location of the system's temporary storage is rather obscure, we heartily encourage users to select a different output location. In this example, we specify `C:\example\output.mln` to store the Mauve output.

3.7.6. Configuring the Alignment Parameters

By default, Mauve configures the alignment parameters so they are appropriate for aligning closely related genomes with moderate to high amounts of genome rearrangement. However, some alignment parameters can be modified to better suit alignments involving more distantly related genomes, or with a lower amount of rearrangement. For example, when aligning more divergent genomes, the seed size can be reduced to find additional alignment anchors and achieve greater alignment coverage over the genomes. Another option disables the full alignment process, allowing Mauve to quickly generate a simple comparative picture of genome organization. For this example, we start with the default parameters, and later use the interactive LCB weight slider (*see* **Fig. 2**) to determine an LCB weight that excludes most spurious rearrangements (1500). We then recomputed the alignment with the LCB weight set to 1564.

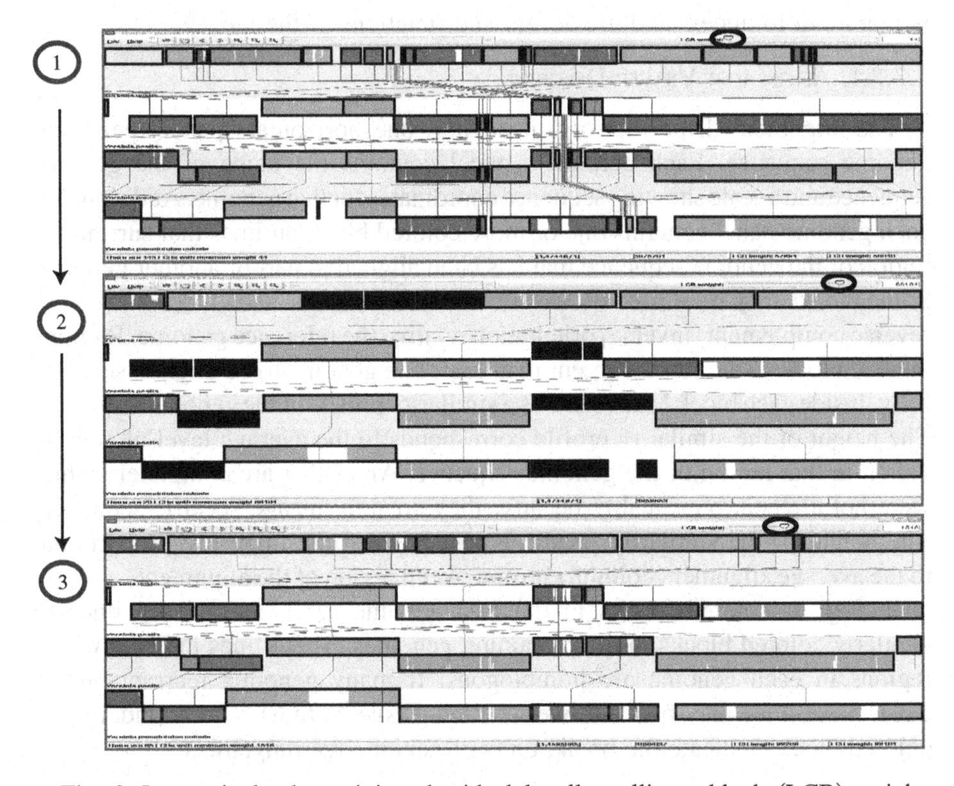

Fig. 2. Interactively determining the ideal locally collinear block (LCB) weight. Starting with an alignment based on the default LCB weight parameter (**1**), we move the interactive LCB weight slider to the right (**2**) to exclude spurious rearrangements. The LCB weight of 66,104 in **2** is too large as it deletes many large segmental homologies that appear to be valid—shown as dark gray blocks in panel **2**, or as bright orange in the Mauve viewer. We thus move the interactive LCB weight slider to the left again (**3**) to arrive at a weight of 1564 that excludes most spurious small matches, and retains all valid large matches. We can now recompute a full alignment with the weight set to 1564.

3.7.7. The Interactive Mauve Alignment Viewer

To start the genome alignment, click on the "Align..." button. Once Mauve finishes its global alignment of the four *Yersinia* genomes, we are ready to interactively inspect the results. Should the alignment fail, please check the console log (*see* **Note 4**). The Mauve Alignment viewer enables manual evaluation of both the proposed global homology and the nucleotide level alignment in the context of genome annotation. *See* **Note 5** for details on manual editing of the genome alignment. To further analyze the *Yersinia* alignment results, it is important to understand the design and functions of the viewer.

3.7.7.1. Alignment Viewer Description

The alignment display is organized into one horizontal "panel" per input genome sequence. Each genome's panel contains the name of the genome sequence and a scale showing sequence coordinates for that genome. Additionally, each genome panel contains one or more colored block outlines that surround a region of the genome sequence that has been aligned to part of another genome. Some blocks may be shifted downward relative to others; such blocks are in the reverse complement (inverse) orientation relative the reference genome. Regions outside blocks were too divergent in at least one genome to be aligned successfully. Inside each block Mauve draws a similarity profile of the genome sequence. The height of the similarity profile corresponds to the average level of conservation in that region of the genome sequence. Areas that are completely white were not aligned, presumably because they contain lineage specific sequence. The height of the similarity profile is calculated to be inversely proportional to the average alignment column entropy over a region of the alignment.

In **Fig. 1**, colored blocks in the first genome are connected by lines to similarly colored blocks in the remaining genomes. These lines indicate which regions in each genome are homologous. If many genomic rearrangements exist, these lines may become overwhelming (*see* **Note 6**). The boundaries of colored blocks indicate the breakpoints of genome rearrangement.

3.7.7.1.1. Navigation The alignment display is interactive, providing the ability to zoom in on particular regions and shift the display to the left and right. Navigating through the alignment visualization can be accomplished by using the control buttons on the toolbar immediately above the display. Alternatively, keyboard shortcuts allow rapid movement through the alignment display. The keystrokes `Ctrl+up arrow` and `Ctrl+down arrow` zoom the display in and out, whereas `Ctrl+left arrow` and `Ctrl+right arrow` shift left and right, respectively. When moving the mouse over the alignment display

Mauve will highlight the aligned regions of each genome with a black vertical bar, and clicking the mouse will vertically align the display on the selected orthologous site.

3.7.7.2. Viewing Annotated Features

If the aligned genome sequences were in GenBank files containing annotated features Mauve will display the annotated features immediately below the sequence similarity profiles. Annotated CDS features show up as white boxes, tRNAs are green, rRNAs are red, and misc_RNA features are blue. Features annotated on the reverse strand are on a track immediately below the forward strand track. Repeat_region features are displayed in red on a third annotation track. Mauve displays the product qualifier when the mouse cursor is held over a feature. When a feature is clicked, Mauve shows a detailed listing of feature qualifiers in a popup window. For performance reasons, the annotated sequence features appear only when the display has been zoomed in to view less than 1 Mbp of genome sequence.

3.7.8. Analyzing/Interpreting the Results

Now that the viewer has been explained in detail, we can use it to analyze the results from the *Yersinia* genome alignment. These four *Yersinia* genomes have a rich and well-studied evolutionary history ***(38–40)***. In **ref. *40*** it was reported that transmission by fleabite is a recent evolutionary adaptation that distinguishes *Y. pestis*, the agent of plague, from *Y. pseudotuberculosis* and all other enteric bacteria. The high level of sequence similarity between *Y. pestis* and *Y. pseudotuberculosis* implies that only a few minor genetic changes were needed to induce flea-borne transmission. The question is thus; can we identify these changes using the Mauve genome alignment system? From the global view presented in **Fig. 3** it is difficult to see individual nucleotide substitutions and deletions, so we need to exploit some of the advanced features of the interactive viewer, such as the zoom and gene annotation. **Figure 4** shows a zoomed in view of the one gene responsible for adhesion to the gut of host organisms. The variability in conservation of this gene could contribute to the differences in pathogenicity between *Y. pestis* and *Y. pseudotuberculosis* reported in **ref. *(40)***. Additionally, using the Mauve alignment viewer we can see that the plasmid *pMT1*, which mediates infection of the plague flea vector, is present in all of the *Y. pestis* genomes, but not in *Y. pseudotuberculosis*.

A number of recent studies indicate that microbial genomes evolve and adapt by integrating novel genetic elements through lateral transfer ***(41–43)***. Such

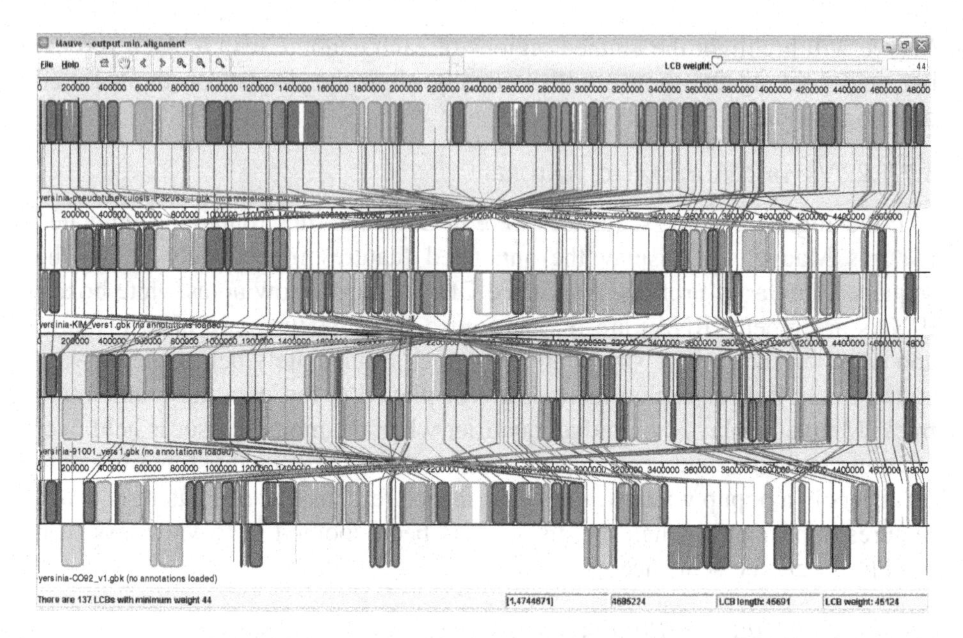

Fig. 3. A Mauve alignment of *Yersinia pseudotuberculosis* IP32953, *Yersinia pestis* KIM, *Y. pestis* 910001, and *Y. pestis* CO92. Notice how inverted regions among the four genomes are clearly depicted as blocks below a genome's center line. The crossing locally collinear block connecting lines give an initial look into the complicated rearrangement landscape among these four related genomes.

novel regions are commonly referred to as genomic islands (GIs) and appear as large gaps in the genome sequence alignment. Mauve identifies large alignment gaps that correspond to putative islands and saves their sequence coordinates in a ".islands" file. Similarly, regions conserved among all genomes are frequently referred to as backbone, and appear as large regions of the alignment that contain only small gaps. Mauve saves the sequence coordinates of backbone segments in a ".backbone" file. Despite their recent speciation, *Y. pestis* and *Y. pseudotuberculosis* contain a number of GIs.

3.7.9. Using the Mauve Command-Line Tool to Align Large Genomes: D. melanogaster *and* D. yakuba

In contrast to the previous example, where each genome contained around 5 million nucleotides, the two *Drosophila* genomes involved in this comparison are each over 100 million nucleotides in size. Although Mauve can efficiently analyze smaller microbial genome comparisons in memory, genomes as large

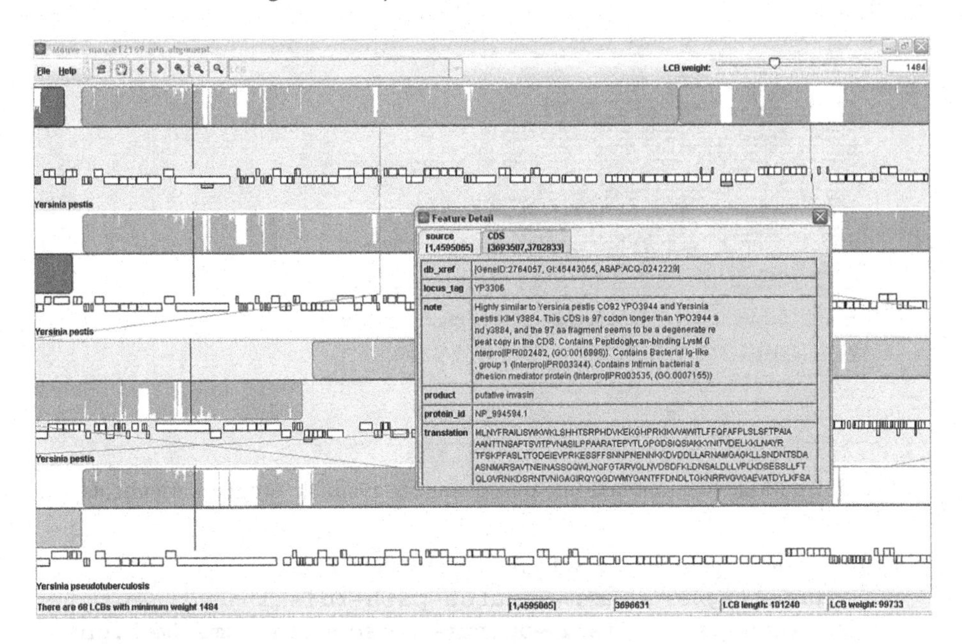

Fig. 4. A closer look at the alignment among the four *Yersinia*. In the regions indicated by the black vertical bar we see a putative orthologous gene with varying levels of conservation among the four genomes. The large white region in the *Yersinia pseudotuberculosis* genome indicates lineage-specific content—either because of ancestral deletion from the *Yersinia pestis* genome or an insertion in the *Y. pseudotuberculosis* lineage. A third possible explanation could be lateral transfer of a homologous allele from an unknown strain. The "Feature Detail" window gives a detailed listing of feature qualifiers for annotated genes which can be displayed by right-clicking a gene and selecting "view GenBank annotation" from the pop-up menu.

as Drosophila would require 2–3 GB of RAM to align in memory. Fortunately, Mauve can utilize two or more scratch disks to reduce memory requirements during generation of unique local alignments—the most memory-intensive part of the alignment process. Thanks to this feature, aligning the two complete *Drosophila* genomes using the Mauve Aligner can be accomplished in four steps:

1. Create sorted mer lists (.sml) for each genome using the scratch-path parameter to identify two or more disks that can be used. The same command must be run once to create an SML file for each genome and once to generate the local alignments (MUMs).

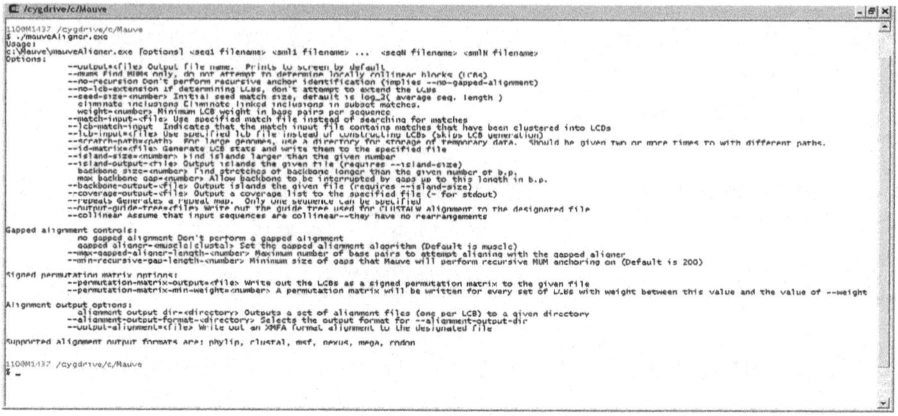

Fig. 5. The Mauve command-line tool listing its available program arguments.

a. Command to generate the SML file for *D. melanogaster*:
   >*mauveAligner* —mums —scratch-path=/disk1 —scratch-path=/disk2 —output=drosophila.mums dmel.gbk dmel.gbk.sml dyak.fas dyak.fas.sml
b. Run again to generate the SML file for *D. yakuba*:
   >*mauveAligner* —mums —scratch-path=/disk1 —scratch-path=/disk2 —output=drosophila.mums dmel.gbk dmel.gbk.sml dyak.fas dyak.fas.sml
c. And one more time to generate the local alignments (MUMs):
   >*mauveAligner* —mums —scratch-path=/disk1 —scratch-path=/disk2 —output=drosophila.mums dmel.gbk dmel.gbk.sml dyak.fas dyak.fas.sml

2. Generate initial alignment anchors without actually aligning:
   >*mauveAligner* —match-input=drosophila.mums —weight=10000 —no-gapped-alignment —no-recursion —no-lcb-extension —output=drosophila_lcb.mums dmel.gbk dmel.gbk.sml dyak.fas dyak.fas.sml
3. Configure MUSCLE alignment parameters to reduce memory usage and speed up the gapped alignment process:
   >*mauveAligner* —match-input=drosophila_lcb.mums —lcb-match-input —muscle-args="-stable —maxiters 1 —diags1 -sv" —output-alignment=drosophila.xmfa dmel.gbk dmel.gbk.sml dyak.fas dyak.fas.sml
4. View the output files using the Mauve alignment display (*see* **Note 3**).
   >*java —Xmx1200m —jar Mauve.jar drosophila.xmfa*

Analyzing the results with the Mauve alignment display, we can see evidence of the two main lineages of gypA and gypB gypsy elements most likely resulting from multiple lateral transfer events reported in **ref.** ***(44)***. *See* **Note 7** for converting the alignment output file into a signed gene-order permutation matrix for use with systems for rearrangement phylogeny such as the GRIMM/MGR server or BADGER. Additional command-line options are shown by running mauveAligner with no arguments, *see* **Fig. 5**.

4. Notes

1. When aligning divergent genomes, the seed size parameter can be reduced to between 9 and 13.
2. If alignment is unacceptably slow or using too much memory, try using ClustalW instead of MUSCLE.
3. If encountering a java.lang.OutOfMemoryError: Java heap space error, try increasing the heap space by running Mauve with the –Xmx command or closing any unused Mauve alignment windows.
4. If the console window is not present or has been closed, it can be reopened in the Help menu->show console.
5. Similar to all existing genome alignment systems, it is possible that Mauve may generate inaccurate alignments. Alignments can be manually edited using the Cinema-MX *(45)* alignment editor that has been incorporated into the Mauve interface. To access this feature, right click on the suspect LCB and then select "Edit this LCB." Then, the Cinema-MX alignment editor window will appear and allow for dynamic alignment correction, and when finished the Mauve interface will update with the adjusted alignment.
6. In the Mauve alignment viewer the LCB connecting lines can be hidden (or made visible again) by typing Shift+L (pressing shift and L simultaneously).
7. The LCBs generated by Mauve in the alignment output file can be transformed into a signed gene-order permutation matrix by supplying the –permutation-matrix-output=<directory> command-line argument. The permutation matrix makes suitable input to systems for rearrangement phylogeny such as the GRIMM/MGR server or BADGER.

Acknowledgments

This work was funded in part by National Institutes of Health grant GM62994-02. A.E.D. was supported by NLM grant 5T15LM007359-05. T.J.T. was supported by Spanish Ministry MECD research grant TIN2004-03382.

References

1. Needleman, S. B. and Wunsch, C. D. (1970) A general method applicable to the search for similarities in the amino acid sequence of two proteins. *J. Mol. Biol.* **48,** 443–453.
2. Smith, T. F. and Waterman, M. S. (1981) Identification of common molecular subsequences. *J. Mol. Biol.* **147,** 195–197.
3. Higgins, D. G. and Sharp, P. M. (1988) CLUSTAL: a package for performing multiple sequence alignment on a microcomputer. *Gene* **73,** 237–244.
4. Notredame, C., Higgins, D. G., and Heringa, J. (2000) T-Coffee: a novel method for fast and accurate multiple sequence alignment. *J. Mol. Biol.* **302,** 205–217.
5. Edgar, R. C. (2004) MUSCLE: multiple sequence alignment with high accuracy and high throughput. *Nucleic Acids Res.* **32,** 1792–1797.
6. Lee, C., Grasso, C., and Sharlow, M. F. (2002) Multiple sequence alignment using partial order graphs. *Bioinformatics* **18,** 452–464.
7. Abouelhoda, M. I. and Ohlebusch, E. (2003) A local chaining algorithm and its applications in comparative genomics. *Algorithms in Bioinformatics, Proceedings* **2812,** 1–16.
8. Haas, B. J., Delcher, A. L., Wortman, J. R., and Salzberg, S. L. (2004) DAGchainer: a tool for mining segmental genome duplications and synteny. *Bioinformatics* **20,** 3643–3646.
9. Hampson, S. E., Gaut, B. S., and Baldi, P. (2005) Statistical detection of chromosomal homology using shared-gene density alone. *Bioinformatics* **21,** 1339–1348.
10. Hampson, S., McLysaght, A., Gaut, B., and Baldi, P. (2003) LineUp: statistical detection of chromosomal homology with application to plant comparative genomics. *Genome Res.* **13,** 999–1010.
11. Tesler, G. (2002) GRIMM: genome rearrangements web server. *Bioinformatics* **18,** 492–493.
12. Spang, R., Rehmsmeier, M., and Stoye, J. (2002) A novel approach to remote homology detection: Jumping alignments. *Journal of Computational Biology* **9,** 747–760.
13. Calabrese, P. P., Chakravarty, S., and Vision, T. J. (2003) Fast identification and statistical evaluation of segmental homologies in comparative maps. *Bioinformatics* **19,** i74–i80.
14. Darling, A. E., Mau, B., Blattner, F. R., and Perna, N. T. (2004) GRIL: genome rearrangement and inversion locator. *Bioinformatics* **20,** 122–124.
15. Durbin, R. (1998) *Biological Sequence Analysis: Probabalistic Models of Proteins and Nucleic Acids*. Cambridge University Press, Cambridge, UK, pp. xi, 356.
16. Fitch, W. M. (2000) Homology a personal view on some of the problems. *Trends Genet.* **16,** 227–231.
17. Larget, B., Kadane, J. B., and Simon, D. L. (2005) A Bayesian approach to the estimation of ancestral genome arrangements. *Mol. Phylogenet. Evol.* **36,** 214–223.

18. Bourque, G. and Pevzner, P. A. (2002) Genome-scale evolution: reconstructing gene orders in the ancestral species. *Genome Res.* **12,** 26–36.
19. Wu, S. and Gu, X. (2003) Algorithms for multiple genome rearrangement by signed reversals. *Pac. Symp. Biocomput.* 363–374.
20. Lu, C. L., Wang, T. C., Lin, Y. C., and Tang, C. Y. (2005) ROBIN: a tool for genome rearrangement of block-interchanges. *Bioinformatics* **21,** 2780–2782.
21. Yancopoulos, S., Attie, O., and Friedberg, R. (2005) Efficient sorting of genomic permutations by translocation, inversion and block interchange. *Bioinformatics* **21,** 3340–3346.
22. Holder, M. and Lewis, P. O. (2003) Phylogeny estimation: traditional and Bayesian approaches. *Nat. Rev. Genet.* **4,** 275–284.
23. Yang, Z., Ro, S., and Rannala, B. (2003) Likelihood models of somatic mutation and codon substitution in cancer genes. *Genetics* **165,** 695–705.
24. Lunter, G., Ponting, C. P., and Hein, J. (2006) Genome-Wide Identification of Human Functional DNA Using a Neutral Indel Model. *PLoS Comput. Biol.* **2,** e5.
25. Darling, A. C., Mau, B., Blattner, F. R., and Perna, N. T. (2004) Mauve: multiple alignment of conserved genomic sequence with rearrangements. *Genome Res.* **14,** 1394–1403.
26. Brudno, M., Malde, S., Poliakov, A., et al. (2003) Glocal alignment: finding rearrangements during alignment. *Bioinformatics* **19,** i54–i62.
27. Ovcharenko, I., Loots, G. G., Giardine, B. M., et al. (2005) Mulan: multiple-sequence local alignment and visualization for studying function and evolution. *Genome Res.* **15,** 184–194.
28. Treangen, T. J. and Messeguer, X. (2006) M-GCAT: interactively and efficiency constructing large-scale multiple genome comparision frameworks in closely related species. *BMC Bioinformatics* **7,** 433.
29. Kurtz, S., Phillippy, A., Delcher, A. L., et al. (2004) Versatile and open software for comparing large genomes. *Genome Biol.* **5,** R12.
30. Hohl, M., Kurtz, S., and Ohlebusch, E. (2002) Efficient multiple genome alignment. *Bioinformatics* **18,** S312–S320.
31. Blanchette, M., Kent, W. J., Riemer, C., et al. (2004) Aligning multiple genomic sequences with the threaded blockset aligner. *Genome Res.* **14,** 708–715.
32. Bray, N. and Pachter, L. (2004) MAVID: constrained ancestral alignment of multiple sequences. *Genome Res.* **14,** 693–699.
33. Brudno, M., Do, C. B., Cooper, G. M., et al. (2003) LAGAN and Multi-LAGAN: efficient tools for large-scale multiple alignment of genomic DNA. *Genome Res.* **13,** 721–731.
34. Choi, K. P., Zeng, F., and Zhang, L. (2004) Good spaced seeds for homology search. *Bioinformatics* **20,** 1053–1059.
35. Henz, S. R., Huson, D. H., Auch, A. F., Nieselt-Struwe, K., and Schuster, S. C. (2005) Whole-genome prokaryotic phylogeny. *Bioinformatics* **21,** 2329–2335.

36. Saitou, N. and Nei, M. (1987) The neighbor-joining method: a new method for reconstructing phylogenetic trees. *Mol. Biol. Evol.* **4,** 406–425.
37. Cleri, D. J., Vernaleo, J. R., Lombardi, L. J., et al. (1997) Plague pneumonia disease caused by Yersinia pestis. *Semin. Respir. Infect.* **12,** 12–23.
38. Carniel, E. (2003) Evolution of pathogenic Yersinia, some lights in the dark. *Adv. Exp. Med. Biol.* **529,** 3–12.
39. Chain, P. S., Carniel, E., Larimer, F. W., et al. (2004) Insights into the evolution of Yersinia pestis through whole-genome comparison with Yersinia pseudotuberculosis. *Proc. Natl. Acad. Sci. USA* **101,** 13,826–13,831.
40. Hinnebusch, B. J. (2005) The evolution of flea-borne transmission in Yersinia pestis. *Curr. Issues Mol. Biol.* **7,** 197–212.
41. Perna, N. T., Plunkett, G., 3rd, Burland, V., et al. (2001) Genome sequence of enterohaemorrhagic Escherichia coli O157:H7. *Nature* **409,** 529–533.
42. Hsiao, W. W., Ung, K., Aeschliman, D., Bryan, J., Finlay, B. B., and Brinkman, F. S. (2005) Evidence of a large novel gene pool associated with prokaryotic genomic islands. *PLoS Genet.* **1,** e62.
43. Tettelin, H., Masignani, V., Cieslewicz, M. J., et al. (2005) Genome analysis of multiple pathogenic isolates of Streptococcus agalactiae: implications for the microbial "pan-genome". *Proc. Natl. Acad. Sci. USA* **102,** 13,950–13,955.
44. Terzian, C., Ferraz, C., Demaille, J., and Bucheton, A. 2000) Evolution of the Gypsy endogenous retrovirus in the Drosophila melanogaster subgroup. *Mol. Biol. Evol.* **17,** 908–914.
45. Lord, P. W., Selley, J. N., and Attwood, T. K. (2002) CINEMA-MX: a modular multiple alignment editor. *Bioinformatics* **18,** 1402–1403.

11

Visualization of Syntenic Relationships With SynBrowse

Volker Brendel, Stefan Kurtz, and Xioakang Pan

Summary

Synteny is the preserved order of genes between related species. To detect syntenic regions one usually first applies sequence comparison methods to the genomic sequences of the considered species. Sequence similarities detected in this way often require manual curation to finally reveal the—in many cases—hidden syntenies. The open source software SynBrowse provides a convenient interface to visualize syntenies on a genomic scale. SynBrowse is based on the well-known GBrowse software. In this chapter, we describe the basic concepts of SynBrowse and show how to apply it to different kinds of data sets. Our exposition includes the description of software pipelines for the complete process of synteny detection: (1) applying standard software to compute sequence similarities, (2) parsing, combining, and storing detected similarities in a standard database, (3) installing, configuring, and using SynBrowse. The complete set of programs making up these pipelines as well as the data sets used are available on the SynBrowse homepage.

Key Words: Genome alignment; genome annotation; synteny blocks.

1. Introduction

The availability of complete or near-complete genome sequences of various species sampled from different clades of molecular phylogenies provides great opportunities for comparative and evolutionary genomics studies. The underlying assumption in these studies is that conserved sequences in generally diverged species are retained because of selective constraints, and, thus, such conserved sequences are candidate functional elements in the genomes. Detailed

From: *Methods in Molecular Biology, vol. 396: Comparative Genomics, Volume 2*
Edited by: N. H. Bergman © Humana Press Inc., Totowa, NJ

analysis of conserved regions will typically lead to the identification of protein-coding and noncoding RNA genes as well as regulatory elements involved in gene expression. One major focus of comparative and evolutionary genomics is the search for "synteny," a term literally meaning "same thread" that is used in sequence analysis to mean a set of conserved genes and other features that share the same relative ordering on the chromosomes of two species or between duplicated chromosome segments within one species. Comparisons of synteny regions provide important clues to evolutionary events that underlie species diversification *(1)*. In addition, the identification of syntenic regions is critical to studying chromosome evolution, in particular providing evidence for genome polyploidization and potential subsequent diploidization events *(2–5)*.

Visual displays of the extent of sequence conservation and chromosomal order of conserved elements are an important technical aide in comparative genomics studies. In this chapter we describe the use of SynBrowse, a generic and portable web-based synteny browser for comparative sequence analysis *(6)*. SynBrowse provides scalable views of conserved sequence elements, which are supplied to the software in standardized format, thus allowing output of any sequence-level comparison tools to be displayed. Typical use of SynBrowse involves evaluation of a "query" (or "focus") sequence relative to a "reference" sequence. Often, but not necessarily, the reference sequence will be well-annotated, whereas the query sequence represents a novel sequence. In this case, the overlay of sequence conservation with annotated features in the reference sequence (which are displayed on different tracks of the SynBrowse view) suggests tentative annotations for the query sequence. We discuss installation and general configuration of SynBrowse and illustrate the use of the software with a few typical examples.

2. Materials

2.1. Overview

SynBrowse is a GBrowse *(7)* family software tool that is written in Perl and builds on open source BioPerl modules such as Bio::Graphics and Bio::DB::GFF *(8)*. It extends GBrowse, a generic single genome browser, to a synteny browser for sequence comparisons. The software consists of a web-based front end and a back end comprising a minimum of two relational databases that contain annotation and alignment data for the query and reference species, respectively. Each species database may contain data for multiple sequences, for example all the chromosomes of a species or a collection of bacterial artificial chromosome (BAC) sequences. Either protein-level or nucleotide-level sequence alignments can serve as the key comparison feature.

In addition, a display option can be set to view only blocks of synteny to give an overall view of sequence conservation. Zooming functions allow detailed viewing of the conserved elements within each synteny block.

2.2. Installation of SynBrowse and Input Format Specification

SynBrowse extends GBrowse, which must be installed before installing SynBrowse. Both software distributions can be downloaded from the Generic Model Organism Database Project homepage (http://www.gmod.org). Additional scripts associated with the SynBrowse project are available from the SynBrowse homepage (http://www.synbrowse.org/). Installation instructions are provided with the software distribution. Installation steps are similar to those for GBrowse (*see* **Note 1**), except that SynBrowse requires the set up of the two backend relational databases. At this point, MySQL has been most frequently used as database backend for SynBrowse (*see* **Note 2**), although other relational databases can also easily be adopted (*see* **Note 3**). The MySQL databases can be populated with data provided in GFF files (http://www.sanger.ac.uk/Software/formats/GFF/; http://song.sourceforge.net/gff3.shtml; *see* **Note 4**) using BioPerl scripts. For example, for each species database, all sequences of the given species are identified by records as illustrated in **Fig. 1**. The protein- and nucleotide-level alignments are similarly specified as described in the following sections.

2.3. Configuration

Settings applicable to the entire SynBrowse installation are set in a general configuration file. In this file, the system administrator must specify the set of

```
XooMAFF  chromosome  Sequence  1 4940217   .  +  .  Sequence XooMAFF

AY632360 segment     Sequence  1 135862    .  +  .  Sequence AY632360
```

Fig. 1. Sample GFF format records for the chromosomes or sequence segments to be made accessible from SynBrowse. The first nine columns are separated by a "tab," and the last column is separated by a space. The first column should be chromosome or segment id (e.g., XooMAFF identifies the entire chromosome of *Xanthomonas oryzae*, and AY632360 refers to cotton BAC AY632360). The second column in each record must be "chromosome" or "segment" and the third column (GFF class name) must be "Sequence." The fourth and fifth columns are the start and end positions of a chromosome or segment. The sixth and eighth columns with "." represent null values for GFF columns "score" and "phase," respectively. The seventh column indicates the DNA strand. The last tab-delimited column is the GFF "group" designation and in this case contains a repeat of the class name followed by the sequence identifier.

```
# comparison = qry_sequence ref_sequence alignment_track1 alignment_track2
comparison1  = Cotton       Cotton_ref   Protein_Align    DNA_Align
comparison2  = XooMAFF      XccATCC      Protein_Align
```

Fig. 2. Arrangements of comparative species and their alignments in the General.conf configuration file. The values of # in the "comparison#" lines represent the order of the compared species. The first column after the equation mark "=" is the query or focus species. The second column is the reference species. The third and forth columns are the track names of protein or nucleotide alignments, respectively.

allowable species comparisons, the tracks of the key comparative alignments, and other global parameters. An example is given in **Fig. 2**. Similarly, species-specific configuration files identify the back end relational databases and control the track displays of the comparison features for each species according to the GBrowse configuration format (*see* **Note 5**).

3. Methods

Application of SynBrowse involves the following steps:

1. Generation of annotation and alignment data.
2. Formatting of annotation and alignment data in GFF.
3. Identification of synteny blocks.
4. Formatting of synteny block information in GFF.
5. Upload of data.
6. Adjustment of configuration parameters.

Annotation data should be supplied in the same format as specified for GBrowse (*see* **Notes 4** and **5**). In the following sections we discuss examples of generating nucleotide-level alignment data, protein-level alignment data, and identification of synteny blocks.

3.1. Nucleotide-Level Alignments

Comparisons displayed with SynBrowse can be based on nucleotide level alignments. We illustrate this with a comparison of two homoeologous BACs derived from the A_T and D_T genomes of *Gossypium hirsutum* (cotton), respectively. The two BACs correspond to GenBank accessions AY632360 (our query sequence) and AY632359 (our reference sequence).

1. The two BACs were compared at the nucleotide level with the BlastZ *(9)*, a whole genome nucleotide alignment program. The BlastZ parameters C, K, and H were set to 2, 2500, and 2200, respectively (*see* **Note 6**).
2. The BlastZ-to-SynBrowse-parser.pl, available at http://www.synbrowse.org/pub_source/, was used to generate a GFF representation of the alignment data.

```
AY632360  conserved50  similarity  16519  16546  57  +  .  Target "Sequence:AY632359" 5782 5809
AY632360  conserved90  similarity  16549  16614  92  +  .  Target "Sequence:AY632359" 5810 5875
......
AY632360  conserved90  similarity  18420  18482  90  +  .  Target "Sequence:AY632359" 7636 7698
AY632360  align80      similarity  16519  18482  80  +  .  Target "Sequence:AY632359" 5782 7698
AY632360  conserved60  similarity  18508  18515  63  +  .  Target "Sequence:AY632359" 7699 7706
AY632360  conserved60  similarity  18533  18550  67  +  .  Target "Sequence:AY632359" 7707 7724
......
AY632360  conserved90  similarity  19935  20153  94  +  .  Target "Sequence:AY632359" 9073 9291
AY632360  align80      similarity  18508  20153  80  +  .  Target "Sequence:AY632359" 7699 9291


AY632359  conserved50  similarity  5782   5809   57  +  .  Target "Sequence:AY632360" 16519 16546
AY632359  conserved90  similarity  5810   5875   92  +  .  Target "Sequence:AY632360" 16549 16614
......
AY632359  conserved80  similarity  7636   7698   88  +  .  Target "Sequence:AY632360" 18420 18482
AY632359  align80      similarity  5782   7698   80  +  .  Target "Sequence:AY632360" 16519 18482
AY632359  conserved70  similarity  7699   7706   78  +  .  Target "Sequence:AY632360" 18508 18515
AY632359  conserved90  similarity  7707   7724   95  +  .  Target "Sequence:AY632360" 18533 18550
......
AY632359  conserved90  similarity  9073   9291   94  +  .  Target "Sequence:AY632360" 19935 20153
AY632359  align80      similarity  5782   9291   80  +  .  Target "Sequence:AY632360" 16519 20153
```

Fig. 3. Nucleotide level alignment positions in GFF format. The upper panel shows data uploaded into the database that contains data for sequence AY632360, whereas the lower panel shows data uploaded into the database which contains data for sequence AY632359. Alignment records in upper panel are reciprocal to the alignment records in the lower panel. The first nine columns are separated by a "tab," whereas entries in the last tab-delimited column are separated by spaces. The first column is the BAC identifier. The second column in each record must be "align" or "conserved" appended with an identity percentage in deciles, and the third column must be "similarity." The source column keyword "conserved" represents a single gap-free matching region (e.g., a conserved exon), whereas the "align" keyword indicates an alignment typically comprised of several successive gap-free regions (such as a conserved gene spanning several conserved exons). The fourth and fifth columns are the start and end positions of an alignment on the query BAC. The sixth column gives the identity percentage of the match. The seventh column represents the query sequence strand of the match. The eighth column is populated with null values ".", as it is reserved for phase information for coding sequence features which is not applicable here. The ninth column is in the format Target "Sequence:BAC id" appended with the start and end positions of the alignment matches on the reference BAC.

Partial output is shown in **Fig. 3**. Each alignment is required to enter reciprocally in two GFF files used in the next step for each of the two compared sequences.

3. Two MySQL databases were created. One is for AY632360-related gene annotations and the alignment data in which AY632360 is considered the query sequence. The other is for AY632359-related gene annotations and the alignment data in which AY632359 is considered the query sequence.
4. The bp_bulk_load_gff.pl and bp_fast_load_gff.pl scripts distributed with BioPerl were used to upload the GFF files into the databases.
5. **Figure 4** shows a screen shot of the related alignment visualization in SynBrowse after appropriate editing of the configuration files (*see* **Subheading 2.3.**).

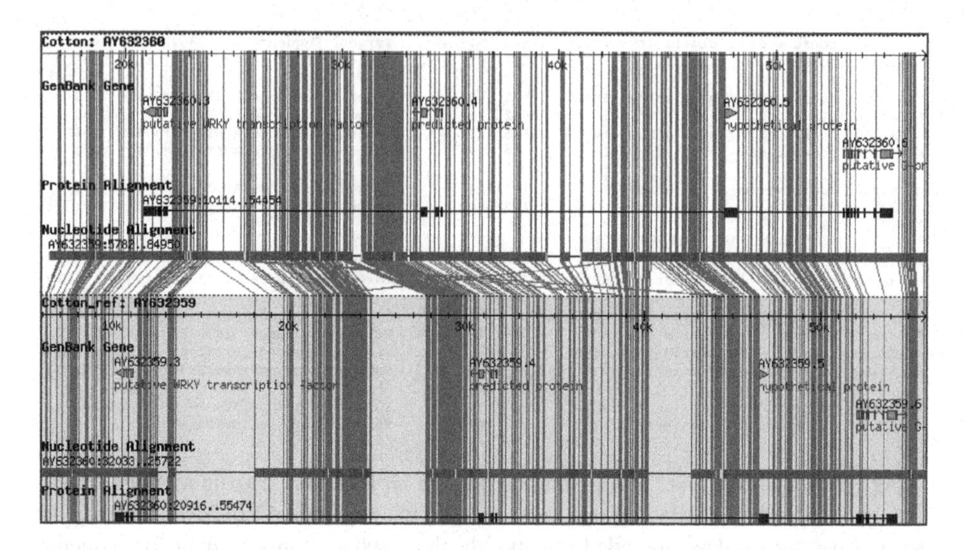

Fig. 4. Using nucleotide alignment as key comparative feature to show extensive synteny between CesA1 homoeologous BACs, A_T and D_T, two coresident genomes of cotton species *Gossypium hirsutum*. The conserved regions cover not only genetic segments but also intergenic areas.

3.2. Protein-Level Alignments

Alignment data can also be generated on the protein level and displayed on a protein alignment track, provided the coordinates of the matching segments are given relative to the chromosomal sequences being compared. We illustrate this option with a comparison of the two cotton BACs on the protein level.

1. Protein alignments were generated by aligning annotated coding sequences of the reference sequences to the query sequence using the GenomeThreader spliced alignment tool ***(10)***. GenomeThreader produces similar output to that of the GeneSeqer spliced alignment tool ***(11)***, which is extensively used for nucleotide-level spliced alignments (***[12]***; *see* **Note 7**).
2. The GeneSeqer-to-SynBrowse-parser.pl, available at http://www.synbrowse.org/pub_source/, was used to generate a GFF representation of the alignment data. Partial output is shown in **Fig. 5**. Each alignment is required to enter reciprocally in two GFF files used in the next step for each of the two compared sequences.
3. The bp_bulk_load_gff.pl and bp_fast_load_gff.pl scripts distributed with BioPerl were used to upload the GFF files into the databases.
4. **Figure 6** shows a screen shot of the related alignment visualization in SynBrowse after appropriate editing of the configuration files (*see* **Subheading 2.3.**).

```
AY632360  coding90   similarity  21800 21940   0.970   -  .  Target "Sequence:AY632359" 10987 11126
AY632360  coding100  similarity  21548 21694   1.000   -  .  Target "Sequence:AY632359" 10743 10889
AY632360  coding90   similarity  20916 21446   0.950   -  .  Target "Sequence:AY632359" 10114 10645
AY632360  gene90     similarity  20916 21939   0.962   -  .  Target "Sequence:AY632359" 10114 11126
......
AY632360  coding100  similarity  53102 53197   1.000   +  .  Target "Sequence:AY632359" 52079 52174
......
AY632360  coding90   similarity  55388 55474   0.973   +  .  Target "Sequence:AY632359" 54365 54454
AY632360  gene90     similarity  53102 55477   0.967   +  .  Target "Sequence:AY632359" 52079 54454


AY632359  coding90   similarity  10987 11126   0.970   -  .  Target "Sequence:AY632360" 21800 21940
AY632359  coding100  similarity  10743 10889   1.000   -  .  Target "Sequence:AY632360" 21548 21694
AY632359  coding90   similarity  10114 10645   0.950   -  .  Target "Sequence:AY632360" 20916 21446
AY632359  gene90     similarity  10114 11126   0.962   -  .  Target "Sequence:AY632360" 20916 21939
......
AY632359  coding100  similarity  52079 52174   1.000   +  .  Target "Sequence:AY632360" 53102 53197
......
AY632359  coding90   similarity  54365 54454   0.973   +  .  Target "Sequence:AY632360" 55388 55474
AY632359  gene90     similarity  52079 54454   0.967   +  .  Target "Sequence:AY632360" 53102 55477
```

Fig. 5. Protein level alignments in GFF format. The upper panel shows sample alignment data for the database that contains data for sequence AY632360, whereas the lower panel shows the corresponding entries for the database which contains data for sequence AY632359. Alignment records in upper panel are reciprocal to the alignment records in lower panel. Columns are spaced and populated as explained in the legend to **Fig. 3**. The second column in each record must be "gene" or "coding" appended with an identity percentage in deciles. The source keyword "gene" indicates entirely conserved coding gene regions, whereas the keyword "coding" represents matching coding exons.

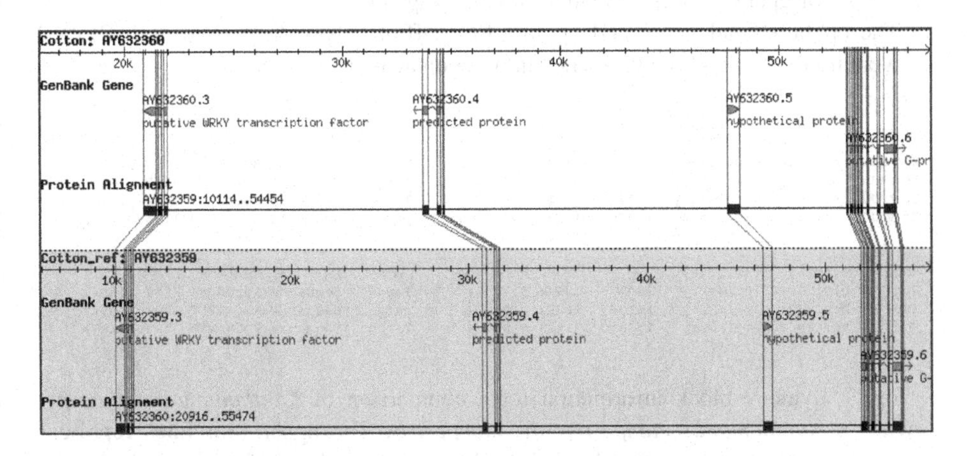

Fig. 6. Using protein alignments as key comparative feature to show conserved gene order between CesA1 homoeologous BACs, A_T and D_T, two coresident genomes of cotton species *Gossypium hirsutum*. Sequence conservation is high in genic regions, with four conserved genes detected in the two compared sequences.

3.3. Identification of Synteny Blocks

We illustrate identification and display of synteny blocks with a prokaryotic genome annotation example. In this case, we use *Xanthomonas campestris pv. campestris* strain ATCC 33913 (GenBank accession NC_ 003902; our abbreviation: XccATCC) as the reference genome and explore synteny-based gene annotation for *Xanthomonas oryzae pv. oryzae* MAFF 311018 (GenBank accession NC_007705; our abbreviation: XooMAFF).

1. We used the GenBank XccATCC.gff file (obtained from the National Center for Biotechnology Information ftp site ftp://ftp.ncbi.nih.gov/genomes/Bacteria/Xanthomonas_campestris/ as a template for uploading the reference genome annotation. Some reformatting was necessary (*see* **Note 3**).
2. We generated all open reading frames of lengths at least 100 codons in all six frames for both sequences. These open reading frames were then pairwise compared using BlastP ***(13)***. The BlastP output was piped through the MuSeqBox output parser ***(14)*** and reformatted to generate both GFF files for upload into the SynBrowse databases and input for the synteny identification step.
3. Synteny blocks were identified using the gpcluster program (available at http://www.synbrowse.org/pub_source/). This program reads files specifying "gene locations" and pairs of "duplicated genes." It clusters two duplicated genes, if their chromosomal locations are close together relative to some user-supplied distance. Single-linkage clusters are considered syntenic blocks.
4. Output of gpcluster was formatted in GFF (**Fig. 7**).
5. The bp_bulk_load_gff.pl and bp_fast_load_gff.pl scripts distributed with BioPerl were used to upload the GFF files into the databases.

```
XooMAFF  block60  similarity  2      2744   0.63  +  .  Target "Sequence:XccATCC" 3 2743
XooMAFF  block70  similarity  3957   13584  0.72  +  .  Target "Sequence:XccATCC" 3507 12956
XooMAFF  block80  similarity  14042  15913  0.82  +  .  Target "Sequence:XccATCC" 14104 16461
XooMAFF  block80  similarity  17076  21873  0.88  +  .  Target "Sequence:XccATCC" 116600 121321

XccATCC block60  similarity   3          2743 0.63   +  .  Target "Sequence:XooMAFF" 2 2744
XccATCC block70  similarity   3507      12956 0.72   +  .  Target "Sequence:XooMAFF" 3957 13584
XccATCC block80  similarity   14104     16461 0.82   +  .  Target "Sequence:XooMAFF" 14042 15913
XccATCC block80  similarity   116600 121321 0.88   +  .  Target "Sequence:XooMAFF" 17076 21873
```

Fig. 7. Synteny block information for a comparison of *Xanthomonas oryzae pv. oryzae* vs *Xanthomonas campestris pv. campestris*. The upper four lines represent partial GFF input into the XooMAFF database, and the lower four lines depict the corresponding entries uploaded into the XccATCC database. The GFF method column (third) must be specified as "similarity," and the second column must be specified as "block" for protein-level alignments ("region" for nucleotide-level alignments), appended by an identity percentage in deciles.

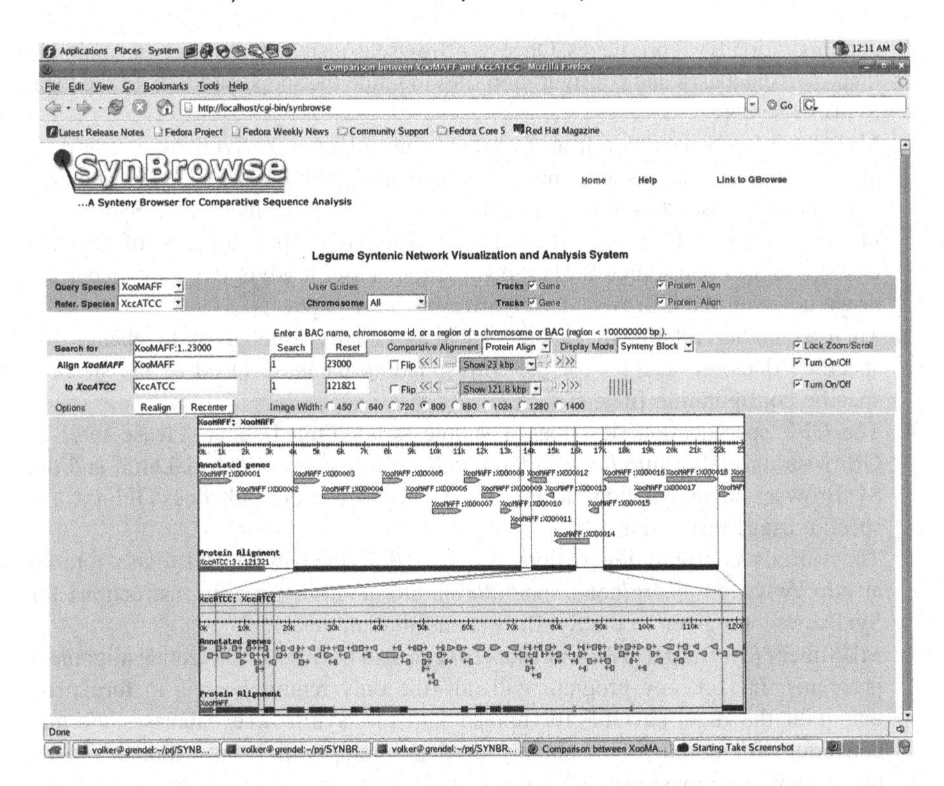

Fig. 8. Syntenic blocks for the *Xanthomonas oryzae pv. oryzae* vs *Xanthomonas campestris pv. campestris* comparison in the first 23 Kb of the query species. The gpcluster gap parameter was set to 1000. Note that the first three blocks map to the beginning of the *Xanthomonas campestris pv. campestris* chromosome, whereas the fourth block appears to have been shuffled into a different location. Users can manipulate the display by shifting, realigning, zooming, or flipping the region. Clicking on a syntenic block, a gene or another alignment on the upper panel will zoom in to a page showing a detailed view of the clicked object.

6. **Figure 8** shows a screen shot of the related alignment visualization in SynBrowse after appropriate editing of the configuration files (*see* **Subheading 2.3.**).

4. Notes

1. Both Gbrowse and SynBrowse are distributed as source code (binary packages available for some platforms) and come with detailed installation instructions. The code is written entirely in Perl (http://www.perl.org). Several Perl modules will have to be installed as prerequisites. Initial setup typically requires system

administration level privileges. Once SynBrowse is installed, any user can maintain their own databases and configuration files to guide the scope and look of the local SynBrowse copy.

2. MySQL is freely available from http://www.mysql.com/. A SynBrowse user does not have to learn the ins and outs of a relational database system. A few standard commands are used to set up a particular database, as described in the SynBrowse installation notes. Data are ingested using BioPerl scripts. Because of frequent changes in the open source BioPerl distribution, a user is advised to look up current usage notes on the GBrowse and SynBrowse homepages.
3. To use another relational database such as Oracle, users need to change the specified database adaptor for the corresponding database in all of the species-specific configuration files.
4. The GFF specification describes a simple tab-delimited input. Please refer to GBrowse tutorial at http://www.gmod.org/nondrupal/tutorial/tutorial.html and the SynBrowse documentation at http://www.synbrowse.org/pub_docs/GFF.txt for specific usage instructions.
5. The various configuration options are best described in the GBrowse tutorial at http://www.gmod.org/nondrupal/tutorial/tutorial.html. Specific instructions for SynBrowse are given in the SynBrowse installation notes.
6. MUMmer ***(15)***, BlastN ***(13)***, and Blat ***(16)*** can also be used as nucleotide alignment programs. In fact, any program will do—the only requirement is to format its output in the required GFF for upload into the SynBrowse databases. Format converters for output of commonly used programs will be distributed via the http://www.synbrowse.org/pub_source/ site.
7. BlastX and BlastP can also be used for the comparisons of both prokaryotic and eukaryotic genomes at the protein level. BlastX output can be parsed into GFF for input to SynBrowse using the BlastX-to-SynBrowse-parser.pl script, available at the http://www.synbrowse.org/pub_source/ site. The exon positions in the query species are predicted based upon the conserved coding regions related to annotated exons in the reference species.

Acknowledgments

This work was supported in part by the Specific Cooperative Agreement no. 58-3625-5-124 of the USDA-ARS to V.B.

References

1. Frazer, K. A., Elnitski, L., Church, D. M., Dubchak, I., and Hardison, R. C. (2003) Cross-species sequence comparisons: a review of methods and available resources. *Genome Res.* **13,** 1–12.
2. Wendel, J. F. (2000) Genome evolution in polyploids. *Plant Mol. Biol.* **42,** 225–249.

3. Guiliano, D. B., Hall, N., Jones, S. J. M., et al. (2002) Conservation of long-range synteny and microsynteny between the genomes of two distantly related nematodes. *Genome Biology* **3,** 1–14.
4. Bowers, J. E., Chapman, B. A., Rong, J., and Paterson, A. H. (2003) Unravelling angiosperm genome evolution by phylogenetics analysis of chromosomal duplication events. *Nature* **422,** 433–438.
5. Cannon, S. B., McCombie, W. R., Sato, S., et al. (2003) Evolution and microsynteny of apyrase gene family in three legume genomes. *Mol. Gen. Genomics* **270,** 347–361.
6. Pan, X., Stein, L., and Brendel, V. (2005) SynBrowse: a synteny browser for comparative sequence analysis. *Bioinformatics* **21,** 3461–3468.
7. Stein, L. D., Mungall, C., Shu, S., et al. (2002) The generic genome browser: A building block for a model organism system database. *Genome Res.* **12,** 1599–1610.
8. Stajich, J. E., Block, D., Boulez, K., et al. (2002) The BioPerl toolkit: Perl modules for the life sciences. *Genome Res.* **12,** 1611–1618.
9. Schwartz, S., Kent, W. J., Smit, A., et al. (2003) Human-mouse alignments with BLASTZ. *Genome Res.* **13,** 103–107.
10. Gremme, G., Brendel, V., Sparks, M. E., and Kurtz, S. (2005) Engineering a software tool for gene prediction in higher organisms. *Information and Software Technol.* **47,** 965–978.
11. Brendel, V., Xing, L., and Zhu, W. (2004) Gene structure prediction from consensus spliced alignment of multiple ESTs matching the same genomic locus. *Bioinformatics* **20,** 1157–1169.
12. Dong, Q., Lawrence, C. J., Schlueter, S. D., et al. (2005) Comparative plant genomics resources at PlantGDB. *Plant Physiol.* **139,** 610–618.
13. Altschul, S. F., Madden, T. L., Schäffer, A. A., et al. (1997) Gapped BLAST and PSI-BLAST: a new generation of protein database search programs. *Nucleic Acids Res.* **25,** 3389–3402.
14. Xing, L. and Brendel, V. (2001) Multi-query sequence BLAST output examination with MuSeqBox. *Bioinformatics* **17,** 744–745.
15. Kurtz, S., Phillippy, A., Delcher, A. L., et al. (2004) Versatile and open software for comparing large genomes. *Genome Biology* **5,** R12.
16. Kent, W. J. (2002) BLAT: the BLAST-like alignment tool. *Genome Res.* **12,** 656–664.

12

Gecko and GhostFam

Rigorous and Efficient Gene Cluster Detection in Prokaryotic Genomes

Thomas Schmidt and Jens Stoye

Summary

A popular approach in comparative genomics is to locate groups or clusters of orthologous genes in multiple genomes and to postulate functional association between the genes contained in such clusters. For a rigorous and efficient detection in multiple genomes, it is essential to have an appropriate model of gene clusters accompanied by efficient algorithms locating them. The Gecko method described herein was designed to serve as a basic tool for the detection and visualization of gene cluster data in prokaryotic genomes founded on a formal string-based gene cluster model.

Key Words: Comparative genomics; gene cluster; Gecko; GhostFam; common intervals.

1. Introduction

The comparative analysis of gene order in completely sequenced prokaryotes showed that they share many gene clusters, which are sets of genes in close proximity to each other, but not necessarily contiguous nor in the same order. The existence of such gene clusters has been explained in different ways: by functional selection *(1)*, operon formation *(2,3)*, and other processes in evolution which affect the gene order and content *(4)*. Consequently, the conservation of gene order is a source of information for many fields in genomic research, like the prediction of groups of functionally associated genes *(5–7)*

From: *Methods in Molecular Biology, vol. 396: Comparative Genomics, Volume 2*
Edited by: N. H. Bergman © Humana Press Inc., Totowa, NJ

or the extraction of evolutionary information *(8)* to develop distance measures allowing the reconstruction of phylogenetic trees *(9)*.

1.1. The Challenge of Exhaustive Gene Cluster Analysis

Primarily, the main task of a computer program for gene cluster analysis is the detection of all gene clusters of a given set of genomes above a defined level of significance. The subsequent analysis of the cluster frequency, length distribution, and gene content then gives a first idea of the evolutionary relatedness of the selected genomes. A more detailed analysis of single clusters points at possible functional associations and might elucidate so far hidden functions of encoded proteins. Usually, these functions and associations will then be further investigated by other methods including lab experiments. Therefore, an automated method for gene cluster detection is useful to generate well founded hypotheses for interacting genes, reducing the search space for further time and cost intensive experiments.

A method for a systematic search for conserved genomic regions on a set of multiple genomes must fulfill the following requirements. (1) Speed: to find all gene clusters appearing in at least two of a hundred or more genomes, each containing up to 10,000 genes, the time and space complexity must scale at most quadratic in the number of genes and genomes. (2) Flexibility: a meaningful gene cluster model not only must be able to recognize clusters appearing in all, but also those appearing in a few genomes, and also if they are only partially conserved. (3) Completeness: to allow the detection of so far uncharacterized gene clusters, it is necessary to output all occurring gene clusters, not only those sharing a particular query gene. (4) Interactive visualization: the large amount of cluster data requires an overview for easy navigation, as well as the option to select and analyze single clusters.

The method Gecko described herein fulfills these criteria as follows. The cluster detection part of Gecko uses an efficient implementation of the fastest known algorithm to find gene clusters based on the model of common intervals over strings *(10)*. Then, the gene clusters found in all pairwise genome comparisons are used as seeds to detect also partially conserved clusters in further genomes. Because the detection algorithm is not a heuristic, it guarantees to find all gene clusters according to the user-defined parameters: minimum cluster size, minimum number of genomes a cluster occurs in, and minimum rate of matching genes in partially conserved clusters. The output of the detection algorithm is given by the complete list of all located clusters, sorted by the number of genes in the cluster. From this list, the user can select single clusters for more detailed graphical representation of the contained genes together with their functional annotation. The stand-alone version of Gecko was written in JAVA and can be obtained free of charge for noncommercial use from http://bibiserv.techfak.uni-bielefeld.de/gecko together with some sample data and documentation. A further description of the underlying model, the detailed algorithm and an evaluation on real data can be found in **ref. *11***.

1.2. The Formal Gene Cluster Model in Brief

In the last decade, several approaches have been developed for the analysis of gene neighborhoods in prokaryotic genomes, *see* **ref. *12*** for a survey. However, so far none of these practical approaches was based on a formal definition of a gene cluster, such that the results were not easily comparable among each other. To solve this problem, in recent years formal models for gene clusters have been developed ***(10,13–16)***. The method Gecko described herein is based on such a model. The basic ideas are given in the following, before the tool is described in detail in the next two sections.

A genome is modeled as a sequence (or *string*) of numbers, representing the genes. If the same number appears more than once in the sequence, these are paralogous copies of a gene that we can not (or do not want to) distinguish from each other. Given such a string S, a *substring* that starts with the ith and ends with the jth character of S is denoted $S[i, j]$. The set of all genes contained in a substring $S[i, j]$ is called its *character set* $CS(S[i, j]) := \{S[k] | i \leq k \leq j\}$. Note that the order and the number of occurrences of genes in a character set is irrelevant.

Given a set of genes C, a CS-location of C in S is a contiguous region of S that contains exactly the genes in $C = CS(S[i, j])$. The CS-location is *maximal* if $S[i-1]$ and $S[j+1]$ are not elements of C.

In the comparison of multiple genomes, we consider a collection of k strings $\sigma = (S_1, S_2, \ldots, S_k)$ such that, whenever the same number occurs in different strings, this denotes orthologous genes. A set of genes C that has at least one CS-location in each S_i is called a *common CS-factor* of σ. The set $L(C)$ of all maximal CS-locations of a common CS-factor C is called its *location set*. Common CS-factors correspond to gene clusters that are present in all of the considered genomes.

We study the following problem: given a collection of k genomes, each of length at most n, find all its common CS-factors, and for each of these report its location set.

For $k = 2$ strings, the problem was well studied in recent years. Using an efficient coding (fingerprint) of the subalphabets of substrings, Amir et al. ***(17)*** developed an algorithm that takes $O(n^2 \log^2 n)$ time where we assume that the number of different genes is in $\Theta(n)$. Didier *(15)* improved upon this result presenting an $O(n^2\log n)$ time algorithm using a tree-like data structure. For the Gecko program we developed the worst-case optimal $O(n^2)$ algorithm "Connecting Intervals" (CI) based on elementary data structures *(10)*.

An additional advantage of Algorithm CI is that it can easily be generalized to more than two strings. The general idea is that a set of genes C is a common CS-factor of $\sigma = (S_1, S_2, \ldots, S_k)$ if and only if it is a (pairwise) common CS-factor of a fixed string (e.g., S_1) and all other strings in σ. Therefore, Algorithm CI is applied to each pair of input strings (S_1, S_r) with $S_r \in \sigma$ and $1 \leq r \leq k$, where for $r = 1$ we locate paralogous copies of gene clusters in S_1.

In practice, the probability of finding a cluster of genes that is conserved in all input genomes decreases rapidly with an increasing number of genomes. Therefore, it is even more interesting to find clusters that appear in only a subset of k' out of the k given genomes. This is possible by another simple extension of Algorithm CI, running in time $O(k(k-k'+1)n^2)$.

A disadvantage of the gene cluster model as described so far is its strictness with respect to deviation of gene content between the genomes: two common intervals must always contain exactly the same set of genes. That is why in Gecko a postprocessing step is performed in which clusters that share a certain amount of their genes are joined together into larger clusters that no longer need to contain exactly the same set of genes in each genome.

2. Materials

As previously described, the input for the gene cluster detection method Gecko are genomes, represented by sequences of numbers, representing the genes. Genes that are assumed to be homologous get the same number.

One possible way to obtain the homology information required for this representation is the use of the COG (Clusters of Orthologous Groups of proteins, http://www.ncbi.nlm.nih.gov/COG) database *(18)*. In this case, genes from the same COG cluster are considered homologous and are represented by the number of their COG cluster.

Genes belonging to no cluster are represented by the identifier "0". A disadvantage using the COG database is that it does not contain all completely sequenced prokaryotic genomes, and from those genomes, which are in the database, many genes do not belong to any COG cluster. On the other hand, the database also contains some too "inclusive" clusters (e.g., for the ABC transport system: 36 different periplasmic components, 54 different ATPase components, 67 different permease components), making it difficult to assign genes, especially from new genomes, to their correct corresponding cluster. For this reason, we developed the GhostFam method as a part of Gecko to allow a more flexible definition of groups of putative homologs.

The data flow in GhostFam and Gecko can be divided into the following steps (*see also* **Fig. 1**):

1. Preparation of the input data for Gecko.

   a. Create the homology information using GhostFam.

      i. Create a file containing the list of input genomes.
      ii. Collect the annotation for the coding regions.
      iii. Collect the coding regions of genomes and perform an all-against-all TBLASTN comparison.
      iv. Use the created files as input for the GhostFam procedure. Build families of putative homologs using the manual editor provided with the method, analyze and if necessary modify the given results.
      v. Export the final gene family information.

   or

   b. Use a data file containing sequences of COG numbers.

2. Run Gecko to compute the classification of genes into conserved clusters.
3. Analyze the displayed result in the graphical browser and use the export function to report your results.

In the next section, these steps are explained in more detail along with a running example.

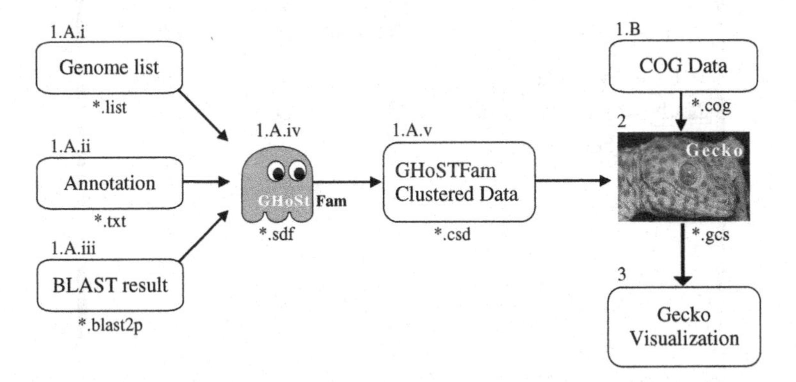

Fig. 1. Data flow. The figure shows the required data input files for each part of the method together with the typical file extensions. Note that Gecko allows two alternative data sources (.cog or .csd). The extensions displayed next to the applications are the defaults for the saved session files.

3. Methods

Here, we describe how Gecko and GhostFam are applied to detect conserved gene clusters, given a set of bacterial genomes. The subsections refer to the steps of the data flow scheme from the previous section.

3.1. Preparation of the Input Data

Figure 2 shows the possible input data files for Gecko. Export data files created in a GhostFam session have extension .csd, COG data files have extension .cog. Previously saved Gecko sessions have the file extension .gcs.

a. Preparation of gene order data with GhostFam. A GhostFam session starts with the selection of genomes that are available for the comparative genomics approach. Then a number of data files must be created, followed by the semiautomatic clustering into gene families, before the final result can be exported.

   i. The project list file. The project list file summarizes the information about the project: the genome names together with the short and unique tags and some source information (*see* **Fig. 3**).

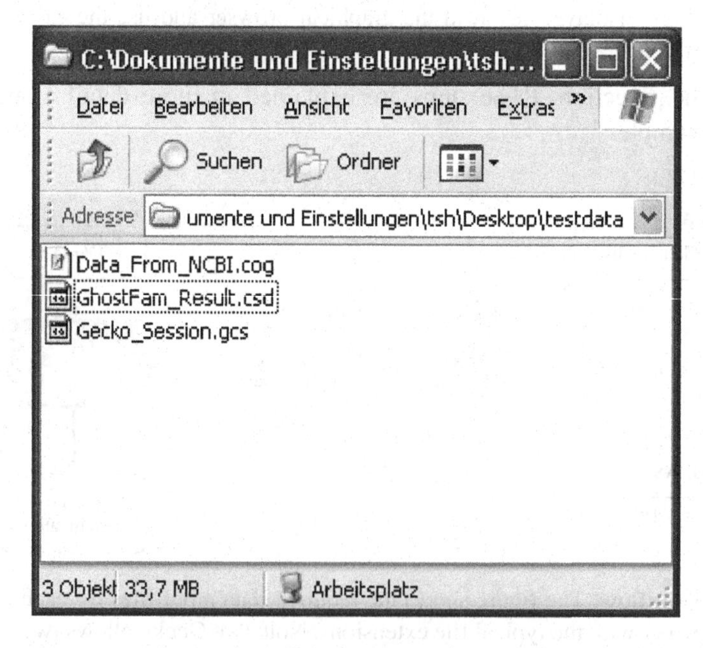

Fig. 2. Input files for the Gecko method.

```
EMBL_Genomes.list - WordPad
Datei  Bearbeiten  Ansicht  Einfügen  Format  ?

C.efficiens       Coref   ce   /vol/biodb/genomes/Bacteria/cefficiens_YS314/BA000035.embl
C.glutamicum      Corgl   cg   /vol/biodb/genomes/Bacteria/cglutamicum_ATCC13032/BX927147.embl
E.coli K12        Ecoli   ec   /vol/biodb/genomes/Bacteria/ecoli_K12/U00096.embl
L.lactis          Lacla   ll   /vol/biodb/genomes/Bacteria/llactis_IL1403/AE005176.embl
L.xyli            Leixy   lx   /vol/coryne/Cg_Blastdatenbanken/Bacterial_Genomes/Leixy_AE016822.embl
M.avium           Mycav   ma   /vol/biodb/genomes/Bacteria/mavium/AE016958.embl
M.leprae          Mycle   ml   /vol/biodb/genomes/Bacteria/mleprae_TN/AL450380.embl
M.tuberculosis    Myctu   mt   /vol/biodb/genomes/Bacteria/mtuberculosis/AL123456.embl
N.farcinica       Nocfa   nf   /vol/coryne/Cg_Blastdatenbanken/Bacterial_Genomes/Nocfa_AP006618.embl
P.acnes           Proac   pb   /vol/coryne/Cg_Blastdatenbanken/Bacterial_Genomes/Pacnes_AE017283.embl
P.aeruginosa      Pseae   pa   /vol/biodb/genomes/Bacteria/paeruginosa/AE004091.embl
P.putida          Psepu   pp   /vol/biodb/genomes/Bacteria/pputida_KT2440/AE015451.embl
S.avermitilis     Strav   sa   /vol/biodb/genomes/Bacteria/savermitilis/BA000030.embl
S.coelicolor      Strco   sc   /vol/biodb/genomes/Bacteria/scoelicolor/AL645882.embl
S.thermophilum    Symth   st   /vol/coryne/Cg_Blastdatenbanken/Bacterial_Genomes/Symth_AP006840.embl
T.whipplei        Trowh   tw   /vol/biodb/genomes/Bacteria/twhipplei_Twist/AE014184.embl
W.succinogenes    Wolsu   ws   /vol/biodb/genomes/Bacteria/wsuccinogenes/BX571656.embl

Drücken Sie F1, um die Hilfe aufzurufen.
```

Fig. 3. Project list file for GhostFam.

ii. The annotation files. To load gene annotation information into GhostFam, for each genome an annotation file can be created, containing for each ORF in the genome information about the coding direction, identification numbers in different databases, and the gene name and functional annotation if available. The structure of such a file is shown in **Fig. 4**.

```
Cordi_annotation.txt - WordPad
Datei  Bearbeiten  Ansicht  Einfügen  Format  ?

Cordi_00001  cd0001  + 553  19..1677                   CAE48512.1  DIP0001 dnaA  chromosomal replication initiator protein Similar t
Cordi_00002  cd0002  + 406  2286..3503                 CAE48513.1  DIP0002 dnaN  DNA polymerase III, beta chain      Similar to Myco
Cordi_00003  cd0003  + 398  3542..4735                 CAE48514.1  DIP0003 recF  DNA replication and repair protein  Similar to Myco
Cordi_00004  cd0004  + 184  4725..5276                 CAE48515.1  DIP0004 DIP0004     Conserved hypothetical protein       Similar t
Cordi_00005  cd0005  + 686  5379..7436                 CAE48516.1  DIP0005 gyrB  DNA gyrase subunit B    Similar to Mycobacterium sm
Cordi_00006  cd0006  - 146  complement(7503..7940)     CAE48517.1  DIP0006 DIP0006     Hypothetical protein     Doubtful CDS. No sign
Cordi_00007  cd0007  - 91   complement(8201..8473)     CAE48518.1  DIP0007 DIP0007     Hypothetical protein     Doubtful CDS. No sign
Cordi_00008  cd0008  - 68   complement(8478..8681)     CAE48519.1  DIP0008 DIP0008     Conserved hypothetical protein       Similar t
Cordi_00009  cd0009  + 857  8792..11362                CAE48520.1  DIP0009 gyrA  DNA gyrase subunit A    Similar to Mycobacterium sm
Cordi_00010  cd0010  + 115  11362..11706               CAE48521.1  DIP0010 DIP0010     Putative membrane protein      Similar in its
Cordi_00011  cd0011  - 243  complement(12176..12904)   CAE48522.1  DIP0011 DIP0011     Putative regulatory protein    Similar to Esch
Cordi_00012  cd0012  - 557  complement(12992..14662)   CAE48523.1  DIP0012 lctP  L-lactate permease      Similar to Escherichia coli
Cordi_00013  cd0013  + 158  15159..15632               CAE48524.1  DIP0013 DIP0013     Hypothetical protein    No significant databa
Cordi_00014  cd0014  + 548  15721..17364               CAE48525.1  DIP0014 DIP0014     Putative membrane protein      Low similarity
Cordi_00015  cd0015  + 549  17354..19000               CAE48526.1  DIP0015 DIP0015     Putative transport system, ATP-binding protei
Cordi_00016  cd0016  + 107  <19447..19767              CAE48527.1  DIP0016 DIP0016     Putative membrane protein      No database mat
Cordi_00017  cd0017  + 55   19865..20029               CAE48528.1  DIP0017 DIP0017     Hypothetical protein     Doubtful CDS. No stro
Cordi_00018  cd0018  + 66   20524..20721               CAE48529.1  DIP0018 DIP0018     Hypothetical protein     Doubtful CDS. No stro
Cordi_00019  cd0019  - 82   complement(20713..20958)   CAE48530.1  DIP0019 DIP0019     Hypothetical protein     Doubtful CDS. No stro
Cordi_00020  cd0020  - 77   complement(20955..21185)   CAE48531.1  DIP0020 DIP0020     Hypothetical protein     Doubtful CDS. No stro
Cordi_00021  cd0021  + 333  21381..22379               CAE48532.1  DIP0021 DIP0021     Putative solute-binding lipoprotein Similar t
Cordi_00022  cd0022  + 344  22376..23407               CAE48533.1  DIP0022 DIP0022     Putative ABC transport protein, membranecompo
Cordi_00023  cd0023  + 254  23408..24169               CAE48534.1  DIP0023 DIP0023     Putative ABC transport protein, ATP-bindingpr
Cordi_00024  cd0024  - 203  complement(24217..24825)   CAE48535.1  DIP0024 DIP0024     Putative membrane protein      Possible membra
Cordi_00025  cd0025  + 181  24888..25430               CAE48536.1  DIP0025 ppiA  probable peptidyl-prolyl cis-trans isomerase A  Sim

Drücken Sie F1, um die Hilfe aufzurufen.
```

Fig. 4. Annotation file for GhostFam. The file contains the unique long-tag in the first column followed by the short-tag, coding direction, length in amino acids, location on the genome, TrEMBL-ID, Locus-Tag, gene name, and the functional annotation.

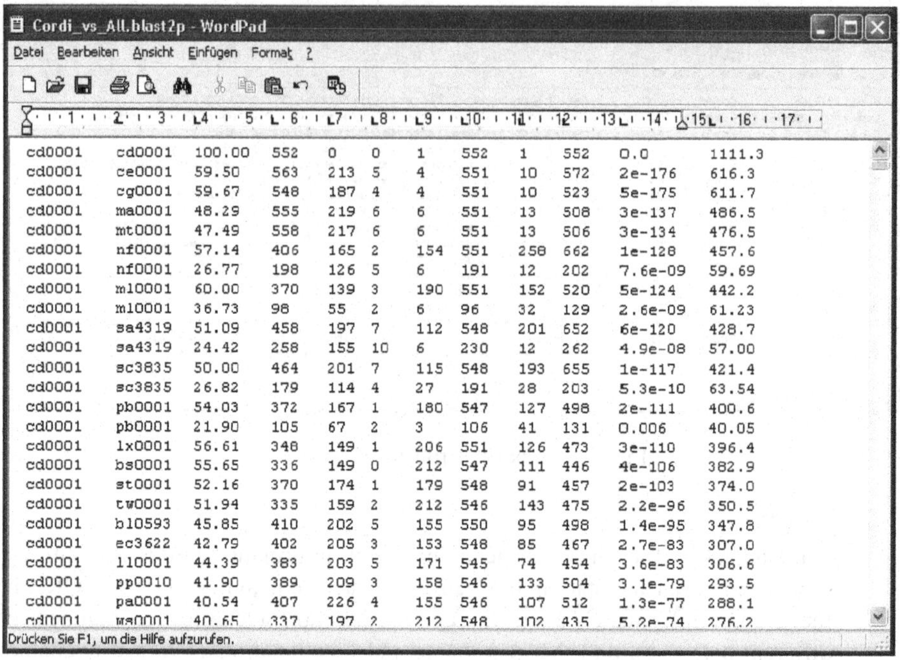

Cordi_vs_All.blast2p - WordPad

Datei Bearbeiten Ansicht Einfügen Format ?

```
cd0001   cd0001   100.00  552  0    0   1    552  1    552  0.0       1111.3
cd0001   ce0001   59.50   563  213  5   4    551  10   572  2e-176    616.3
cd0001   cg0001   59.67   548  187  4   4    551  10   523  5e-175    611.7
cd0001   ma0001   48.29   555  219  6   6    551  13   508  3e-137    486.5
cd0001   mt0001   47.49   558  217  6   6    551  13   506  3e-134    476.5
cd0001   nf0001   57.14   406  165  2   154  551  258  662  1e-128    457.6
cd0001   nf0001   26.77   198  126  5   6    191  12   202  7.6e-09   59.69
cd0001   ml0001   60.00   370  139  3   190  551  152  520  5e-124    442.2
cd0001   ml0001   36.73   98   55   2   6    96   32   129  2.6e-09   61.23
cd0001   sa4319   51.09   458  197  7   112  548  201  652  6e-120    428.7
cd0001   sa4319   24.42   258  155  10  6    230  12   262  4.9e-08   57.00
cd0001   sc3835   50.00   464  201  7   115  548  193  655  1e-117    421.4
cd0001   sc3835   26.82   179  114  4   27   191  28   203  5.3e-10   63.54
cd0001   pb0001   54.03   372  167  1   180  547  127  498  2e-111    400.6
cd0001   pb0001   21.90   105  67   2   3    106  41   131  0.006     40.05
cd0001   lx0001   56.61   348  149  1   206  551  126  473  3e-110    396.4
cd0001   bs0001   55.65   336  149  0   212  547  111  446  4e-106    382.9
cd0001   st0001   52.16   370  174  1   179  548  91   457  2e-103    374.0
cd0001   tw0001   51.94   335  159  2   212  546  143  475  2.2e-96   350.5
cd0001   bl0593   45.85   410  202  5   155  550  95   498  1.4e-95   347.8
cd0001   ec3622   42.79   402  205  3   153  548  85   467  2.7e-83   307.0
cd0001   ll0001   44.39   383  203  5   171  545  74   454  3.6e-83   306.6
cd0001   pp0010   41.90   389  209  3   158  546  133  504  3.1e-79   293.5
cd0001   pa0001   40.54   407  226  4   155  546  107  512  1.3e-77   288.1
cd0001   ms0001   40.65   337  197  2   212  548  102  435  5.2e-74   276.2
```

Drücken Sie F1, um die Hilfe aufzurufen.

Fig. 5. Structure of an input file obtained from a *BLAST2P* comparison with the output option "-*m*8," during the comparison of *C. diphtheriae* with 19 other genomes.

iii. The BLAST data files. The last type of input data files for GhostFam contain the results of an all-against-all BLAST comparison in which each ORF of each of the genomes is compared with all other ORFs in the same and in all other genomes. The results are stored in a separate file for each genome, as shown in **Fig. 5**. The relevant columns for GhostFam are the rate of matching amino acids (column 3) and the *e*-value (column 11). Columns 7–10 containing the start and end positions of the alignment in the sequences are used to calculate the length of the alignment compared to the length of the ORF.

iv. Semiautomatic family determination. After all files previously described are created (*see* **Fig. 6** for a complete list), a new GhostFam session can be started, in which families of putative paralogous and orthologous genes are determined. The clustering into families is performed in two steps. In the first step, families of putative paralogs are computed for each genome separately. In the second step, for each family of paralogs of each genome, corresponding families in the other genomes are searched.

After each of these two steps, the result of the procedure is presented to the user in form of different tables and a graphical cluster representation. This allows

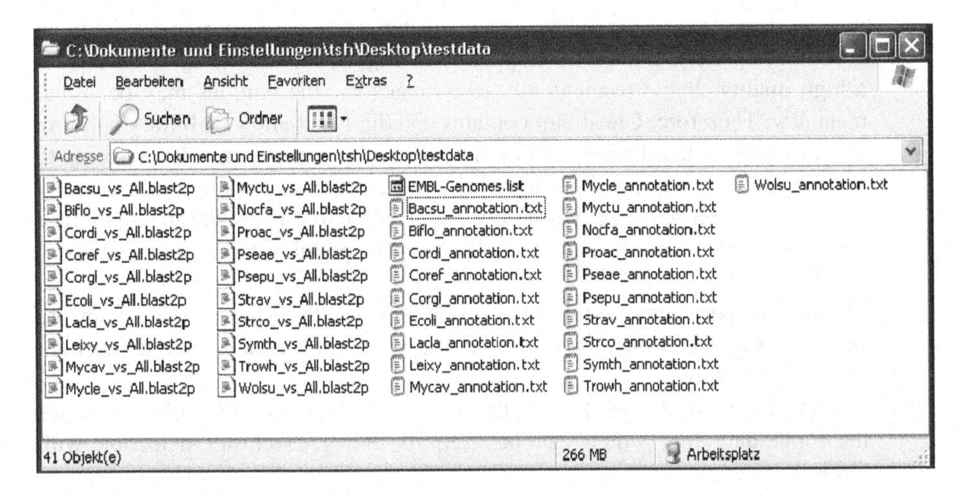

Fig. 6. Overview of the files required for a GhostFam run with 20 genomes.

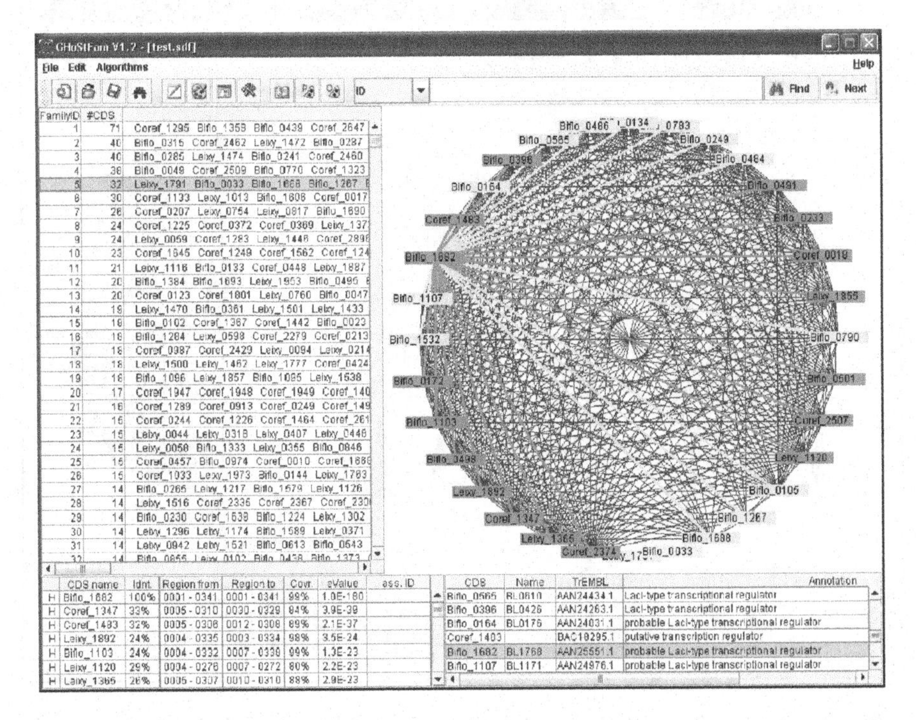

Fig. 7. An almost perfectly connected family of orthologs. Thin black connections indicate a high similarity between two genes. A light gray line is drawn between two genes if they are not highly similar but fulfill the similarity criteria.

manual inspection of the quality of the clustering. If the clustering results in an almost completely connected graph like in **Fig. 7**, the clustering is usually of a high quality. If the result is not yet satisfactory, the families may be edited manually. Therefore, GhostFam contains an editor that allows arbitrary splitting and merging of families. For example, the family depicted in **Fig. 8** seems to imply that two groups of genes have been merged into one family by a relatively weak connection between two highly connected parts. Here, the manual editor could be used to separate the family into two or possibly more groups (**Figs. 9** and **10**).

The default parameter settings for the automatic clustering of GhostFam are shown in **Fig. 11**. Some hints for an optimization of the parameter settings are discussed in **Note 1**.

v. Export of the family data. Finally the data export function of GhostFam creates the input data file, which can be used for the gene cluster detection with Gecko.

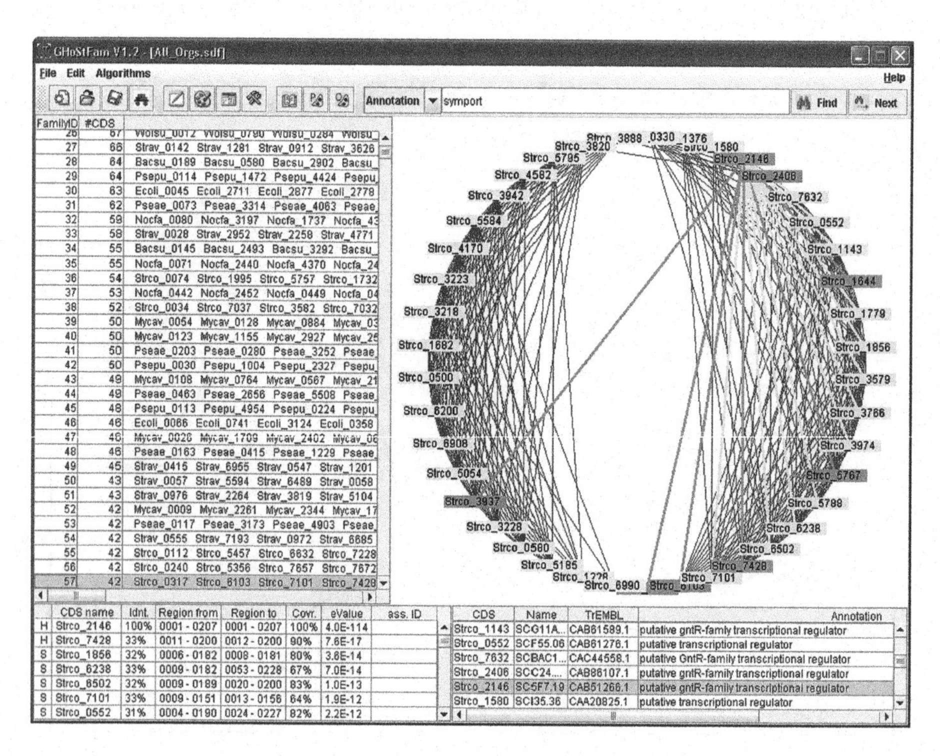

Fig. 8. Two families of orthologs falsely connected by an accidental high similarity of two ORFs.

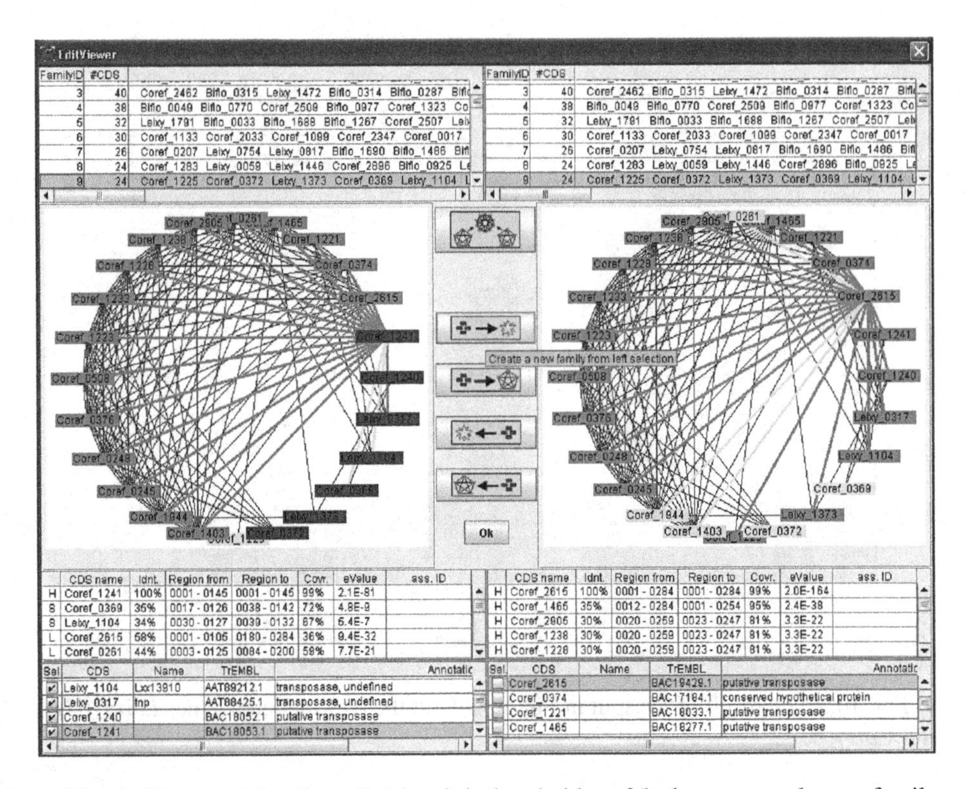

Fig. 9. The manual editor. On the right-hand side a falsely connected gene family is depicted. The same family is shown on the left-hand side with seven marked genes (dark gray boxes), which are manually selected for being split into a separate family. The result is shown in the next figure.

b. Import of gene order data from COG. The extraction of the genome data from the National Center for Biotechnology Information database to create the input file for Gecko is straight forward and depicted in **Fig. 12**.

3.2. Using Gecko

Independent of the source of the input data, after its loading into Gecko, one can select which of the included genomes should be used for gene cluster detection (**Fig. 13**). Three parameters are chosen: the minimal number of sequences that a gene cluster has to be contained in, the minimal number of genes a cluster has to contain, and the threshold for minimal cluster similarity

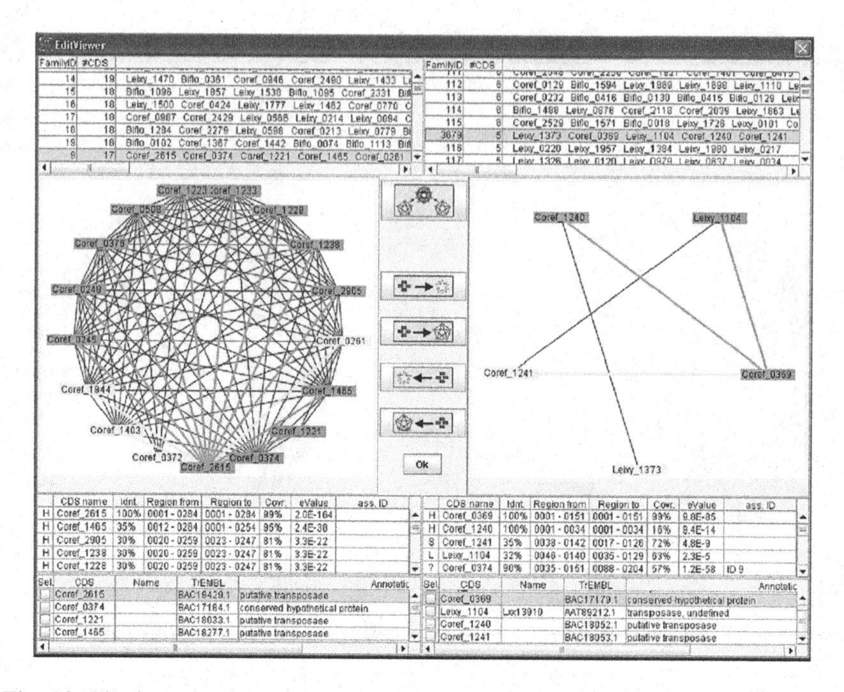

Fig. 10. Final result of the split in **Fig. 9** without the two singleton genes *Leixy_0317* and *Coref_0372*.

in the postprocessing. When Gecko finishes the cluster detection, all found clusters are displayed in a summary table ordered by their size (*see* **Fig. 14**, top left). Some hints for an optimization of the quality and speed are discussed in **Notes 2–4.**

3.3. Analyzing the Results

From the cluster summary table one can select a cluster for more detailed inspection, which is then shown in a graphical representation below the table (**Fig. 14**, bottom left). Gecko moreover provides easy access to the annotation of any gene displayed in form of a tool-tip with information about its position in the genome, its name, and annotation. By clicking on a gene or interleaving region (italic blue numbers), all the available information about the selected location is displayed in an additional window.

For the export of results from Gecko or GhostFam, both tools allow saving the cluster graphics in common file formats.

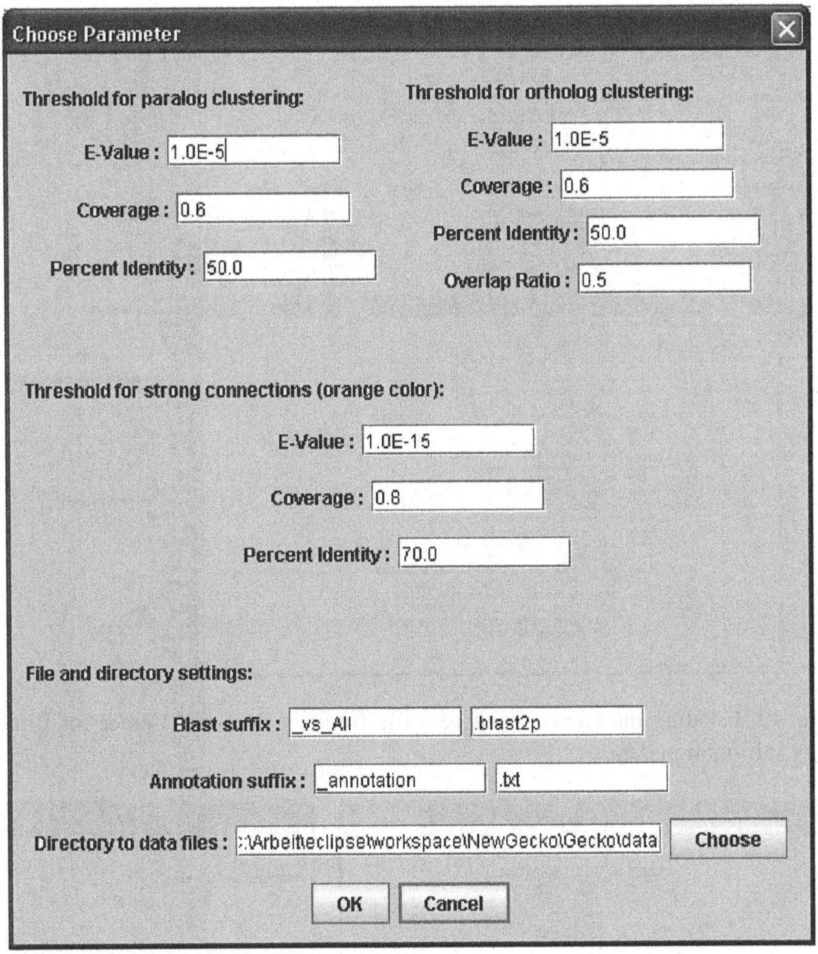

Fig. 11. The default parameter setting for the GhostFam method.

4. Notes

In this section, we give some hints on how to achieve the best results using GhostFam (*see* **Note 1**) and Gecko (*see* **Notes 2–4**).

1. Improving the results of GhostFam. Essential for obtaining good results with the GhostFam method is the selection of optimal parameters for the classification of putative paralogs and orthologs. The default parameter settings given in **Fig. 11** are a first suggestion extracted from our initial experiments with the method. A rule of thumb is to use similar values in separation of paralogs and of orthologs.

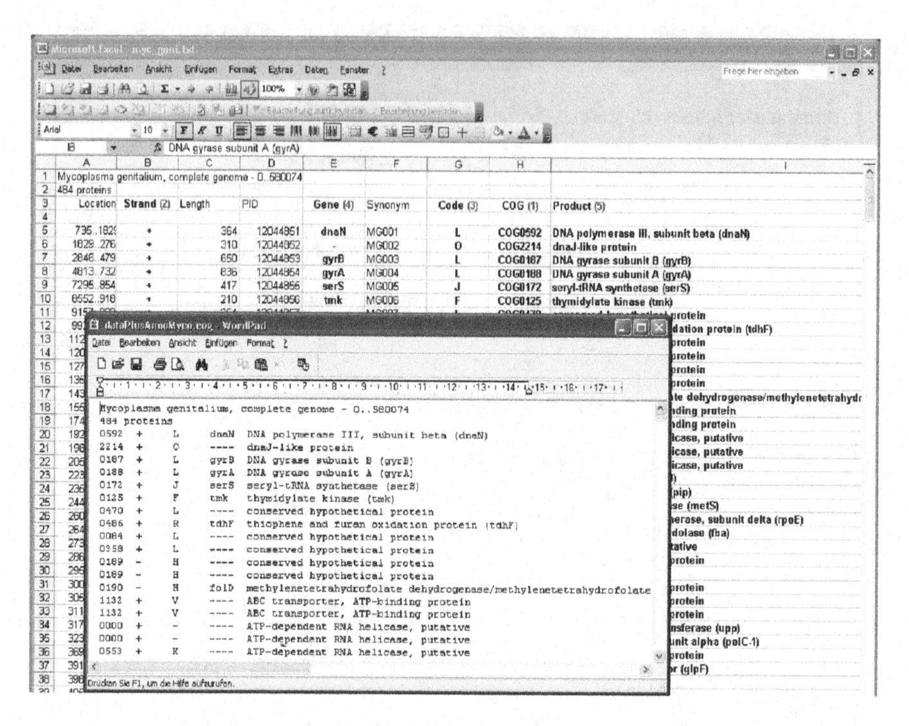

Fig. 12. Creating the Gecko input data file from the National Center for Biotechnology Information data.

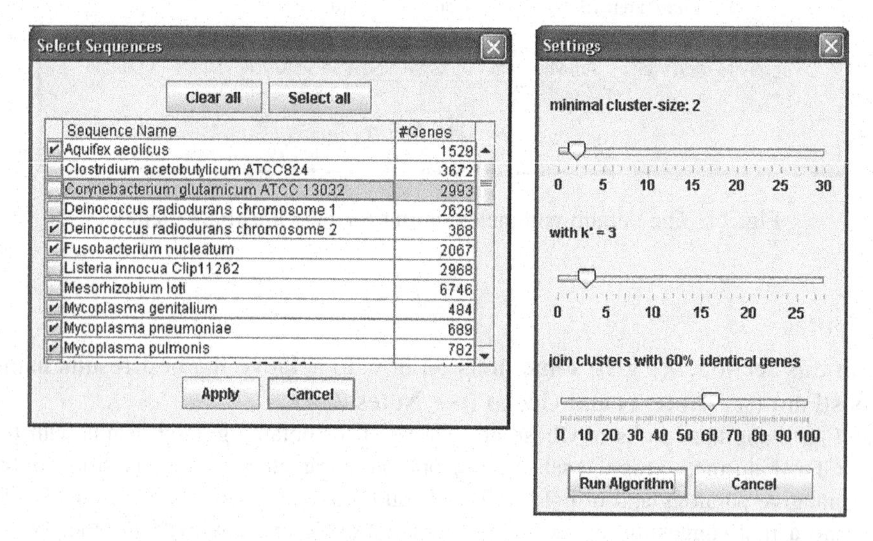

Fig. 13. Sequence and parameter selection in Gecko.

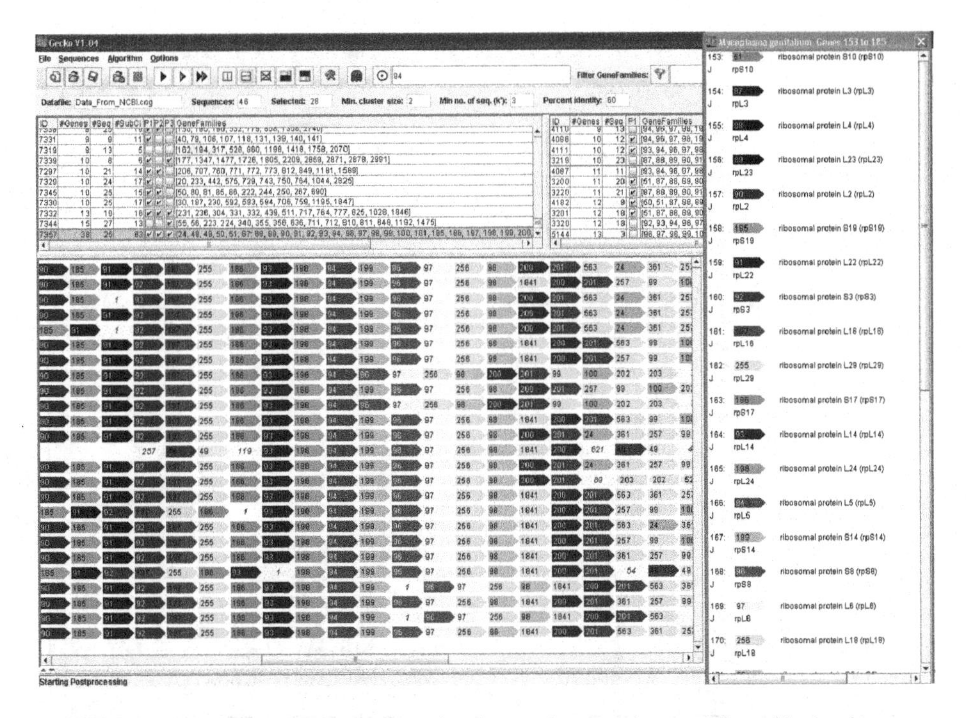

Fig. 14. Graphical representation of a detected gene cluster. The table in the upper left region shows the list of all detected gene clusters. The window below shows the selected cluster graphically. Each arrow represents a gene, where the same color (also the same number) denotes genes from the same family.

Otherwise results like that in **Fig. 15** are likely, where a loose definition of paralogs together with too strict a setting for orthologs led to a somehow random grouping.

2. Selecting the set of genomes. The choice of sequences to be used in the cluster detection is a crucial point for the evaluation of the reported results, because the conservation of gene order depends on the evolutionary distances of the analyzed species. On the one hand, if the aim of the cluster detection is to find groups of genes that are probably functionally related, it is important to choose genomes of a sufficient phylogenetic distance. On the other hand, if the detected gene clusters are used to analyze the evolutionary development of a genomic region in a particular lineage, the computed results become more significant the more genomes from this lineage are used in the cluster detection.

   Additionally, in the selection dialog for each genome the number of contained genes is listed. This number can give a first hint indicating whether an organism

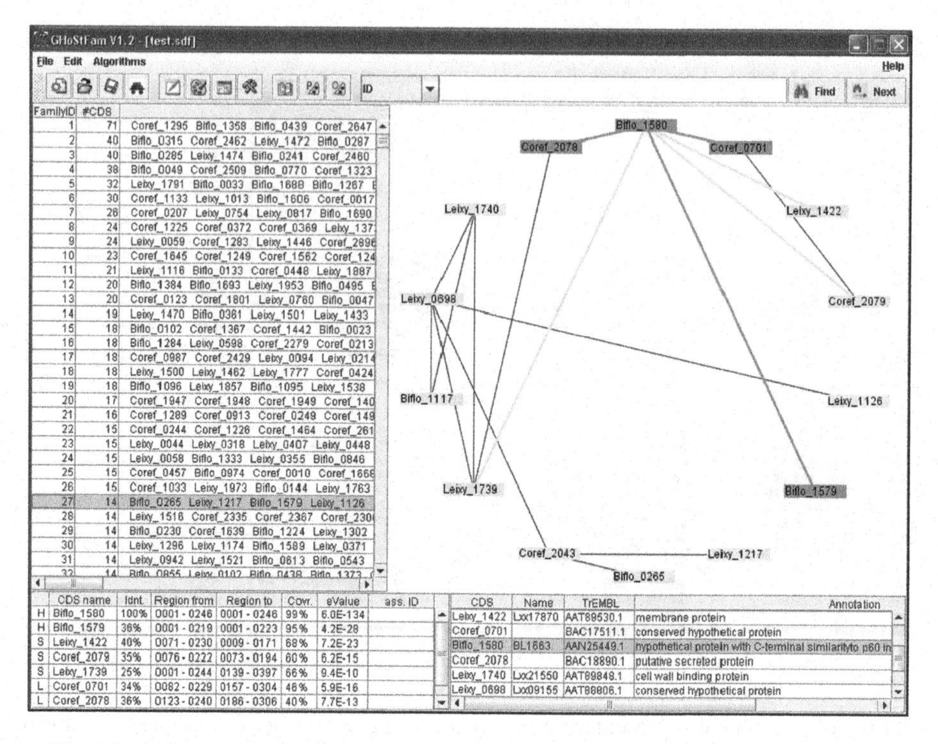

Fig. 15. A disrupted gene cluster caused by too different criteria for putative paralogs and orthologs.

is able to synthesize all its required products for the housekeeping functions by itself, or depends on an environment providing these metabolites. For the analysis of presence or absence of particular gene clusters encoding elementary functions of the organism, the total number of genes in a genome also provides a certain estimation about the number of gene clusters to be expected.

3. Avoiding runtime and quality problems. The running time of Gecko not only depends on the number and size of genomes considered, but also on the parameter "identity rate" of the postprocessing phase. Here, it turned out that the best reconstructions resulted from values between 55 and 75 %. Within this range most of the clusters were reconstructed correctly without too many false-positives. Values less than 50 % usually cause a significant increase in runtime while getting less meaningful results. On the other hand, values more than 80 % often result in an incomplete clustering, leaving many conserved fragments in the set of all conserved clusters.
4. Memory problems. The speed of Gecko is partially paid with the price of increased space consumption. Therefore, in some cases it could be practical to perform the

computation of the gene clusters on a powerful compute server with larger memory and save the session for later continuation on a laptop. Opening an existing session usually requires less than a 10th of the memory needed for the cluster computation.

Acknowledgments

The authors wish to thank Christian Rückert and Jörn Kalinowski for their helpful discussions on the topic of gene clusters and their valuable feedback during the development of GhostFam and Gecko.

References

1. Overbeek, R., Fonstein, M., D'Souza, M., Pusch, G. D., and Maltsev, N. (1999) The use of gene clusters to infer functional coupling. *Proc. Natl. Acad. Sci. USA* **96,** 2896–2901.
2. Bork, P., Snel, B., Lehmann, G., et al. (2000) Comparative genome analysis: exploiting the context of genes to infer evolution and predict function, in *Comparative Genomics,* (Sankoff, D. and Nadeau, J. H., eds.), Kluwer Academic Publishers, pp. 281–294.
3. Lathe, W. C., III, Snel, B., and Bork, P. (2000) Gene context conservation of a higher order than operons. *Trends Biochem. Sci.* **25,** 474–479.
4. Tamames, J., Casari, G., Ouzounis, C., and Valencia, A. (1997) Conserved clusters of functionally related genes in two bacterial genomes. *J. Mol. Evol.* **44,** 66–73.
5. Dandekar, T., Snel, B., Huynen, M., and Bork, P. (1998) Conservation of gene order: a fingerprint of proteins that physically interact. *Trends Biochem. Sci.* **23,** 324–328.
6. Yanai, I. and DeLisi, C. (2002) The society of genes: networks of functional links between genes from comparative genomics. *Genome Biol.* **3,** 1–12.
7. Kolesov, G., Mewes, H. W., and Frishman, D. (2002) Snapper: gene order predicts gene function. *Bioinformatics* **18,** 1017–1019.
8. Tamames, J. (2001) Evolution of gene order conservation in prokaryotes. *Genome Biol.* **2,** 1–11.
9. Korbel, J. O., Snel, B., Huynen, M. A., and Bork, P. (2002) Shot: a web server for the construction of genome phylogenies. *Trends Genet.* **18,** 158–162.
10. Schmidt, T., and Stoye, J. (2004) Quadratic time algorithms for finding common intervals in two and more sequences, in *Proceedings of the 15th Annual Symposium on Combinatorial Pattern Matching, CPM 2004,* volume 3109 of *LNCS,* Springer Verlag, pp. 347–358.
11. Schmidt, T. (2005) *Efficient Algorithms for Gene Cluster Detection in Prokaryotic Genomes.* Dissertation, Technische Fakultät der Universität Bielefeld, Bielefeld, 2005. Available at http://bieson.ub.uni-bielefeld.de/volltexte/2005/749/.
12. Rogozin, I. B., Makarova, K. S., Wolf, Y. I., and Koonin, E. V. (2004) Computational approaches for the analysis of gene neighborhoods in prokaryotic genomes. *Briefings in Bioinformatics* **5,** 131–149.

13. Uno, T. and Yagiura, M. (2000) Fast algorithms to enumerate all common intervals of two permutations. *Algorithmica* **26,** 290–309.
14. Heber, S. and Stoye, J. (2001) Finding all common intervals of k permutations, in *Proceedings of the 12th Annual Symposium on Combinatorial Pattern Matching, CPM 2001*, pp. 207–218.
15. Didier, G., Schmidt, T., Stoye, J., and Tsur, D. (2007) Character sets of strings. *J. Discr. Alg.* **5,** 330–340.
16. Bergeron, A., Corteel, S., and Raffinot, M. (2002) The algorithmic of gene teams, in *Proceedings of the Second International Workshop on Algorithms in BioInformatics, WABI2002*, pp. 464–476.
17. Amir, A., Apostolico, A., Landau, G. M., and Satta, G. (2003) Efficient text fingerprinting via parikh mapping. *J. Discr. Alg.* **26,** 1–13.
18. Tatusov, R. L., Natale, D. A., Garkavtsev, I. V., et al. (2001) The COG database: new developments in phylogenetic classification of proteins from complete genomes. *Nucleic Acids Res.* **29,** 22–28.

III

Experimental Analysis of Whole Genomes: *Analysis of Copy Number and Sequence Polymorphisms*

13

Genome-wide Copy Number Analysis on GeneChip® Platform Using Copy Number Analyzer for Affymetrix GeneChip 2.0 Software

Seishi Ogawa, Yasuhito Nanya, and Go Yamamoto

Summary

Genome-wide copy number detection using microarray technologies has been one of the recent topics in cancer genetics and also in the research on large-scale variations in human genomes. This chapter describes methods to analyze copy number alterations in cancer genomes using Affymetrix GeneChip®, high-density oligonucleotide microarrays originally developed for large-scale single nucleotide polymorphism (SNP) typing. Combined with the large numbers of SNP-specific probes, the robust algorithms developed for analyzing raw array signals (Copy Number Analyzer for Affymetrix GeneChip v2.0) enable not only accurate and high resolution copy number estimations, but also allelic imbalances commonly found in cancer genomes, which provides a powerful platform to explore the complexities of cancer genomes.

Key Words: GeneChip; SNP; copy number detection; allelic imbalance; CNAG2.0.

1. Introduction

Cancer genomes undergo a variety of changes and modifications during neoplastic development. Because tumorigenesis should be ultimately ascribed to the alterations of the genetic programs caused by these changes and modifications, it is of particular importance to establish a technology to analyze such genetic and epigenetic changes in a genome-wide manner, to get better understanding of human cancers. In this respect, recent development of a number of genome-wide copy number detection systems using array technologies

From: *Methods in Molecular Biology, vol. 396: Comparative Genomics, Volume 2*
Edited by: N. H. Bergman © Humana Press Inc., Totowa, NJ

represents one of the major advances in the field of cancer genetics ***(1–3)***. Complementing conventional cytogenetics and FISH analysis, or replacing chromosome-based CGH techniques, microarray-based copy number detection methods are now widely used to analyze complex cancer genomes as well as to study large-scale copy number variations found in normal human populations ***(4,5)***.

Among different copy number detection array systems now commercially available, the Affymetrix GeneChip® system has a unique feature that it can detect copy number alterations in an allele-specific manner using a set of single nucleotide polymorphisms (SNP)-specific probes. In fact, currently, this is one of the most sensitive systems to detect allelic imbalances or loss of heterozygosity (LOH) in cancer genomes using allele-specific copy number determination. In addition, the total number of probes on a single array now exceeds 250K, which allows very high resolution copy number detection (http://www.affymetrix.com/products/arrays/specific/500k.affx) ***(6)***. A main drawback about these SNP arrays arises from PCR amplification before microarray analysis, which biases copy numbers as measured from raw signal ratios between test and reference experiments. However, well-designed experiments and a set of algorithms to compensate the biases which we developed in the software tool Copy Number Analyzer for Affymetrix GeneChip (CNAG) could overcome the drawback and produce satisfactory results in most situations *(7)*.

In this chapter, we describe the basics of genome-wide copy number/LOH analysis using CNAG2.0 on the GeneChip platform, focusing on the practical use of the software with the underlying principles of the algorithms CNAG2.0 uses. Before going on to the next section, some key issues in the "wet" part of the analysis are discussed in **Note 1** for those who are unfamiliar with copy number analysis using Affymetrix platform, but these issues are also essential to get high-quality results and the readers are recommended to see the section in advance.

2. Materials

2.1. Cell Lines

NCI-H2171, H1437, and H2009, and NCI-BL2171, BL1437, BL2009 are lung cancer cell lines and immortalized lymphoblastoid cell lines established from the corresponding patients, which were obtained from American Type Culture Collection (Rockville, MD; http://www.atcc.org/common/cultures/NavByApp.cfm). A glioma cell line U251 was distributed from Riken (Tsukuba, Japan). Peripheral blood samples and primary tumor specimens were collected from 96 normal volunteers and 33 patients with a variety of malignancies, and subjected to DNA extraction (*see* **Note 1**). Microarray experiments were performed completely according to the manufacturer's protocols, using Affymetrix GeneChip 50K *Xba*I, 50K *Hind*III, 250K *Sty*I, and 250K *Nsp*I arrays.

2.2. Microarray Data Files

CNAG2.0 uses both .cel and .chp files for the copy number calculations and LOH inference. These files are created by GCOS and GTYPE programs, respectively, when array signals are scanned (.cel files) and analyzed genotypes (.chp files). The wet part details are not provided in this article, because they are essentially the same as what is described in the manuals of GeneChip. See the Affymetrix manuals about how to do array experiments and create these files (http://www.affymetrix.com/products/application/whole_genome.affx). Note that they need to be placed within the same directory that is accessible from the PC in which CNAG2.0 is installed.

2.3. CNAG2.0 Software

CNAG2.0 is distributed from our website (http://www.genome.umin.ac.jp). After downloading the installer, set up the program according to the manual. By default, it is installed under the directory, \C\Program Files\Genome Lab\CNAG. Requirements for machines are more than 500 Mb RAM and more than 1 GHz Pentium processors. Recommended speck is, however, more than 1 Mb RAM and more than 3 GHz processors. Portable PC performs well but desktop PC equipped with enough RAM and faster processor is recommended when a large number of 500 K samples are analyzed.

2.4. Chip Description Files

CNAG2.0 uses Chip Description (CDF) files to extract signal data from .cel files. CDF files should be downloaded from the Affimetrix website (http://www.affymetrix.com/support/technical/libraryfilesmain.affx).

3. Methods

3.1. Starting Up CNAG2.0 Software and Setting Up the Environment

1. Launch CNAG2.0 from Desktop CNAG icon or from within the Windows Startup Menu->Programs->CNAG->CNAG.
2. Parameter->Default Setting to open the "setting" dialog, where all the settings of CNAG2.0 are specified (**Fig. 1**).

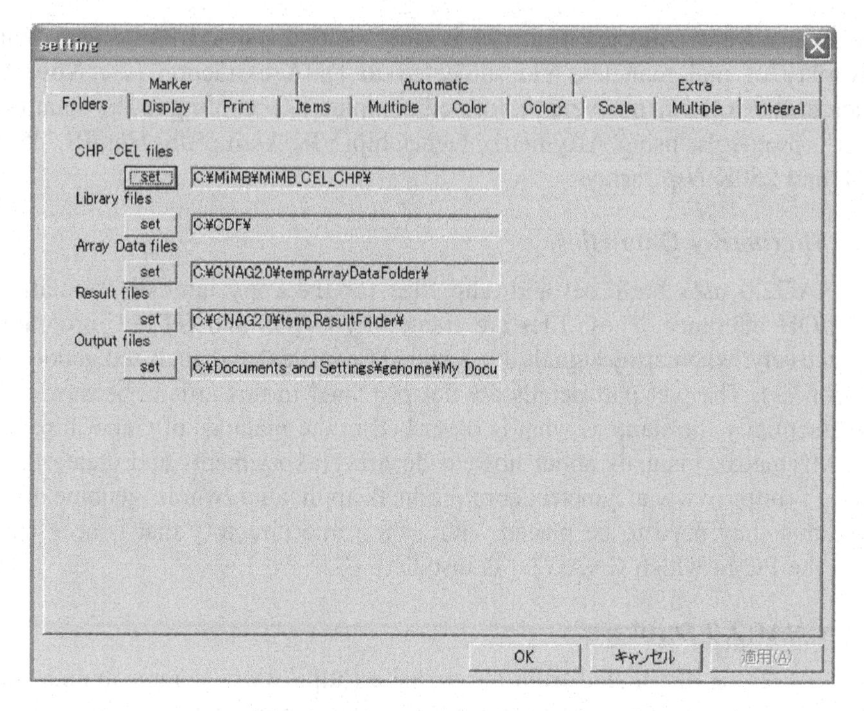

Fig. 1. The dialog for folder setting.

3. Choose "Folders" tab and set indicated directories as follows.
4. Set "CHP_CEL files" folder to that directory where all the .cel and .chp files are stored to be analyzed.
5. Set "Library files" folder to that directory where the CDF files are installed. .cdf files describe probe information in .cel files and prepared for individual array types. They usually come with Affymetrix programs or are available from the Affymetrix website. CDF files are essential for CNAG to extract data from .cel files.
6. Set "Array data files" folder to that directory where you want to save .cfh files in which CNAG2.0 stores the signal and genotype data extracted from .cel and .chp files as well as other information used for copy number calculations. Any directory can be used for this purpose but we recommend creating a new folder named, for example, "array_data," under CNAG directory.
7. Similarly, create another folder to save the analyzed results, for example, "Results_folder" and set "Results files" folder to that directory. CNAG holds all the results within this folder.
8. Set "Output files" folder, where CNAG output copy number as well as other data in text files. After all the folders are properly set, close the dialog by clicking "OK."

3.2. Extracting Array Signals and Genotyping Data From .cel and .chp Files

1. Data-> Extract Data to open a dialog for extracting data from .cel and .chp files (**Fig. 2**).
2. Place each sample that appears in the middle window in one of the four categories, by choosing each sample name and click "Add" on the appropriate window on the right. Paired tumor and reference files are put in the "Paired TEST" and "Paired REF" windows. Otherwise, non-paired samples are put into either the "non-Paired TEST" or "non-Paired REF" window. Each paired samples should appear at the same level in the corresponding windows.
3. After all samples are properly placed, click "OK" to start extraction. Through this step, CNAG2.0 looks into each .cel file to get array information and separately calculate signal sum of all the perfect match probes for each SNP type for all SNP loci, whereas genotyping and the gender information are extracted from the

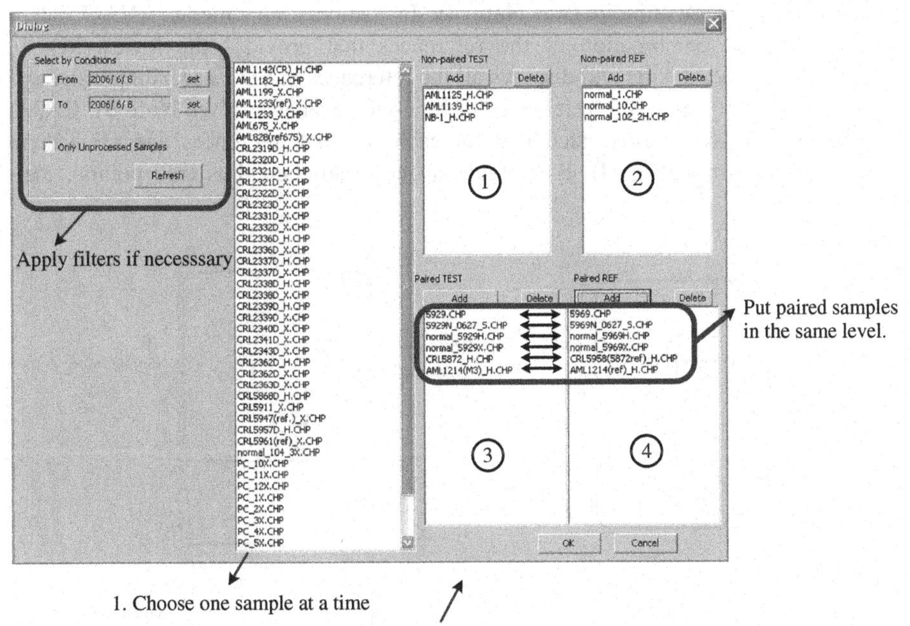

Fig. 2. Dialog for extracting array data. Copy Number Analyzer for Affymetrix GeneChip v2.0 extracts array signals and genotype information directly from .cel and .chp files using Extract Data function and store the data into .cfh files in the Array Data Files folder.

corresponding .cfh file. Extracted data is then stored in the .cfh file, which is used for all the CNAG analyses hereafter.

3.3. Nonallele-Specific Copy Number Analysis Using Unrelated References

Two modes of copy number analysis are available in CNAG2.0. In nonallele-specific analysis, array signals for both SNP types are combined and compared between tumor and reference files.

1. Data->Sample Manager to launch Sample Manager (**Fig. 3**).
2. Apply appropriate filters based on the date of array experiments, array types, sample types (test or reference), and other user-defined attributes.
3. Copy number analysis in CNAG is iteratively performed. First, choose the sample to be analyzed and the mode of analysis, and do "non-self analysis" (*see* **Note 2**). For every available reference, CNAG takes log2 ratios of signals between tumor and reference (measured copy numbers) for all SNP loci, applies regressions to these log2 ratios (*see* **Note 3**), and calculates their standard deviation assuming all autosomes are diploid (*see* **Note 4**). In "automation" mode, CNAG calculates the best combination of the references that provides the lowest SD. In "manual" mode, it sorts and displays all the references in SD's ascending order and prompts one to select the references to be used for analysis (*see* **Note 5**). CNAG averages the log2 ratios calculated for each selected references and plots them on chromatogram (**Fig. 4**). By default, all copy number plots (red points), their

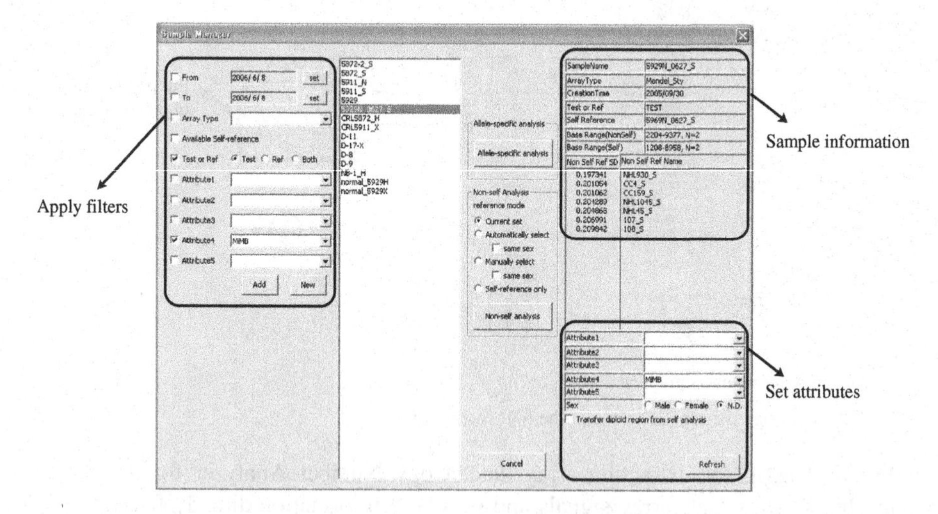

Fig. 3. Sample manager. Sample Manager dialog is used to control the mode of analysis, to modify attributes of samples, and to show sample information.

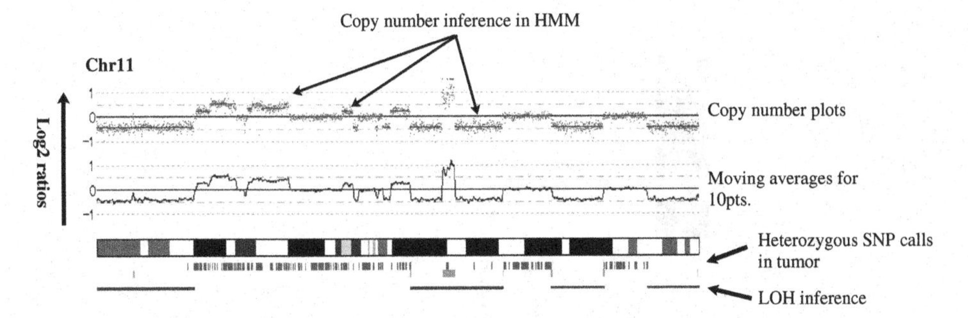

Fig. 4. Nonallele-specific copy number analysis in chromosome view. Copy Number Analyzer for Affymetrix GeneChip v2.0 calculates log2 ratios of signals for all single nucleotide polymorphism (SNP) loci and plots them on chromatograms. It also displays inferred copy numbers (brown lines), moving averages for 10 adjacent SNPs (blue lines), heterozygous SNP calls (green bars), and loss of heterozygosity prediction (light blue lines). The horizontal axis is interchangeably depicted in SNP order with uniform intervals or in proportion to real genetic distance (not shown).

moving average for 10 loci (blue lines), and heterozygous SNP calls (green bars) in tumor samples are displayed. One may change items to be displayed and their appearances.

4. In the next step, do Parameter-> Set range of SD calculation to select a region to be assigned as diploid by moving the mouse. CNAG recalculate copy numbers and applies regressions assuming the region as diploid (*see* **Note 6**).

3.4. Copy Number Inference From log2 Ratios

By default, CNAG infers real copy numbers, which take integer values, from the observed log2 ratios based on the HMM and overlays the inferred values on the log2 plots. When processing multiple samples, some kind of automation is required to extract copy number abnormalities. Because the default parameters of the HMM is optimized for pure tumor cells, they should be adjusted for fresh tumor samples having significant degrees of normal tissue components (**Fig. 5**) (*see* **Note 7**). To do this:

1. Parameter->ContaminationRate to select a region used for setting the parameters. It should have copy number 1 (loss) or 3 (gain).
2. Assign a copy number for that region (1 or 3).
3. CNAG recalculates the real copy numbers based on the adjusted parameters.

In addition to the inference of real copy numbers, CNAG predicts the presence of LOH based on the regional reduction of heterozygous SNP calls.

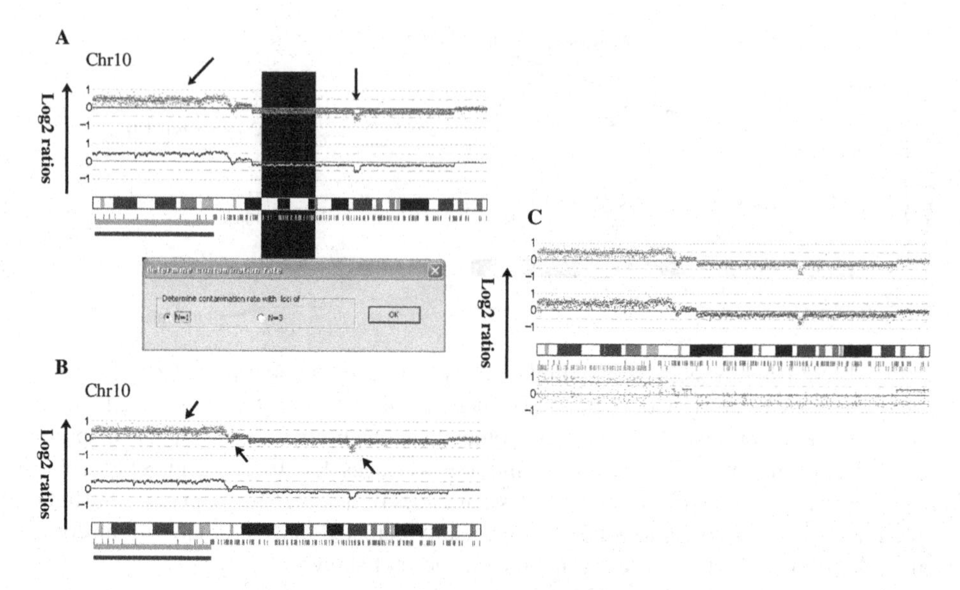

Fig. 5. Adjustment of parameters for HMM. In primary tumor samples suffering from contamination of normal tissues, the copy number inference algorithm based on HMM may not work correctly because of reduced amplitude of copy number shifts for a given gain or loss (**A**). After adjustment of the parameters of HMM, copy number inference becomes stable and sensitively detects the homozygous deletion and other deletions that escaped detection under the default parameters (**B**). Copy number inference on HMM is enabled in allele-specific analysis, but it fails to detect allelic deletion in the short arm, where the DNA fragments from the highly amplified allele cross-hybridize to the probes for the other polymorphic allele, which is actually lost in this sample (**C**).

Putative LOH regions are indicated by blue bars, which are drawn in varying height that is proportional to the logarithm of LOH likelihood (**Fig. 4**) (*see* **Note 8**).

3.5. Allele-Specific Copy Number Analysis Using the Paired Reference

For those samples where paired normal DNA is available, copy numbers in log2 ratios can be calculated in an allele-specific manner (*see* **Note 9**).

1. Choose a tumor sample in Sample Manager as described in **Subheading 3.3., steps 1** and **2**. When the sample has a paired reference, "Allele-specific analysis" becomes enabled.
2. Do "Allele-specific analysis."
3. Assign a diploid region just in the same manner as **Subheading 3.3., step 4**.

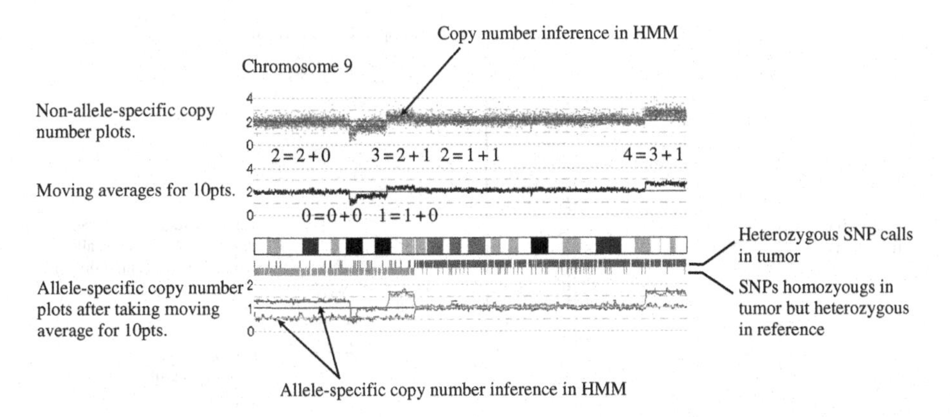

Fig. 6. Allele-specific analysis on StyI 250K array. Allele-specific copy number analysis on chromosome 9 reveals allelic imbalances in a lung cancer cell line (NCI-H1437) chromosome 9, where signal ratios are separately calculated for different SNPs at heterozygous SNP loci in paired normal DNA (bottom panel). Allele-specific copy number inferences are provided by dark red and light green lines. Combined signal ratios are plotted for all SNP loci together with inferred copy numbers in HMM (upper panel). Moving averages for 10 adjacent SNPs and heterozygous SNP calls in the tumor sample are also depicted.

In this mode, CNAG displays allele-specific copy numbers by separately plotting larger and smaller log2 ratios for each heterozygous SNP locus (**Fig. 6**). The parameters of allele-specific copy number inference by HMM should be adjusted as described in **Subheading 3.4.** (*see* **Note 10**).

The latest CNAG version (CNAG 3.0) enables allele-specific copy number analysis without depending on the availability of paired normal DNA, but using a few anonymous references (AsCNAR algorithm) *(12)*.

3.6. Automated Processing of Multiple Samples

Multiple samples can be analyzed in background, although diploid assignment is the responsibility of researchers and cannot be automated.

1. Automation-> Batch Analysis to open a dialog for automated analysis (**Fig. 7**).
2. Apply filters to show candidate samples to be analyzed.
3. Check samples subjected to automated analysis.
4. Check the preferences with regard to diploid assignment and gender of references.

For paired samples, both allele-specific analysis and nonallele-specific analysis are performed. Also within this dialog, attributes of multiple samples can be changed at one time using "Change attributes."

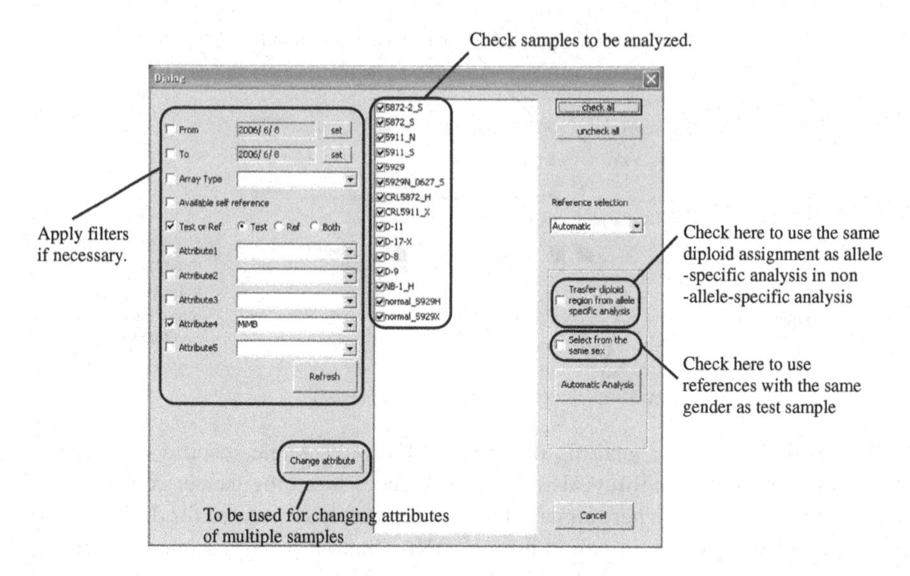

Fig. 7. The dialog for automated analysis. Background analyses are enabled by the automation function.

3.7. *Exploring Copy Number Abnormalities and Allelic Imbalances*

In CNAG2.0, capabilities of exploring the results of copy number analysis are greatly extended from the previous version. Descriptions of the basic components of the functionalities implemented in CNAG2.0 are provided here.

3.7.1. Open "Display Samples" Dialog and Select Samples to be Explored

1. Data->Display Samples to open a dialog for displaying the results (**Fig. 8**).
2. Apply filters to show up the samples.
3. Check samples to be explored.

3.7.2. Select the Mode of Displaying Results

1. Single display to open a pop up window that shows selected samples. Choose one and CNAG display the sample.
2. Multi display to explore multiple samples (**Fig. 9**).
3. Integral view to summarize the results of multiple samples. "Set the Integration Parameters" dialog is open to prompt the input of the required parameters (**Fig. 10**). Then do "OK."

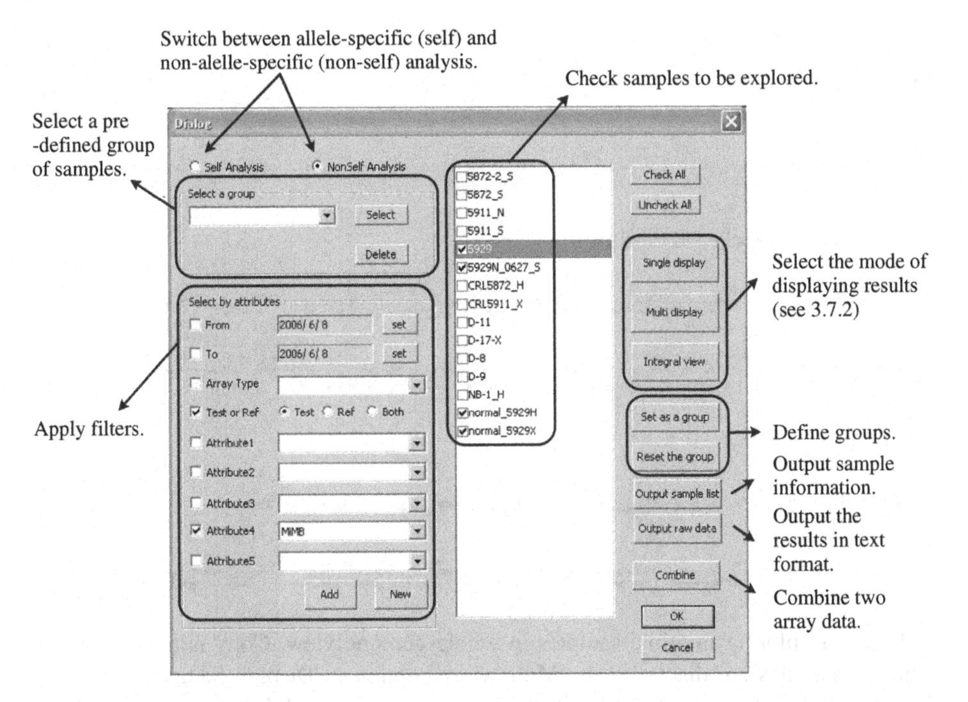

Fig. 8. The dialog for display samples. The "Display Sample" supports a variety of analyses, including simple display of samples, combined data from different array types, output raw data into other file formats, and integration of multiple data, which is chosen by its dialog.

3.7.3. Combining Paired Microarray Data

Copy number data from different array types can be analyzed in combination. This is possible between 50K *Xba* and 50K *Hind* arrays, and 250K *Sty*I and 250K *Nsp*I arrays (*see* **Note 11**).

1. Data Display Samples to open a dialog for displaying the results.
2. Apply filters to display specific samples.
3. Check the paired samples to be combined.
4. Combine. The combined result is displayed and stored in a file, named A_and_B.CFS(CFN) for a combination of A.CSF(CFN) and B.CFS(CFN) files.

3.7.4. Output Sample Information and Copy Number Data in Text Format (*see* **Note 12**)

1. Data Display Samples to open a dialog for displaying the results.
2. Apply filters to show up the samples.

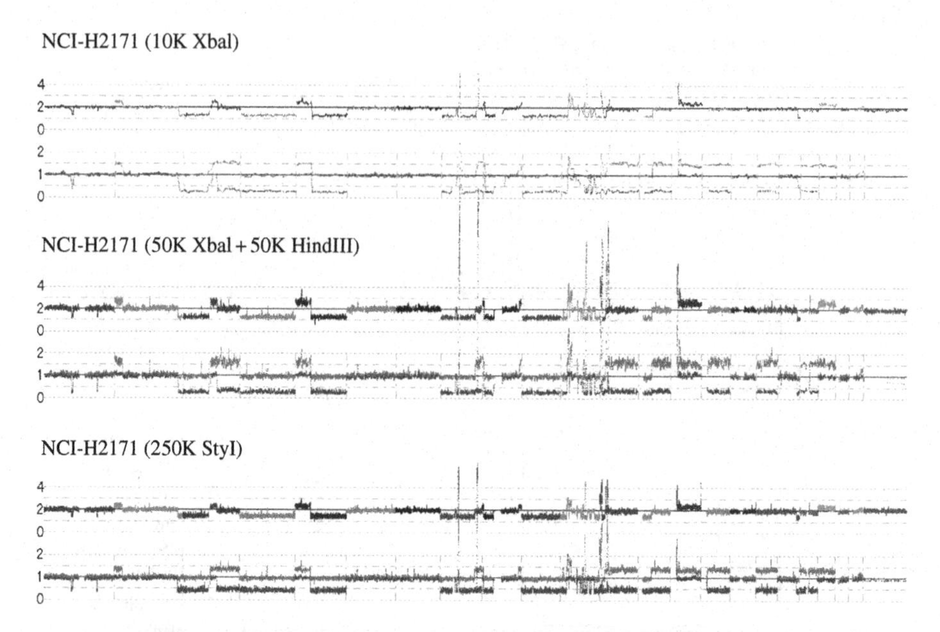

Fig. 9. Exploring multiple samples in whole genome view. Copy number plots of multiple samples are displayed in "Multi view" option in Display Samples function. Both whole genome view and chromosome view are supported. In this example, whole genome views of a lung cancer genome (NCI-H2171) analyzed by 10K *Xba*I, 50K *Xba*I+50K *Hind*III, and 250K *Sty*I arrays are all displayed at one time. The order of samples can be changed by drag and drop. The items to be displayed are indicated in the setting menu (Parameter-> Default Setting-> Multiple).

3. Check the paired samples whose data is going to be output.
4. Output raw data and "Output raw data" dialog will open.
5. Check the data to be output.
6. Also check "combined to one file for each array type" if all the data should be output into a single text file. Otherwise, data of each sample goes into a separate file.

3.7.5. Getting Information From the Region Under Interest

1. Display the sample and move the chromosome to be explored.
2. Parameter measure.
3. Assign the region under interest by drugging the mouse.
4. "Statistics" dialog will open to show the information and statistics about the assigned region **(Fig. 11)**, including the start and end position and corresponding SNP

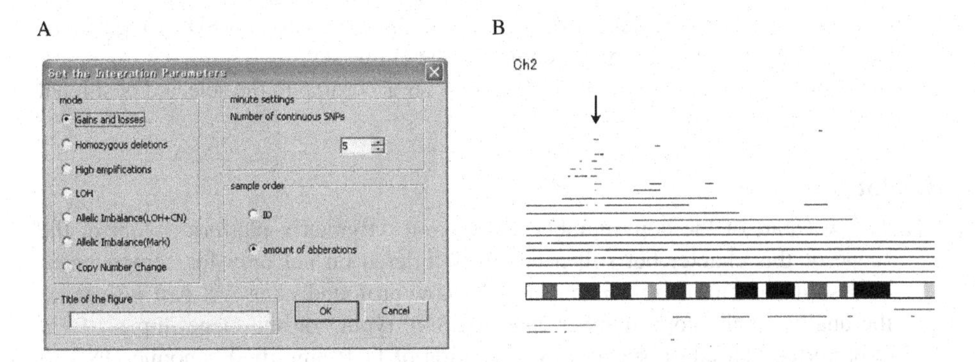

Fig. 10. Integration of data from multiple samples. Copy Number Analyzer for Affymetrix GeneChip v2.0 integrates the data from multiple samples and displays the integrated results on chromatograms. The type of abnormality and the minimum number of the abnormal points to be integrated is indicated in the setting dialog **(A)**. Integration of high-Grade amplifications in 25 neuroblastoma cell lines **(B)** discloses a common region containing the *MycN* gene on the short arm of chromosome 2 (arrow).

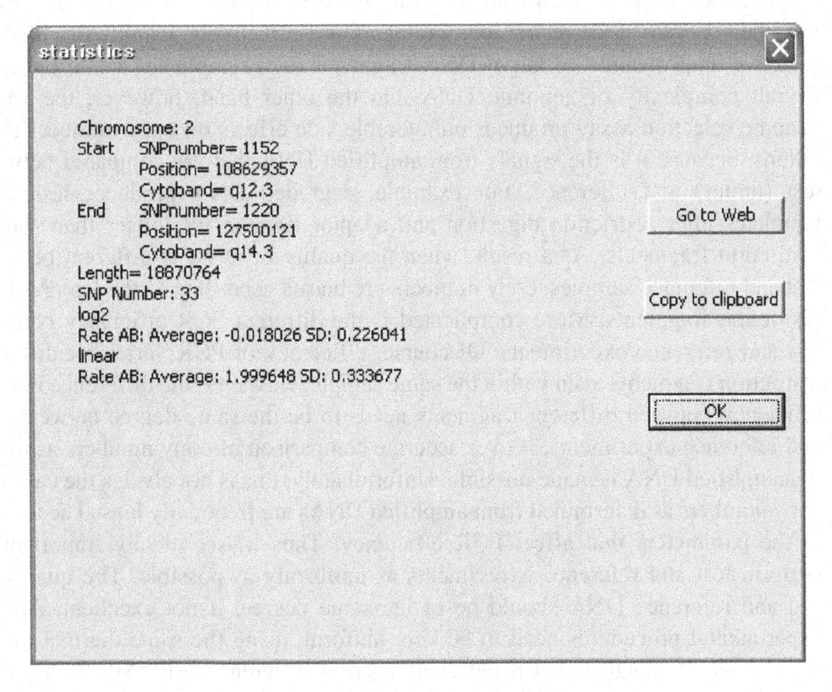

Fig. 11. "statistics" dialog.

numbers and cytobands, length of the region, and the mean and SD of log2 ratios therein. The data can be directly copied to the clipboard.

5. Click "go to web" to jump to the UCSC browser according to the selected positional parameters.

4. Notes

1. The wet part of the copy number analysis on Affymetrix platform is out of the scope of this chapter but still needs to be briefed on not only for completeness of description but also for understanding how profoundly the wet part influences the quality of the copy number analysis. SNP typing on the GeneChip platform is performed based on specific hybridization of PCR-amplified genomic DNA to large numbers ($\sim$10, 000 to $\sim$250, 000 depending on array types) of SNP-specific probe sets chemically synthesized on microarrays *(8,9)*. Each probe set consists of varying numbers of 25-mer oligonucleotide probes perfectly matched to each SNP type together with the same number of mismatch base-introduced probes. The key step on this platform is the whole genome selection assay. Briefly, 250 ng genomic DNA is digested with an indicated restriction enzyme, attached with a common primer, and PCR-amplified using a single primer *(8)*. The PCR products are subjected to ultrafiltration for desalting followed by DNaseI fragmentation, and biotinylated DNA fragments are hybridized to microarrays. Through this assay, less than 1.5- to 2-kb restriction fragments are effectively amplified and thus, selective amplification of target DNA fragments is accomplished while reducing overall complexity of genomic DNA. On the other hand, however, the whole genome selection assay produces unfavorable side effects on copy number calculations, because it is the signals from amplified DNA that are compared between test (tumor) and reference. For example, degraded DNA produces less PCR templates after restriction digestion and adaptor ligation for longer than shorter restriction fragments. As a result, when the quality of DNA is different between test and reference samples, copy numbers are biased according to the length of the restriction fragments. More complicated is the different PCR efficiency between test and reference experiments. Of course, efficiency of PCR varies for different restriction fragments even within the same sample. However, the difference in PCR efficiency between different fragments needs to be the same degree between test and reference experiments, if ever accurate comparison of copy numbers as to the nonamplified DNA is made possible. Unfortunately, this is not always the case and copy numbers as determined from amplified DNAs are frequently biased according to the parameters that affect PCR efficiency. Thus it is critically important to perform test and reference experiments as uniformly as possible. The quality of test and reference DNA should be of the same degree, if not excellent. All the experimental procedures need to be kept uniform, using the same thermal cycler that can achieve uniform thermal cycles across different wells. All the reagents must be prepared as premix. Performance of polymerase is another parameter.

According to our experience for the 100K protocol, TAKARA LA Taq outperforms the Platinum Taq recommended by Affymetrix. It is more difficult to keep experimental conditions uniform when the experiments are performed on different occasion. As discussed in the following sections, it is possible to compensate the effects from different PCR conditions using computations to a certain extent, but keep in mind that these compensations never replace well-performed experiments.

2. As described next, CNAG uses multiple references to improve copy number calculations. However the random selection of references from both sex prevents the accurate copy number calculations in X chromosome. When the "same sex" box is checked, CNAG only uses the references having the same sex as the tumor sample, and therefore correctly detect copy number gains and losses in X chromosome. Note that in this mode, however, the quality of over all analysis may be affected, because availability of references is restricted.
3. On any array based platform, copy numbers are calculated as relative values to a reference sample, usually from normal individual. Given that the PCR amplification preserves relative copy numbers across different restriction fragments and that the array signals are proportional to the amount of PCR products for each fragment, the copy number of the ith SNP fragment in log2 ratio is expressed as:

$$C^i = K \times \frac{\sum(PA^i_{tum} + PB^i_{tum})}{\sum(PA^i_{ref} + PB^i_{ref})} \tag{1}$$

or

$$\log_2 C^i = \log_2 K + \log_2 \frac{\sum(PA^i_{tum} + PB^i_{tum})}{\sum(PA^i_{ref} + PB^i_{ref})}, \tag{1'}$$

where PA^i and PB^i are signal intensities for type A and type B SNPs at the ith SNP probe, respectively, and lower suffixes are to distinguish tumor (*tum*) and reference (*ref*) values. Signal sums are taken for all the perfect match probes for the ith SNP. K is a constant determined from the mean array signals of tumor and reference samples and also from the mean ploidy of tumor sample. CNAG separately calculates $\sum PA^i$ and $\sum PB^i$ and stores them into the .cfh file at the data extraction time and at the initial step of non-self copy number analysis, it preliminary set the K-value so that the mean intensities of the two arrays are the same and mean ploidy for all autosomes is **Eq. 2**, or,

$$K_0 = \frac{\sum_i \sum(PA^i_{ref} + PB^i_{ref})}{\sum_i \sum(PA^i_{tum} + PB^i_{tum})} \tag{2}$$

where sum is taken for all autosomal SNPs.

4. As discussed in **Note 1**, the K-value is not a constant but varies for different SNP fragments, and copy numbers calculated as **Eq. 1** are deviated from the real values

according to the variable K_i-value, which is determined by the base composition of the ith fragment and the difference in experimental parameters between test and reference experiments. In fact it can be demonstrated, empirically though, that the K_i explicitly depends on the length and GC content of the ith SNP fragment, when $\log_2 C^i$s in expression **Eq. 1'** under a constant K given by **Eq. 2** are plotted against these parameters **(Fig. 12A)**. Based on this finding, we obtain a new expression using the second order approximation for $\log_2 K_i$:

$$\bar{\Lambda}^i = \log_2 \frac{\sum(PA^i_{tum} + PB^i_{tum})}{\sum(PA^i_{ref} + PB^i_{ref})} + \log_2 K_0 - \sum_{j=0}^{2}(a_j x_i^j + b_j y_i^j) \quad (3)$$

where x_i and y_i are the length and GC content of the ith SNP fragment and a_j and b_j are determined by the quadratic regression from the measured PAs and PBs for the SNPs in a given diploid ($C_i = 1$) region **(Fig. 12B,C)**. Different DNA quality between both samples also affects k-value but the effect is already incorporated in expression **(Eq. 3)** because it is related to the length of the fragment. After a series of quadratic regressions, the biases from different PCR conditions and

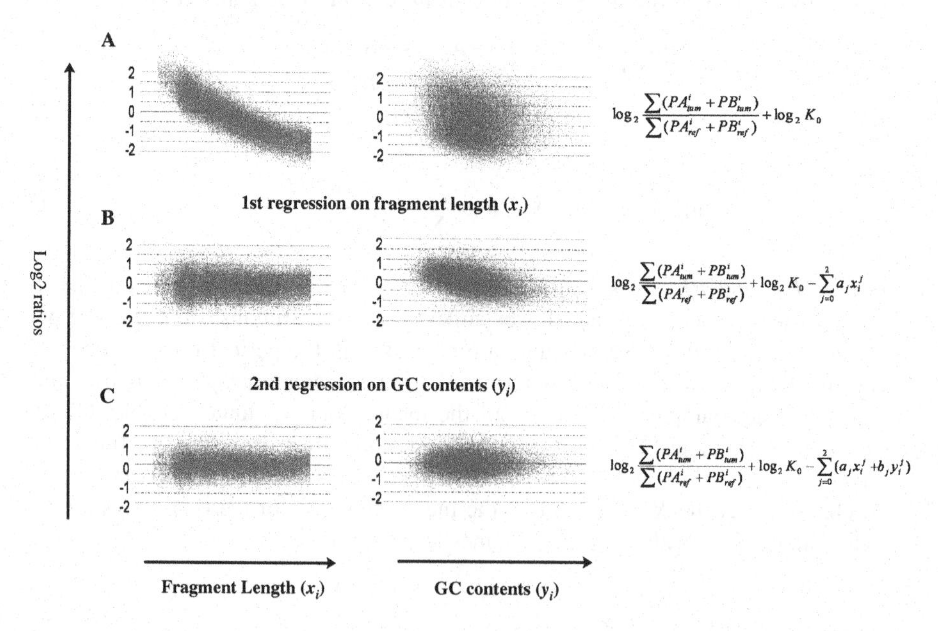

Fig. 12. Dependence of Log2 ratios on fragment length and GC content. Log2 ratios of normalized array signals were calculated between two normal samples and plotted against the two parameters, fragment length (left) and GC content (right). Biased log2 ratios **(A)** are serially compensated by a couple of quadratic regressions with regard to both parameters **(B,C)**, which contributes to improved SD value.

DNA quality are partially compensated and better inference of copy numbers is obtained. CNAG initially calculates the coefficients of the regression (a_js and b_js) using all the autosomal SNPs and in the next step recalculates these coefficients as well as K_0 using the SNPs within the region assigned as diploid and revises all the copy numbers according to the recalculated parameters.

5. Quality of the copy number analysis strongly depends on the selection of the reference, or "control," to be used for copy number calculations, even after a series of regressions are applied, and accordingly it is of particular importance to choose appropriate references or to do well-controlled experiments (*see* **Note 1**). CNAG measures the performance of a reference in terms of standard deviation (SD) of the observed copy number values for a given diploid region, which is defined by:

$$SD = \sqrt{\frac{1}{n-1}\sum_{i}^{n}(\log_2 C^i)^2}$$

where the sum is taken for all n SNPs within a given diploid region. In **Fig. 13**, SD is improved from 0.365 to 0.114 after a series of quadratic regressions regarding length and GC content of SNP fragments. CNAG calculates SD values against every reference available and in the next step, instead of choosing single best reference showing the lowest SD, it tries several combinations of references to achieve the lower SD value by averaging the copy numbers separately calculated using each of the references. More mathematically,

$$\tilde{\Lambda}^i = \frac{1}{m}\sum_{k=1}^{m}\log_2 \bar{\Lambda}^k$$

is adopted as a revised copy number estimate, when SD value is improved. Unfortunately computing all the combinations should be diverged and instead CNAG tries to serially add a reference with the next lower SD value until the combined SD is no more reduced. Generally, but not always, the use of multiple references successfully improves SD values on average. Two factors could contribute the improvement in SD.

First, random error components in reference experiments are expected be reduced by taking averages. Note, however, that including dissimilar references in terms of SD values is of little use to reduce the overall SD value or even deteriorates the result. We recommend that experiments should be planned to include more than five normal controls to get enough references with similar SD values. The other factor that contributes to improved SD is the fact that the difference in genotype between test and reference tends to be neutralized by taking averages. Because SNP specific probes may show significantly different hybridization kinetics, $\sum(PA^i_{tum} + PB^i_{tum})/\sum(PA^i_{ref} + PB^i_{ref})$ could be considerably affected depending on the combination of genotypes in tumor and reference samples **Fig. 14** demonstrates the effect of comparison between different genotypes.

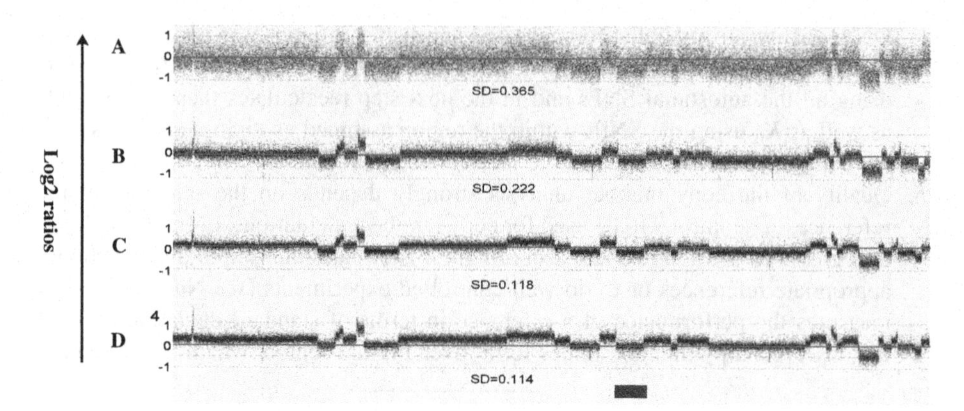

Fig. 13. Improvement of SD after a series of compensations. Genomic DNA from a glioma cell line U251 was analyzed for copy number alterations with or without compensations for different experimental conditions using different compensation algorithms. The raw log 2 ratios in the chromosomal order show widely fluctuated baselines (**A**; SD = 0.365), which are greatly reduced after compensation of experimental conditions and using the best-fit single (**B**; SD = 0.222) and multiple references (**C**; SD = 0.118). Regions of homozygous and hemizygous deletion are indicated by small and large arrows. The SD value is further decreased (SD = 0.114) when the second round compensation procedures are applied by specifying a region of diploid alleles (**D**, red bar). With this approach, diploid regions as well as other regions showing varying ploidy states are correctly assigned to their real ploidy. For proper comparison, SD values are calculated for the SNPs within the diploid range (red bar).

6. To be rigorous, it is impossible to unambiguously determine the K-value based only on array experiments. It is essentially related to the ploidy of the tumor genome, which can be determined only by cell-based analysis, for examples, cytogenetics, FISH, or measurement of DNA content using a flowcytometer. We can only guess the value but unfortunately frequently fail to do that correctly, especially for solid tumor samples, as revealed by FISH confirmation. Also note that assignment of copy number gains (more than two copies) and losses (less than two copies) for a region easily changes depending on the mean ploidy. In allele-specific analysis, purely triploid tumor genome can be distinguished from diploid genome in that the former shows allelic imbalance over the whole genome without showing disappearance of heterozygous SNP calls. However, for clinical samples, it would be difficult to rule out a possibility that the tumor specimen is composed of mostly haploid tumor and considerable components of normal tissues without significant reduction in heterozygous SNP calls. Assigning a region without allelic imbalance as diploid may provide a conservative estimate of ploidy, assuming homozygous deletion of a large genomic region is incompatible with

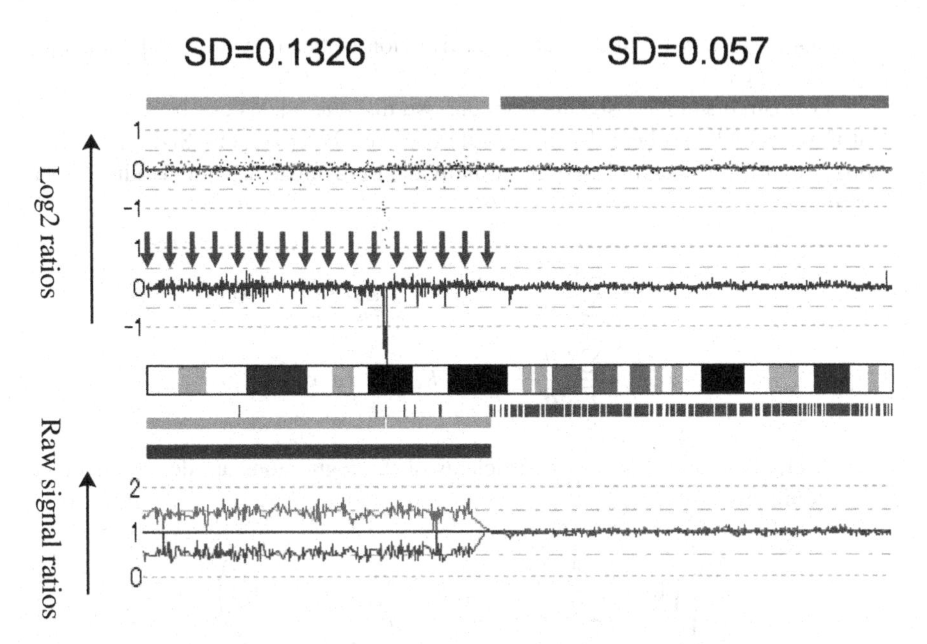

Fig. 14. Effects of difference in genotype on SD. In this example of allele-specific analysis on chromosome 9, signals from test and reference are compared between the identical genotypes on the long arm where there is no allelic imbalance. On the other hand, comparison may be between different genotypes in short arm portion, which is composed of two copies of the identical allele because of uniparental disomy (arrows). As a result, 9p shows a higher SD than 9p, whereas the mean copy number is the same between two.

cell survival, but not infrequently tumors have more ploidy and there remains a possibility of subclones exclusively retaining different alleles. Caution must be taken when mentioning absolute copy numbers of tumor cells.

7. Mathematical details of the hidden Markov model used for copy number inference in CNAG is out of the scope of this paper but a lot of references are available on this issue ***(10)***. It was first introduced for copy number inference on Affymetrix GeneChip platform by Li et al. ***(11)***. In this model, each measured copy number is assumed as an emission from one of the definite copy number states, which is now unknown or hidden. Each SNP locus has its own copy number state and default parameter values for HMM analysis in CNAG were empirically determined from the measurements in tumor cell lines where no normal cells are contaminated. Log2 value for an allelic loss is
8. Prediction of LOH in nonallele-specific analysis is based on a kind of maximum likelihood method, in which LOH regions are locally determined so that the

probability of observing the current distribution of heterozygous SNP calls takes a maximum value.

For paired samples, signals from the two discrete alleles can be discriminated at those SNP loci where the paired reference has heterozygous SNP calls. Thus, copy numbers in each allele can be individually evaluated by separating signals from the different alleles.

$$\Lambda_A^i = \log_2 \frac{\sum PA_{tum}^i}{\sum PA_{ref}^i} + \log_2 K_0 - \sum_{j=0}^{2}(a_j x_i^j + b_j y_i^j)$$

$$\Lambda_B^i = \log_2 \frac{\sum PB_{tum}^i}{\sum PB_{ref}^i} + \log_2 K_0 - \sum_{j=0}^{2}(a_j x_i^j + b_j y_i^j)$$

A ploidy constant K_0 and coefficients of the regressions are determined so as to minimize

$$\sum_{i}^{n}\left\{\left(\log_2 \frac{\sum PA_{tum}^i}{\sum PA_{ref}^i} + \log_2 K_0 - \sum_{j=0}^{2}(a_j x_i^j + b_j y_i^j)\right)^2 + \left(\log_2 \frac{\sum PB_{tum}^i}{\sum PB_{ref}^i} + \log_2 K_0 - \sum_{j=0}^{2}(a_j x_i^j + b_j y_i^j)\right)^2\right\}$$

where sum is taken for all heterozygous SNP fragments within a given diploid region. Two log2 ratios are separated into two groups,

$$L = \{\max(\Lambda_A^i, \Lambda_B^i) | i \in \ hetero\_SNPs\}$$

$$S = \{\min(\Lambda_A^i, \Lambda_B^i) | i \in \ hetero\_SNPs\}$$

which provide allele-specific copy number analysis. CNAG displays allele-specific copy numbers by separately plotting L and S elements.

9. Allele-specific copy number analysis is more sensitive to detecting the presence of LOH than comparison of SNP genotypes between the paired samples. It is one of the most sensitive methods for detecting allelic imbalance. Even when no discordant SNP calls are detected between the two, presence of LOH can be detected by allele-specific copy number analysis (**Fig. 15**).
10. CNAG stores the results of allele specific and non-allele-specific analysis to separate files having extensions .cfs and .cfn, respectively, within the "Results files" folder (**Subheading 3.1.7.**). The diploid region is separately assigned for both analyses, and recorded in the header of .cfh files as well as that of .cfs and .cfn files.

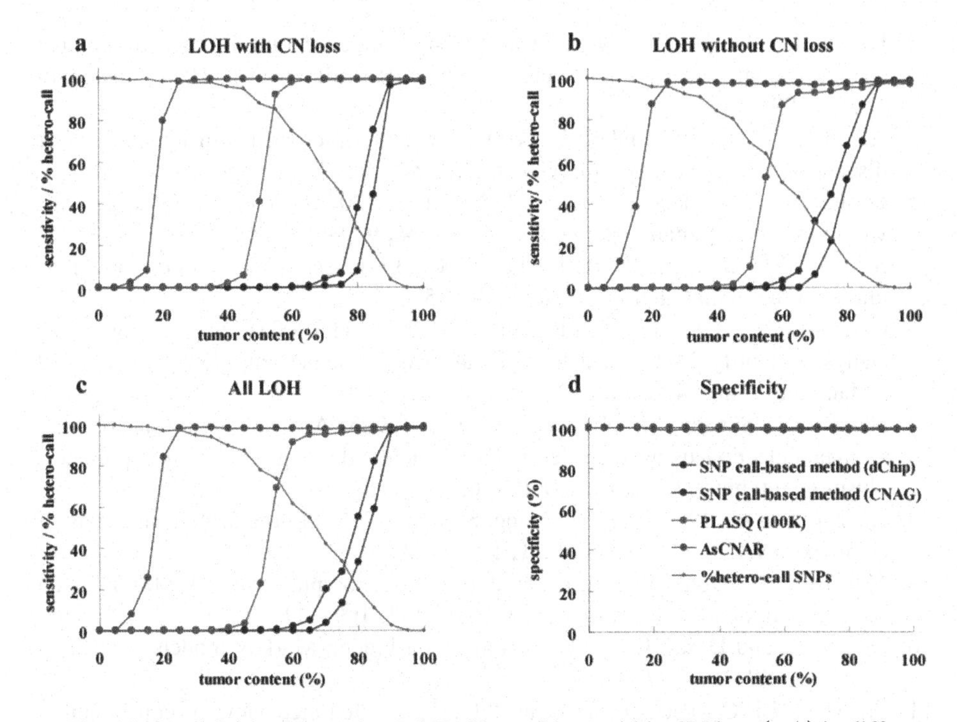

Fig. 15. Sensitivity of detecting LOH either with or within CN loss (a, b) in different algorithms are compared using mixture of tumor (NCI-H2171) and the paired LCL sample (NCI-BL2171) *(see 12)*, where SNPs which are heterozygous in the paired LCL and homozogous in the tumor sample are considered to be truly positive for LOH, and SNPs which are heterozygous both in the paired LCL and the tumor are to be truly negative. SNPs typing were carried out three times and discordant SNPs were exclude from calculation. The result for all LOH regions are shown in c and specificities of LOH detection are depicted in d. Proportions of heterozygous call SNPs remained in LOH regions of each sample are also shown in a-c.

11. CNAG recalculates LOH likelihood and inferred copy numbers in HMM using combined data.
12. CNAG supports data output in Affymetrix IGB format to enable data viewing with IGB. For IGB output, Data->Output IGB files. IGB files are stored in the "Output files" folder.

References

1. Mantripragada, K. K., Buckley, P. G., de Stahl, T. D., and Dumanski, J. P. (2004) Genomic microarrays in the spotlight. *Trends Genet.* **20,** 87–94.

2. Huang, J., Wei, W., Zhang, J., et al. (2004) Whole genome DNA copy number changes identified by high density oligonucleotide arrays. *Hum. Genomics* **1,** 287–299.
3. Albertson, D. G. and Pinkel, D. (2003) Genomic microarrays in human genetic disease and cancer. *Hum. Mol. Genet.* **12,** R145–R152.
4. Iafrate, A. J., Feuk, L., Rivera, M. N., et al. (2004) Detection of large-scale variation in the human genome. *Nat. Genet.* **36,** 949–951.
5. Sebat, J., Lakshmi, B., Troge, J., et al. (2004) Large-scale copy number polymorphism in the human genome. *Science* **305,** 525–528.
6. Garraway, L. A., Widlund, H. R., Rubin, M. A., et al. (2005) Integrative genomic analyses identify MITF as a lineage survival oncogene amplified in malignant melanoma. *Nature* **436,** 117–122.
7. Nannya, Y., Sanada, M., Nakazaki, K., et al. (2005) A robust algorithm for copy number detection using high-density oligonucleotide single nucleotide polymorphism genotyping arrays. *Cancer Res.* **65,** 6071–6079.
8. Kennedy, G. C., Matsuzaki, H., Dong, S., et al. (2003) Large-scale genotyping of complex DNA. *Nat. Biotechnol.* **21,** 1233–1237.
9. Matsuzaki, H., Dong, S., Loi, H., et al. (2004) Genotyping over 100,000 SNPs on a pair of oligonucleotide arrays. *Nat. Methods* **1,** 109–111.
10. Dugad, R. and Desai, U. (1996) A tutorial on hidden Markov models. *Technical report no. SPANN-96.1.*
11. Zhao, X., Li, C., Paez, J. G., et al. (2004) An integrated view of copy number and allelic alterations in the cancer genome using single nucleotide polymorphism arrays. *Cancer Res.* **64,** 3060–3071.
12. Yamamoto, G., Nannya, Y., Kato, M., et al. (2007) Highly sensitive method for genome-wide detection of allelic composition in non-paired, primary tumor specimens using Affymetrix(R) SNP genotyping microarrays. *Am J Hum Genet.* in press.

14

Oligonucleotide Array Comparative Genomic Hybridization

Paul van den IJssel and Bauke Ylstra

Summary

The array CGH technique (array comparative genome hybridization) has been developed to detect chromosomal copy number changes on a genome-wide and/or high-resolution scale. Here, we present validated protocols using in-house spotted oligonucleotide libraries for array CGH. This oligo array CGH platform yields reproducible results and is capable of detecting single copy gains, multicopy amplifications as well as homozygous and heterozygous deletions with high resolution. The protocol allows as little as 300 ng of input DNA, which makes the procedure valuable for small clinical samples and is also functional for DNA samples obtained from archival tissues.

Key Words: Microarray; CGH; comparative genome hybridization; printing; DNA-isolation; DNA-quality; oligonucleotide; amine binding; hybstation; feature extraction.

1. Introduction

Microarray comparative genomic hybridization (array CGH) is now the method of choice for the detection of unbalanced chromosomal changes, such as gains, losses, amplifications, and deletions in tumours and genetic diseases. The fore laying protocols use spotted long oligonucleotides in microarray experiments *(1)*. Oligonucleotides allow a sheer infinite resolution, great flexibility, and are cost-effective *(2)*. They also allow for the generation of microarrays for any organism of which the genome has been sequenced. Using the same oligo

From: *Methods in Molecular Biology, vol. 396: Comparative Genomics, Volume 2*
Edited by: N. H. Bergman © Humana Press Inc., Totowa, NJ

array for CGH and expression profiling allows direct comparison of mRNA expression and DNA copy number ratios.

The presented oligo array CGH protocol provides a highly sensitive and reproducible platform applicable to DNA isolated from both fresh and formalin-fixed paraffin embedded tissue *(1)*. For laboratories that already have oligonucleotide-based expression array platforms implemented or are planning to, we here provide protocols so that array CGH can be implemented in parallel.

Outline: the presented protocols start with commercially obtained oligonucleotide libraries. They describe the whole procedure from printing oligonucleotides onto arrays, isolation of DNA from various tissue sources, assessment of the DNA quality, fluorescent labeling of the DNA, automated hybridization, and array scanning to feature extraction, resulting in array CGH profiles where chromosomal copy number changes are displayed (*see* **Fig. 1**). Downstream array CGH data processing is not included.

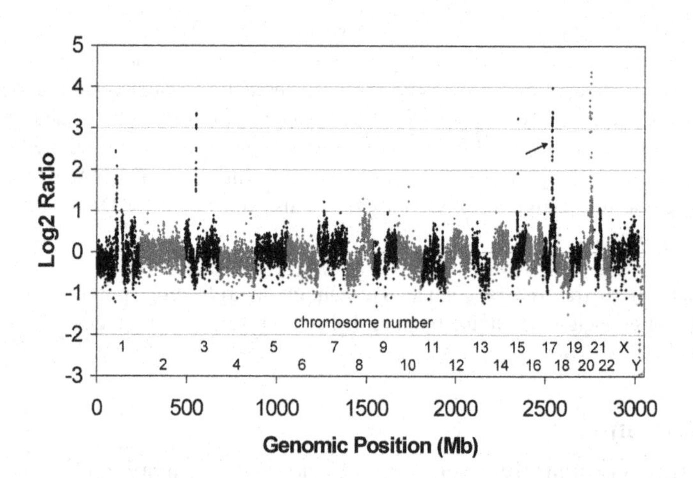

Fig. 1. Example of an oligonucleotide array comparative genome hybridization profile. A genome-wide profile is shown that was obtained from hybridization of DNA obtained from the breast cancer cell line MCF7 in the test (Cy3) channel with human male reference DNA in the reference (Cy5) channel on a 29K human oligonucleotide array according to the protocol described in this chapter. Log2ratios were calculated with a weighted moving average using a window of 250 kb and are displayed as a function of their position in the genome. Log2ratios of the odd and even chromosomes are shown in black and gray, respectively. Chromosome numbers are indicated. Note the excellent coverage of the amplifications, such as the one on chromosome 17 (arrow), as well as the presence of gains (chromosome 8q) and deletions (chromosome 13q). Chromosome 2 does not show aberrations and gives an indication of the noise level.

2. Materials

2.1. Preparation of Oligonucleotide Microarrays

1. Oligonucleotide library consisting of 60- to 70-mer unique 5′-amine-oligonucleotides designed to cover the genome of the target organism with minimal cross-hybridization (i.e., the Human Release 2.0 oligonucleotide library, containing 60-mer oligonucleotides representing 28830 unique genes designed by Compugen (San Jose, CA) and obtained from Sigma-Genosys (*see* **Note 1**). Oligonucleotides should be lyophilized and in 384-well plates (Genetix, cat. no. X7020) (*see* **Note 2**).
2. 50 m*M* Sodium phosphate buffer pH 8.5: prepare from fresh Na_2HPO_4 and NaH_2PO_4 50 m*M* solutions, autoclave and pass through 0.2-μm filter. Store at room temperature (*see* **Note 3** and **4**).
3. PCR-grade microtiter plate seals (i.e., ABgene) to avoid evaporation of the oligonucleotides.
4. Amine-binding slides (CodeLink™, Amersham Biosciences).
5. Microarray printer (i.e., OmniGrid® 100 microarrayer, Genomic Solutions, Ann Arbor, MI).
6. Pins (i.e., SMP3 pins, TeleChem International, Sunnyvale, CA).

2.2. Genomic DNA Isolation

2.2.1. From Cultured Cells

1. Phase lock gel (Eppendorf).
2. Proteinase K: prepare fresh solution of 20 mg/mL in water.
3. Phenol solution (i.e., Sigma, cat. no. P-4557).
4. Chloroform.
5. Isopropanol.
6. Phenol/chloroform: 50% phenol, 50% chloroform.
7. SDS 20% solution: for preparation of 100 mL dissolve 20 g of sodium dodecyl sulfate in 90 mL water, bring final volume to 100 mL.
8. TE: 10 m*M* Tris-Cl, 1 m*M* EDTA, pH 8.0.
9. Saturated NaCl: for preparation of 1 L dissolve 293 g NaCl in 1 L water, keep adding NaCl until it fails to dissolve, usually not more than 58 g of additional NaCl is needed.
10. Ethanol (70% and absolute, ice cold).
11. Sodium acetate (3 *M*, pH 5.2).

2.2.2. From Paraffin-Embedded Tissue

1. Xylene (i.e., Merck-VEL, cat. no. 90380).
2. Methanol.
3. Ethanol (absolute, 96%, 70%).
4. Haematoxylin (i.e., Merck, cat. no. 115938).

5. QIAamp DNA Mini Kit 250 (Qiagen, cat. no. 51306) or QIAamp DNA Micro Kit 50 (Qiagen, cat. no. 56304) if amount of tissue is limited (i.e., biopsy).
6. NaSCN (i.e., Sigma).
7. Proteinase K (i.e., Roche).
8. RNase A (i.e., Roche).
9. PBS.

2.2.3. From Fresh or Frozen Tissue

1. Wizard Genomic DNA Purification Kit (Promega, cat. no. A1120), containing:
   a. EDTA/nuclei lysis solution.
   b. Proteinase K (20 mg/mL).
   c. RNase solution.
   d. Protein precipitation solution.
2. Phase lock gel (Eppendorf).
3. Phenol solution (i.e., Sigma, cat. no. P-4557).
4. Chloroform.
5. Isopropanol.
6. Phenol/chloroform: 50% phenol, 50% chloroform.
7. TE: 10 m*M* Tris-Cl, 1 m*M* EDTA, pH 8.0.
8. Ethanol (70% and absolute, ice-cold).
9. 3 *M* Sodium acetate, pH 5.2.

2.2.4. Back Extraction of DNA From Trizol Samples (After RNA Isolation)

1. Ethanol (70% and absolute, ice cold).
2. 0.1 *M* Sodium citrate in 10% ethanol: for preparation of 500 mL dissolve 14.7 g sodium citrate in 400 mL 10% (v/v) ethanol. Adjust final volume to 500 mL.

2.3. DNA Quality Assessment

1. Spectrophotometer (i.e., Nanodrop ND-1000).
2. Loading buffer: dissolve 10 g sucrose in 40 mL of water, add 100 mg Orange G and adjust volume to 50 mL with water. For use: dilute 1:5.
3. DNA marker solution: 300 ng/μL DNA ladder (i.e., 1 kb and 100 bp DNA ladder, New England Biolabs, or equivalent), in diluted loading buffer.
4. 1% (w/v) Agarose gel and equipment to run the gel.

2.4. DNA Labeling

1. BioPrime DNA labeling system (Invitrogen, cat. no. 18094-011), containing:
   a. 2.5X Random primers solution.

   b. Klenow fragment of DNA polymerase I (40 U/μL, high concentration), keep on ice at all times or preferably use a −20 °C labcooler when taking the Klenow out of the freezer.
2. Cy3 labeled dCTP (i.e., Amersham Biosciences/Perkin Elmer).
3. Cy5 labeled dCTP (i.e., Amersham Biosciences/Perkin Elmer).
4. ProbeQuant G-50 Micro Columns (Amersham Biosciences).
5. dNTP mixture (i.e., Invitrogen), for 200 μL mix:

   a. dATP (100 m*M*) 4 μL
   b. dGTP (100 m*M*) 4 μL
   c. dTTP (100 m*M*) 4 μL
   d. dCTP (100 m*M*) 1 μL
   e. 1 *M* Tris Cl (pH 7.6) 2 μL
   f. 0.5 EDTA (pH 8.0) 0.4 μL
   g. Water 184.6 μL

2.5. Blocking of the Slides

1. Blocking solution: 0.1 *M* Tris, 50 m*M* ethanolamine, pH 9.0: dissolve 6.055 g Trizma Base and 7.88 g Trizma HCl in 900 mL of water. Add 3.05 g (3 mL) ethanolamine and mix thoroughly. Adjust pH to 9.0 using 6 *N* HCl. Adjust final volume to 1 L with water.
2. 10% (w/v) Sodium dodecyl sulfate (SDS).
3. Wash solution: 4X SSC/0.1% SDS: 200 mL 20X SSC, 10 mL 10% SDS, adjust to 1 L with water.

2.6. Hybridization

1. Human Cot-1 DNA (i.e., Invitrogen) (*see* **Note 5**).
2. Yeast tRNA, 100 μg/μL (i.e., Invitrogen).
3. 3 *M* Sodium acetate, pH 5.2.
4. Ethanol absolute.
5. 20X SSC (i.e., Sigma) and dilutions in water (0.2X, 0.1X, and 0.01X SSC).
6. 20% (w/v) SDS solution: for preparation of 100 mL dissolve 20 g of sodium dodecyl sulfate in 90 mL of water. Adjust final volume to 100 mL.
7. Mastermix: 1 g dextran sulfate (USB), 3.5 mL redistilled formamide (Invitrogen, store at −20 °C), 2.5 mL water, 1 mL 20X SSC, pH 7.0, gently shake for several hours to dissolve dextran sulfate, store aliquotted at −20 °C.
8. Washing buffer (formamide/SSC): 50% formamide, 2X SSC, pH 7.0.
9. PN buffer: 0.1 *M* Na_2HPO_4/NaH_2PO_4, pH 8.0, 0.1% (v/v) Igepal CA630 (i.e., Sigma).
10. GeneTAC/HybArray12 hybstation (Genomic Solutions/Perkin Elmer).

2.7. Scanning of the Slides

1. High-resolution laser scanner or imager equipped to detect Cy3 and Cy5 dyes, including software to acquire images (i.e., Microarray Scanner G2505B, Agilent Technologies).

2.8. Feature Extraction and Creation of a Copy Number Profile

1. Feature extraction software (i.e., BlueFuse 3.2, BlueGnome Ltd., UK).
2. Gene Array List (GAL-file or equivalent) is created by the microarray printer software using the oligonucleotide plates content lists provided by the supplier of the oligo library.
3. Position list: a file containing the mapping positions of the oligonucleotides to their genome provided by the supplier of the oligo library or created by mapping the oligonucleotide sequences onto the respective genome.
4. Software that calculates ratios, links the genomic position of the oligonucleotide to the experimental ratio's, and draws a profile (i.e., Microsoft Excel or dedicated software such as BlueFuse).

3. Methods

3.1. Preparation of Oligonucleotide Microarrays

1. Dissolve oligonucleotides at 10 μ*M* concentration in 50 m*M* sodium phosphate buffer pH 8.5. Thoroughly cover plates with a fresh seal and store at 4 °C (*see* **Note 6**).
2. Spot one or more replicates of each oligonucleotide onto the CodeLink slides. Use CodeLink slides immediately after opening the desiccated packaging. Prior to printing allow oligonucleotide plates to equilibrate to room temperature. If small drops of water caused by condensation during storage are visible on the seals, centrifuge plates at 2000–3000g for 2 min, homogenize oligonucleotide solutions by careful but vigorous shaking on a vortex at medium speed and centrifuge again for 1 min to ensure all oligonucleotide solutions are on the bottom of the wells. Otherwise just centrifuge plates at 2000–3000g for 2 min. Settings of the microarray printer should be such that the pins touch the slide with a speed of 1–3 mm/s. Humidity should not exceed 50% during printing. After printing each plate, cover with a fresh seal to prevent evaporation (*see* **Note 7**).
3. After all plates have been printed, incubate slides overnight at 75% humidity to allow cross-linking of the oligonucleotides to the slide. It is sufficient to incubate the slides in a chamber containing saturated NaCl solution at the bottom.
4. Mark slides by inscription with a diamond knife only.
5. Store slides in a desiccator at room temperature (*see* **Note 8**).

3.2. DNA Isolation

3.2.1. From Cultured Cells

1. Trypsinize and pellet the cells.
2. Resuspend the cells in 3 mL of TE per T75 flask.
3. Add 100 μL of 20% SDS and 20 μL of 20 mg/mL Proteinase K and incubate overnight at 55 °C.
4. Add 1 mL of saturated NaCl solution, 10 mL of ethanol absolute and mix gently.
5. Using a yellow tip spool the precipitated DNA mass from the mixture, place in 1.5-mL tube and remove excess liquid using a tissue. If there is not enough DNA to spool then spin it down and decant liquid.
6. Dissolve DNA in an appropriate amount of water. If the DNA solution is cloudy, proceed with a phenol/chloroform extraction followed by a chloroform extraction in a phase lock gel (PLG) tube and an ethanol precipitation. Otherwise proceed with DNA quality assessment (**Subheading 3.3.**).
7. Spin down PLG 2-mL light tube at 12,000–16,000g in microcentrifuge for 20–30 s.
8. Add the DNA-solution to the 2-mL PLG light tube and add an equal volume of phenol-chloroform directly to the tube (if sample volume is larger than 500 μL, use two PLG light tubes).
9. Mix the organic and the aqueous phases thoroughly by inverting (do not vortex).
10. Centrifuge at 12,000–16,000g for 5 min to separate the phases.
11. Add an equal amount of chloroform directly to the tube, do not discard the supernatant.
12. Mix thoroughly by inverting (do not vortex).
13. Centrifuge at 12,000–16,000g for 5 min to separate the phases.
14. Transfer the aqueous solution (above the gel) to a fresh 1.5-mL microcentrifuge tube.
15. Add 0.1X the volume (of the aqueous solution) 3 *M* sodium acetate pH 5.2 and mix.
16. Add 2.5 times the total volume of absolute ethanol (ice cold). Mix and the DNA can be seen coming out of the solution.
17. Spin down the DNA in microcentrifuge at 20,000*g* and 4 °C for 15 min.
18. Discard the supernatant and add 500 μL 70% ethanol (ice cold). Vortex the sample and spin down the DNA in microcentrifuge at 20,000g and 4 °C for 10–15 min.
19. Discard the supernatant, air-dry pellet until no ethanol is visible.
20. Resuspend the pellet in 100 μL TE or water.

3.2.2. From Paraffin Blocks (see **Note 9**)

3.2.2.1. Treatment of 10-μm Coupes Prior to DNA Isolation

1. Cut 1–15 10-μm coupes.
2. Place the coupes on uncoated slides.
3. Dry overnight at room temperature.

4. Deparaffine three times 7 min in Xyleen at room temperature.
5. Wash two times 7 min in methanol at room temperature.
6. Subsequently wash in absolute ethanol, 96% ethanol, 70% ethanol and with from the tap.
7. Color in Haematoxylin for 5 min.
8. Wash for 5 min with water from the tap.
9. Wet the coupe with water, scrape the selected area (by the hand of a representative 4-μm HE-colored coupe) from the slide with a clean knife and transfer the coupe into a new microcentrifuge tube.
10. Spin for 5 min at 14,000*g*, remove the liquid from the tube and discard.
11. Proceed to **Subheading 3.2.2.3.**

3.2.2.2. Treatment of 50-μm Coupes Prior to DNA Isolation

1. Cut at least three 50-μm coupes.
2. Place all coupes of one sample in one microcentrifuge tube.
3. Add 1 mL Xyleen and shake by vortexing.
4. Incubate for 7 min at room temperature.
5. Spin for 5 min at 14,000*g* and remove the supernatant.
6. Repeat Xyleen washing and spinning (**step 3–5**) two times.
7. Add 1 mL of methanol and shake by vortexing.
8. Incubate for 5 min at room temperature.
9. Spin for 5 min at 14,000*g* and remove the supernatant.
10. Repeat methanol washing and spinning (**step 7–9**).
11. Add 1 mL absolute ethanol and shake by vortexing.
12. Spin for 5 min at 14,000*g* and remove the supernatant.
13. Add 1 mL 96% ethanol and shake by vortexing.
14. Spin for 5 min at 14,000*g* and remove the supernatant.
15. Add 1 mL 70% ethanol and shake by vortexing.
16. Spin for 5 min at 14,000*g* and remove the supernatant.
17. Add 1 mL of distilled water and shake by vortexing.
18. Spin for 5 min at 14,000*g* and remove the supernatant.

3.2.2.3. DNA Isolation From Paraffin-Embedded Tissue

1. Add 1 mL 1 *M* NaSCN and shake by vortexing.
2. Incubate overnight at 38–40 °C.
3. Spin for 5 min at 14,000*g* and remove supernatant.
4. Wash pellet three times with 1 mL PBS.
5. Add 200 μL Buffer ATL (QIAamp kit) and 40 μL of Proteinase K, shake by vortexing and incubate at 50–60 °C.
6. If the isolation was started in the morning, then add another 40 μL of Proteinase K at the end of the day.
7. Continue incubation at 50–60 °C overnight.

8. Add 40 μL of Proteinase K and shake by vortexing.
9. Incubate all day at 50–60 °C and vortex regularly.
10. Add 40 μL of Proteinase K add the end of the day and shake by vortexing.
11. Incubate again overnight at 50–60 °C.
12. Add 40 μL RNase A the next morning and shake by vortexing.
13. Incubate for 2 min at room temperature.
14. Add 400 μL Buffer AL (QIAamp kit) and shake by vortexing.
15. Incubate for 10 min at 65–75 °C.
16. Add 420 μL absolute ethanol and shake by vortexing thoroughly.
17. Transfer 600 μL of the solution to a QIAamp column, which is placed in the tube from the QIAamp kit.
18. Spin for 1 min at 2000*g* and discard flow through.
19. Repeat **steps 17** and **18** until the complete sample is applied to the columns.
20. Add 500 μL of Buffer AW1 (QIAamp kit) to the column.
21. Spin for 1 min at 2000*g*, discard flow through.
22. Add 500 μL of Buffer AW2 (QIAamp kit) to the column.
23. Spin for 3 min at 14,000g and discard flow through.
24. Transfer column to a fresh microcentrifuge tube (with lid).
25. Add 75 μL of Buffer AE (QIAamp kit), preheated to 65–75 °C, to the column.
26. Leave at room temperature for 1 min.
27. Spin for 1 min at 2000*g*.
28. Discard column and store DNA at 2–8 °C.

3.2.3. From Fresh or Frozen Tissue

1. Add to 1.5-mL microcentrifuge tube:

   (a) 0.5- to 1-cm^3 Tissue.
   (b) 600 μL EDTA/Nuclei Lysis solution.
   (c) 17.5 μL Proteinase K (20 mg/mL).

2. Incubate overnight at 55 °C with gentle shaking or vortex the sample several times.
3. Add 3 μL of RNase solution to the nuclear lysate, mix the sample by inverting the tube two to five times.
4. Incubate the mixture for 15–30 min at 37 °C, allow the sample to cool to room temperature for 5 min before proceeding.
5. Add 200 μL Protein Precipitation Solution to the sample and vortex vigorously at high speed for 20 s, chill sample on ice for 10 min.
6. Spin 15 min at 20,000*g* to pellet the precipitated protein.
7. Carefully remove the supernatant containing the DNA (leaving the protein pellet behind) and transfer it to a fresh 1.5-mL microcentrifuge tube.
8. Add 600 μL of room temperature isopropanol.
9. Mix the solution by gently inverting until the white thread-like strands of DNA form a visible mass.

10. Centrifuge for 1 min at 20,000*g* at room temperature. The DNA will be visible as a small white pellet. Carefully aspirate supernatant by decanting. Air-dry until no ethanol visible.
11. Add 200 µL TE to resuspend the DNA.
12. Spin down PLG 2 mL light at 12,000–16,000g in microcentrifuge for 20–30 s.
13. Add the 200 µL DNA-containing TE to the 2-mL PLG light tube and add an equal volume of phenol-chloroform directly to the tube.
14. Mix the organic and the aqueous phases thoroughly by inverting (do not vortex).
15. Centrifuge at 12,000–16,000g for 5 min to separate the phases.
16. Add 200 µL of chloroform directly to the tube, do not discard the supernatant.
17. Mix thoroughly by inverting (do not vortex).
18. Centrifuge at 12,000–16,000g for 5 min to separate the phases.
19. Transfer the aqueous solution (above the gel) to a new 1.5-mL microcentrifuge tube.
20. Add 20 µL 3 *M* sodium acetate pH 5.2 and mix.
21. Add 2.5 times the total volume of absolute ethanol (ice cold), after mixing the DNA should come out of solution.
22. Spin down the DNA in microcentrifuge at full speed for 15 min.
23. Discard the supernatant and add 500 µL 70% ethanol (ice cold). Vortex the sample and spin down the DNA in microcentrifuge at 20,000*g* and 4 °C for 10–15 min.
24. Discard the supernatant, air-dry pellet until no ethanol is visible.
25. Resuspend the pellet in 100 µL TE or water.

3.2.4. Back Extraction of DNA From Trizol Samples (After RNA Isolation)

1. Thaw samples if necessary at room temperature.
2. Spin for 5 min at 12,000*g* and 4 °C and remove the remaining aqueous phase overlaying the interphase.
3. Add 0.3 mL of absolute ethanol per milliliter Trizol.
4. Mix well (vortex) and leave at room temperature for 2–3 min.
5. Spin for 10 min at 20,000*g* and 4 °C.
6. Remove supernatant by decanting.
7. Wash pellet with 1 mL 0.1 M sodium citrate in 10% ethanol per milliliter of Trizol, leave for 30 min at room temperature. Vortex several times during the 30 min.
8. Spin for 10 min at 20,000*g* and 4 °C.
9. Repeat washing and spinning (**steps 7** and **8**).
10. Wash pellet with 1.5 mL 70% ethanol per milliliter of Trizol, leave for 20 min at room temperature.
11. Spin for 10 min at 20,000*g* and 4 °C.
12. Remove supernatant completely by pipetting.
13. Air-dry the pellet until no ethanol is visible (approx 5–10 min).
14. Resuspend pellet in 100 µL of water.
15. Leave for 30 min at 65 °C, gently mix several times during the 30 min.

16. Spin for 10 min at 20,000*g* and 4 °C.
17. Transfer the DNA-containing supernatant to a fresh microcentrifuge tube.

3.3. DNA Quantity and Quality Assessment

3.3.1. Spectrophotometric Quantity and Quality Assessment

1. Measure 1 μL of the DNA solution in the Nanodrop.
2. Pure DNA should have A_{260}/A_{280} equal to A_{260}/A_{230} ~2.0. Lower values indicate the presence of proteins (A_{280}) or residual organic contamination (A_{230}), like Trizol and chloroform. An additional phenol/chloroform extraction and ethanol precipitation is then recommended.

3.3.2. Assessment of the Integrity of Chromosomal DNA by Gel Electrophoresis

1. Analyze 200 ng of your DNA sample on a 1% agarose gel next to 3 μL DNA marker solution. The chromosomal DNA is located above the 1-kb band of the marker. A smear across the gel indicates the DNA is broken down and will compromise the array CGH results. Alternatively, a more in depth analysis of DNA quality from paraffin-embedded tissue uses a multiplex PCR approach *(3)*.

3.4. DNA Labeling

1. Mix together in PCR tube 300 ng of genomic DNA and 20 μL 2.5X random primers solution. Adjust volume to 42 μL with water. For paraffin material 600 ng of test and reference DNA samples should be used.
2. Denature the DNA mixture in a PCR machine at 100 °C for 10 min and immediately transfer to an ice/water bath for 2–5 min. Briefly spin the tubes and put back on ice.
3. Add to the mixture while keeping on ice: 5 μL of dNTP mixture, 2 μL of Cy3 (test) or Cy5 (ref) labeled dCTP and 1 μL of Klenow DNA polymerase (40 U/μL).
4. Mix well and incubate at 37 °C (in PCR machine) for 14 h, then keep at 4 °C.
5. Preparation of Probe-Quant G-50 columns for removal of uncoupled dye material:
   (a) Resuspend the resin in the column by vortexing.
   (b) Loosen the cap one-fourth turn and snap off the bottom closure.
   (c) Place the column in a 1.5-mL microcentrifuge tube.
   (d) Prespin the column for 1 min at 735*g*. Start timer and microcentrifuge simultaneously, so total spinning time should not exceed 1 min.
6. Sample application:
   (a) Place the column in a fresh 1.5-mL tube and slowly apply 50 μL of the sample to the top center of the resin, being careful not to disturb the resin bed.
   (b) Spin the column at 735*g* for 2 min. The purified sample is collected at the bottom of the support tube.
7. Discard the column and store the purified and labeled sample in the dark until use on the same day or optionally store at −20 °C for a maximum of 10 d.

3.5. Blocking of the Slides

1. Add 0.01 vol of 10% SDS to blocking solution (final concentration of 0.1% SDS).
2. Prewarm blocking solution at 50 °C.
3. Place the oligonucleotide slides in a slide rack and block residual reactive groups using prewarmed blocking solution at 50 °C for 15 min, extend to 30 min if not warm, do not exceed 1 h.
4. Rinse the slides twice with water.
5. Wash slides with wash solution (4X SSC/0.1% SDS, prewarmed to 50 °C) for 15–60 min, use at least 10 mL per slide.
6. Rinse briefly with water, do not allow slides to dry prior to centrifugation.
7. Place each slide in 50-mL tube and centrifuge at 200*g* for 3 min to dry.
8. Use slides for hybridization within 1 wk.

3.6. Hybridization

3.6.1. Preparation of Hybridization Solution

1. Mix together in a 1.5-mL tube: 50 μL of Cy3-labeled test DNA, 50 μL of Cy5-labeled reference DNA, 10 μL of 1 μg/μL Human Cot-1 DNA (*see* **Note 10**).
2. Add 11 μL of 3 *M* NaAc pH 5.2 (0.1 vol) and 300 μL of ice-cold ethanol absolute, mix the solution by inversion and collect DNA by centrifugation at 20,000*g* for 30 min at 4 °C.
3. Remove the supernatant with a pipet and air-dry the pellet for 5–10 min until no ethanol is visible.
4. Carefully dissolve the pellet (prevent foam formation due to SDS) in 13 μL yeast tRNA (100 μg/μL) and 26 μL 20% SDS. Leave at room temperature for at least 15 min.
5. Add 91 μL of master mix and mix gently.
6. Denature the hybridization solution at 73 °C for 10 min, and incubate at 37 °C for 60 min to allow the Cot-1 DNA to block repetitive sequences.

3.6.2. Programming the Hybstation

On the hybstation store the following program named "CGH.hyb:"

1. Introduce hybridization solution. Temperature 37 °C.
2. Set slide temperature. Temp: 37 °C, time: 38h:00m:00s, agitate: yes.
3. Wash slides (50% formamid/2X SSC). Six cycles, source 1, waste 2 at 36 °C, flow for 10 s, hold for 20 s.
4. Wash slides (PN buffer). Two cycles, source 2, waste 1 at 25 °C, flow for 10 s, hold for 20 s.
5. Wash slides (0,2X SSC). Two cycles, source 3, waste 1 at 25 °C, flow for 10 s, hold for 20 s.
6. Wash slides (0,1X SSC). Two cycles, source 4, waste 1 at 25 °C, flow for 10 s, hold for 20 s.

3.6.3. Hybridization Using the Hybstation

1. Assemble up to six hybridization units (two slides per unit): insert rubber O-rings in the covers and put the slides on the black bottom plate. Make sure slides are in the proper orientation with the printed side up.
2. Introduce the unit into the hybstation, press unit down with one hand while tightening the screw with the other.
3. Insert plugs into the sample ports and the tubes into the corresponding wash bottles.
4. On the touch screen subsequently press: start a run, from floppy, CGH.hyb, load, the positions of the slides you want to use, start, continue (the hybstation starts to warm up the slides).
5. When the hybstation is ready (visible on screen by indication of the module you have to start) apply hyb mix:

   (a) Press Probe to add the hyb mix for the selected slide.
   (b) Check if a mark on the screen appears.
   (c) Take the plug out and inject the hyb mix by pipetting it slowly into the port using a 200 – μL pipet.
   (d) Press the Finished control (check mark) and replace the plug.
   (e) Repeat this for the next slide.
   (f) Press the Finished control for the selected slide.
   (g) Press the Finishes control for the module.
   (h) Repeat this for the selected module(s).

6. Take slides out after 38 h and put them in 0.01X SSC.
7. Place each slide in 50-mL tube and centrifuge at 200*g* for 3 min to dry.
8. Immediately scan slides in a microarray scanner (storage comprises the signal and influences ratios).

3.6.4. Cleaning of the Hybstation

Cleaning the hybstation after each hybridization is essential to maintain proper functioning of the equipment.

1. Reassemble all used hybridization units with dummy slides and introduce them into the hybstation.
2. Insert plugs into all sample ports and place all tubes in a bottle of water (*see* **Note 11**).
3. On the touch screen subsequently press: maintenance, Machine Cleaning Cycle, the positions of the slides used, continue.
4. When cleaning is finished, take out the hybridization units, rinse with water, (never use ethanol) especially the sample port, and dry the unit with the air pistol.

3.7. Scanning of the Slides

Scan slides for both the Cy3 and Cy5 channel taking the following issues into account (settings have been tested for the Microarray Scanner G2505B, Agilent Technologies):

1. Set resolution at 10 μm.
2. Use maximum (100) settings for the photomultiplier (PMT).
3. Make sure lasers are warm before scanning is started. Warm-up time is 15–30 min.
4. Store the results from both channels as separate TIFF images.
5. Check that images from both channels are precisely aligned: pixels in each image obtained from the same location should in fact overlap.

3.8. Feature Extraction and Creation of a Copy Number Profile

To create copy number profiles from TIFF images, follow the steps next. BlueFuse (BF) microarray analysis software takes care of all of these steps as indicated in brackets, but essentially any feature extraction software in combination with a spreadsheet will do.

1. Finding the spots (using the information from the GAL-file, placing the grid is fully automated in BF).
2. Feature extraction (quantification is fully automated in BF and includes background subtraction).
3. Ratio building (automatic in BF).
4. Exclusion of bad spots (exclude spots that have a "confidence value" lower than 0.1 or a "quality flag" lower than 1).
5. Linking ratios to genomic positions (using the information from the position file, this is automated in BF).
6. Normalization (select global mode normalization in BF, avoid Lowess and/or block normalization, because this may compress the profile) (*see* **Note 12**).
7. Drawing the genomic profile (automated in BF): order normalized ratios by chromosomal mapping and display in a graph.

4. Notes

1. Any time we identify a supplier preceded by "i.e." we suggest a possible supplier for that material as currently used in our laboratory. Whenever "i.e." is not written, we believe the material should be obtained from the mentioned supplier to ensure best results.
2. The human oligonucleotide library we have been using is no longer produced by Sigma. We would now recommend the HEEBO (Human Exonic Evidence Based Oligonucleotide) set (Invitrogen) or the human oligonucleotide library

from Operon. Of course the technique can be used with oligonucleotide libraries designed from other species or providers and we believe longer oligonucleotides perform better *(2)*.

3. Unless stated otherwise, all chemicals should be "high grade" or preferably "molecular biology grade."
4. Unless stated otherwise, all solutions should be prepared in water that has a resistivity of 18.2 MΩ-cm and total organic content of less than five parts per billion. This standard is referred to as "water" in this text.
5. Adjust origin of the Cot-1 DNA according to species of the oligonucleotide library.
6. For safety and long-term storage reasons, we recommend to store in batches of 200–300 pmol and if necessary split the library upon arrival to create one or more copies. For this purpose dissolve oligonucleotides in water, pipet part of it to one or more copy plates (preferably using a pipetting robot) and dry down overnight at room temperature. Dissolved stocks can be used for at least 2 yr when stored at 4 °C. Frozen stocks hold for more years. Do not freeze/thaw oligonucleotide solutions.
7. When handling the slides wear nonpowdered gloves at all times.
8. After printing desiccated slides hold for more than 1 yr.
9. The protocol is described for "normal" size tissue samples using the QIAmp Mini kit. In case of small size samples using the QIAmp Micro kit volumes and Proteinase K incubation times may need to be adjusted and the RNase treatment omitted.
10. Test and reference DNA can be labeled with either Cy3 or Cy5.
11. Using 50 °C heated water may result in better cleaning, never use ethanol or other organic solvents.
12. Mode normalization is used to set the "normal" level and is preferred over mean or median normalization. Mode normalization is more accurate because it ignores the ratios generated by gains, amplifications, and deletions. Block normalization is sometimes used to suppress noise. Block normalization may help to suppress noise in samples showing little aberrations, such as cytogenetic samples, but is not recommended for samples with multiple chromosomal aberrations, such as tumour samples.
13. Updated protocols available on www.vumc.nl/microarrays.

References

1. van den IJssel, P., Tijssen, M., Chin, S. F., et al. (2005) Human and mouse oligonucleotide-based array CGH. *Nucleic Acids Res.* **33,** e192.
2. Ylstra, B., van den IJssel, P., Carvalho, B., Brakenhoff, R. H., and Meijer, G. A. (2006) BAC to the future! or oligonucleotides: a perspective for micro array comparative genomic hybridization (array CGH). *Nucleic Acids Res.* **34,** 445–450.
3. van Beers, E. H., Joosse, S. A., Ligtenberg, M. J., et al. (2006) A multiplex PCR predictor for aCGH success of FFPE samples. *Br. J. Cancer* **94,** 333–337.

15

Studying Bacterial Genome Dynamics Using Microarray-Based Comparative Genomic Hybridization

Eduardo N. Taboada, Christian C. Luebbert, and John H. E. Nash

Summary

Genome sequencing has revealed the remarkable amount of genetic diversity that can be encountered in bacterial genomes. In particular, the comparison of genome sequences from closely related strains has uncovered significant differences in gene content, hinting at the dynamic nature of bacterial genomes. The study of these genome dynamics is crucial to leveraging genomic information because the genome sequence of a single bacterial strain may not accurately represent the genome of the species.

The dynamic nature of bacterial genome content has required us to apply the concepts of comparative genomics (CG) at the species level. Although direct genome sequence comparisons are an ideal method of performing CG, one current constraint is the limited availability of multiple genome sequences from a given bacterial species. DNA microarray-based comparative genomic hybridization (MCGH), which can be used to determine the presence or absence of thousands of genes in a single hybridization experiment, provides a powerful alternative for determining genome content and has been successfully used to investigate the genome dynamics of a wide number of bacterial species. Although MCGH-based studies have already provided a new vista on bacterial genome diversity, original methods for MCGH have been limited by the absence of novel gene sequences included in the microarray. New applications of the MCGH platform not only promise to accelerate the pace of novel gene discovery but will also help provide an integrated microarray-based approach to the study of bacterial CG.

Key Words: DNA microarrays; comparative genomics; genomotyping; genome content; phylogenomics; genome evolution; bacteria.

From: *Methods in Molecular Biology, vol. 396: Comparative Genomics, Volume 2*
Edited by: N. H. Bergman © Humana Press Inc., Totowa, NJ

1. Introduction

In the decade since the completion of the first bacterial genome-sequencing project, an ever-increasing amount of genomic information has challenged our definition of a bacterial species and shed light on the dynamic nature of bacterial genomes. Initial results from comparative genomic (CG) analysis of whole-genome sequences from multiple strains of *Helicobacter pylori* ***(1)*** and *Escherichia coli* ***(2)*** revealed that although some genetic variation is in the form of differences in shared conserved gene sequences, the most important source of genetic variation within a bacterial species is at the level of gene content. In the most comprehensive bacterial intraspecies CG study to date, the complete genomes of eight strains of *Streptococcus agalactiae* were recently compared ***(3)***. This study revealed that the *S. agalactiae* species can be described in terms of a "pan-genome," which consists of a "core genome" of genes shared by all isolates and accounting for approx 80% of any single genome, plus a "dispensable genome" of partially shared and strain-specific genes. This confirms results from earlier studies and further affirms that bacterial genomes are in a constant state of flux.

An increasing number of bacterial species have complete genome sequences available for more than one strain **(Fig. 1)** however, as of January 2006, of the 623 bacterial species whose genomes are either fully sequenced or in progress, there exist only 16 CG datasets comprising five or more strains. Of note, these 16 species, and nearly two-thirds of all bacterial species with multiple strain coverage, represent plant or animal pathogens. Thus, although the study of intraspecies CG is important from the standpoint of understanding genome dynamics and evolution, it also has important practical implications. Each species can be thought of being comprised of a spectrum of genetic variants with potential differences in important phenotypes such as host specificity and pathogenicity. The results from CG studies performed thus far have led to an increasing realization that obtaining the complete genome sequence of a single bacterial strain not only fails to address the genetic diversity comprised by the species but also seriously limits the extent to which these genomic data can be exploited for practical applications. For example, in exploiting bacterial genome sequence information toward the development of "broad spectrum" antimicrobial drugs or vaccines, the suitability of a gene target rests in large part on whether the gene is conserved in a large proportion of clinically relevant strains. A proof of concept for this multigenome CG-based approach to vaccine research has recently been described through the development of a universal Group B *Streptococcus* vaccine ***(4)***.

Although obtaining the complete genome sequence of multiple strains per bacterial species is highly desirable, these types of datasets are likely to remain

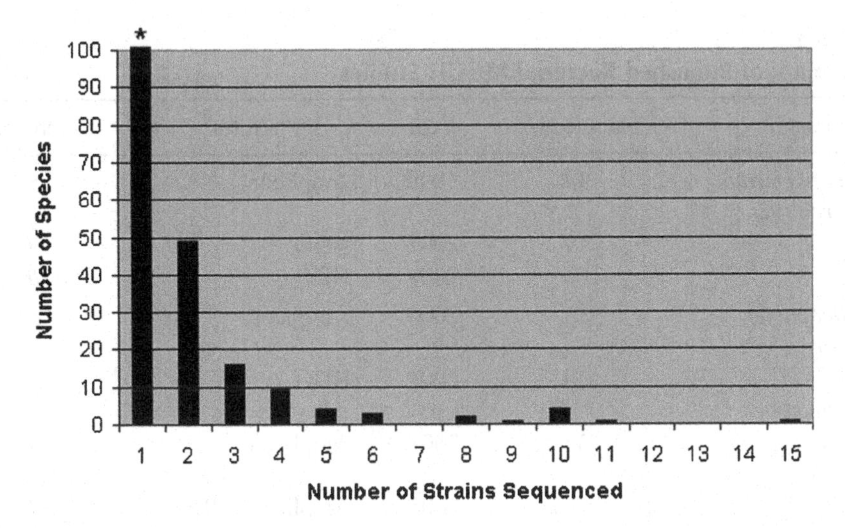

Fig. 1. Multistrain coverage in bacterial sequencing projects. As of January 2006, 847 bacterial genomes (representing 623 different species) have been or are currently being sequenced. Although multiple genome coverage is available for 91 species, only 16 have five or more genomes characterized (*532 species have single genome coverage).

limited in number in the near future. In their absence, the field of CG has benefited greatly from the development and rapid adoption of DNA microarray technology. The development of DNA microarrays comprising every open reading frame (ORF) in a genome strain has become an integral part of nearly every bacterial genomics project, whereas an increasing number of laboratories have acquired the necessary infrastructure to carry out microarray experiments.

Although microarrays were originally applied in gene expression studies (i.e., transcript profiling), studies describing microarray-based comparative genomic hybridization (MCGH) soon followed *(5,6)*. In the years since these initial efforts, MCGH has been used to study the CG of a large number of different bacterial species **(Table 1)** and has shown enormous potential as a cost effective and reasonably high-throughput approach to generating large CG datasets when whole-genome sequence data from multiple strains is lacking.

This chapter will provide an overview of currently published methods for DNA microarray-based bacterial CG, placing a special emphasis on data analysis and visualization issues. It is important to note, however, that this is a newly emerging field and that protocols and analytical methods are in an active state of development.

Table 1
Summary of Published Bacterial MCGH Studies

Organism	Sample size	Year	Platform[a]	Reference
Mycobacterium tuberculosis	14	1999	Amplicon	***5***
	19	2001	HDO	***9***
	100	2004	HDO	***10***
Helicobacter pylori	15	2000	Amplicon	***6***
Streptococcus pneumoniae	20	2001	HDO	***11***
Staphylococcus aureus	36	2001	Amplicon	***53***
	9	2005	Amplicon+HDO	***54***
Campylobacter jejuni	11	2001	Amplicon	***23***
	18	2003	Amplicon	***55***
	26	2004	Amplicon	***56***
	97	2004	Amplicon	***35***
	111	2005	Amplicon	***46***
Vibrio cholerae	10	2002	Amplicon	***40***
Streptococcus (Group A)	36	2002	Amplicon	***57***
Salmonella	22	2002	Amplicon	***58***
	26	2003	Amplicon	***59***
	79	2004	Amplicon	***60***
	40	2005	Amplicon	***61***
Listeria monocytogenes	50	2003	Shotgun/Amplicon	***24***
	52	2004	Shotgun/Amplicon	***25***
	113	2004	Shotgun/Amplicon	***62***
Escherichia coli	26	2003	Amplicon	***63***
	29	2003	Amplicon	***64***
	22	2004	Amplicon	***65***
	10	2005	LDO	***12***
Bacillus thuringiensis, B. cereus and *B. anthracis*	19	2003	Amplicon	***47***

Francisella tularensis	27	2003	Shotgun/Amplicon	***66***
	17	2004	Shotgun/Amplicon	***26***
Pseudomonas aeruginosa	18	2003	HDO	***67***
Yersinia	32	2003	Amplicon	***68***
	18	2004	Amplicon	***69***
Chlamydia trachomatis	14	2004	Amplicon	***70***
Bordetella	42	2004	Amplicon	***71***
Neisseria	38	2005	Shotgun/Amplicon	***72***
Brucella	6	2004	HDO	***19***
Mycobacterium avium	16	2005	Amplicon	***73***
Staphylococcus epidermidis	42	2005	LDO	***13***
Thermotoga maritima	11	2005	Amplicon	***74***

[a] LDO: low-density spotted oligo-array; HDO: high-density *in situ* synthesized oligo-array; Amplicon: amplicon-array; Shotgun/Amplicon: shotgun library inserts amplified by PCR.

1.1. Microarray Platforms Available for Bacterial CGH

1.1.1. Gene-Specific Microarrays

Gene-specific probes for a microarray can be obtained either through synthesis of oligonucleotides to be used directly as probes, or through the PCR amplification of gene-specific DNA fragments.

1.1.1.1. Amplicon-Based Microarrays

The PCR amplicon-based microarray (henceforth referred to as "amplicon-array") platform, pioneered by the Brown laboratory at Stanford *(7,8)* uses DNA fragments amplified by PCR (i.e., amplicons) from cDNA library clones as microarray probes (*see* **Note 1**), which are robotically deposited (printed) by "contact-spotting" as a high-density array onto a derivatized glass microscope slide. This class of microarray, which can be used both for gene expression analysis and for MCGH, has been the most popular platform in bacterial microarray studies.

The first description of bacterial MCGH was by Behr et al. *(5)* in a study in which they examined the genome content of various *Mycobacterium bovis* strains used as vaccines against tuberculosis using an amplicon-array derived from the *M. tuberculosis* H37Rv genome strain. To generate their array, they designed and synthesized PCR primer pairs for every gene to be included in the microarray ("ORFmer sets"), PCR-amplified each probe, and spotted the gene-specific probes onto glass slides. Their basic method for MCGH used differentially labeled targets generated from the genomic DNA of tester and reference strains, which were then cohybridized to the microarray. Once control- and reference-derived signals were compared, genes with differential hybridization signals were identified as likely to show copy number changes and/or sequence divergence between control and tester strains. Using this approach, Behr et al. uncovered a number of genomic deletions in the vaccine strains with respect to the H37Rv genome strain.

1.1.1.2. Oligonucleotide-Based Microarrays

The original Gene Chip® technology developed by Affymetrix relied on *in situ* synthesis of short 25-mer oligonucleotides (oligos) directly on the array surface using a proprietary photolithographic method. Affymetrix arrays have been used in a small number of studies (for examples *see* **refs.** ***9–11***). A number of additional methods have been developed for *in situ* oligo-synthesis (these include digital micromirror-based by NimbleGen™, ink-jet deposition by Agilent Technologies, electrode directed by Nanogen) but they have yet to be widely adopted for MCGH.

Although microarrays made by direct spotting of short oligos perform poorly compared to amplicon arrays, advances in high-throughput synthesis of "long oligos" (i.e., >50 bases), has led to the development of "long-oligo arrays" that perform on par with spotted amplicon arrays with the additional attractive feature that they can be directly printed without the need for a PCR amplification step (for examples, *see* **refs.** ***12*** and ***13***).

1.1.1.3. Advantages and Disadvantages of Amplicon- vs Oligonucleotide-Arrays in MCGH

Even though some commercial sources exist for microarrays (Affymetrix, MGW Biotech), ORFmer sets (Sigma Genosys), or long oligos for spotting (Operon), these are only available for a very select number of species. Thus, most groups will need to design custom-made arrays for their organism of interest. At the present time, spotted arrays have been the platform of choice in bacterial MCGH in large part because the method allows research teams complete autonomy and flexibility in microarray design and production. In addition, the use of nonproprietary fabrication technology can substantially decrease operating costs in the long run.

Although most studies have used amplicon-arrays, oligo-arrays may eventually supersede them as the platform of choice because they are easier to implement because oligo-arrays do not require the additional amplification of probes, amplicon assessment, and amplicon purification steps, which are not a trivial endeavor. An additional attractive feature of oligo-arrays is their increased hybridization specificity as a result of the increased sensitivity of oligos toward probe/target mismatches. Oligonucleotides can be designed to specifically minimize cross-hybridization to homologous (i.e., similar but nonidentical) sequences (for an example, *see* **ref. *14***), which makes discrimination between alleles possible ***(15–17)***. One caveat is that the increased specificity of oligo-arrays can come at the cost of decreased sensitivity in gene detection because unknown alleles of a gene may yield little to no hybridization signal, which can lead to erroneous gene presence/absence calls ***(18)***. Because additional or unknown alleles are generally absent from a microarray, the increased tolerance of amplicon-based arrays toward mismatches may allow for their detection, thus minimizing the likelihood of false gene-absence calls.

Although *in situ* synthesized oligo-arrays have not seen wide use in bacterial MCGH, they merit mentioning because the platforms are likely to play an important role in future CG research. The ultra-high density (*see* **Note 2**) at which oligos can be arrayed with *in situ* synthesis-based platforms (such that multiple probes can be used per gene) and the increased hybridization specificity of oligos can be exploited for high-resolution MCGH, allele determination, and genomic resequencing ***(19–21)***.

1.1.2. Plasmid Clone Microarrays

"Shotgun" clones are derived from genomic libraries that have been generated by cloning randomly sheared genomic DNA fragments (0.5–5 Kb) into a plasmid vector. The approaches described next have used plasmids in microarray construction either directly as probes for spotting or indirectly as templates for amplicon generation.

1.1.2.1. Low-Cost Shotgun Clone Microarrays

Because shotgun library construction and the subsequent plasmid isolation from thousands of individual clones is generally the first step in a typical bacterial genome-sequencing project, low-cost microarrays can be constructed that use these plasmid clones as templates for amplicon probes ***(22)***. The first MCGH application of this approach was described for a *Campylobacter jejuni* microarray made using clones derived from the genome-sequencing project ***(23)***. Although Dorrell et al. first amplified plasmid inserts by PCR to generate their printing stocks, this approach for generating microarrays was relatively inexpensive because a common primer pair was used to generate all of the amplicons, obviating the need for the synthesis of gene-specific primers on a genome-scale. Because of the concurrent characterization of the clones during the sequencing project, the authors were able to select a panel of clones representing the majority of genes in the genome strain. Nevertheless, the main disadvantage of shotgun clone microarrays is that it is generally impossible to completely avoid clones that contain the sequences from adjacent genes, confounding subsequent analysis.

1.1.2.2. Shotgun Library Screening Using Microarrays

Recently, several groups have used the microarray platform to prescreen shotgun-libraries for strain-specific genes (i.e., absent or highly divergent in the genome strain) by directly spotting plasmid DNA from shotgun clones to produce microarrays ***(24–28)***. When hybridized against the genome strain, clones bearing sequences conserved in both strains would be expected to yield significant signal whereas probes derived from clones with "unique" sequences would be expected to yield reduced or no signal. By performing a microarray-based prescreening step, the authors were able to narrow down on the clones bearing unique sequences, enabling them to focus sequencing resources on such clones thereby reducing sequencing costs and speeding up the discovery of novel "strain-specific" genes. This method is likely to play an instrumental role in the development of "pan-genomic" microarrays (i.e., containing nonredundant sets of conserved and strain-specific genes) when additional genome sequence information is lacking (for examples, *see* **refs. *29*** and ***30***).

1.1.3. "Library on a Slide" Microarrays

In conventional microarray platforms, the complete genome of a bacterial strain is represented on a microarray as a collection of probes derived from all of the individual genes contained in its genome. The presence or absence of these genes in a strain of interest is then assayed based on the hybridization pattern on the microarray. Zhang et al. ***(31)*** have described a novel application of microarray technology in which each probe on the array corresponds to whole genomic DNA from a bacterial strain. The array can thus contain thousands of whole genomic DNA samples from different strains arrayed at high density. These arrays can be used to assay for the presence or absence of a specific gene and/or allele in thousands of strains by hybridization to a gene-specific labeled target. Although this method does not allow for whole-genome CG analysis *per se* (only one gene can be assayed at a time), it is well worth mentioning because it enables researchers to assay for the prevalence of a gene of interest (e.g., antibiotic resistance gene, virulence factor, and so on) in very large collections numbering thousands of strains. The method also represents a very efficient way for sharing genomic DNA samples from such large collections with the scientific community.

1.2. Microarray Construction

The successful production of quality microarrays is as much an art as it is a science. Although laboratories exist that own and operate their own printing equipment, most labs will have to resort to either partly or completely outsourcing the production process to dedicated production facilities that have the necessary liquid handling and printing robotics (*see* **Note 3**). In that context, the following sections on microarray construction will focus on the early stages of microarray design and production, prior to the actual microarray printing. These methods will focus primarily on spotted microarrays, as they are the most widely used platform for MCGH (*see* **Note 4**).

1.2.1. Oligonucleotide/PCR Primer Selection

Whether amplicon- or oligo-arrays are used, gene-specific primer-pairs or long-oligos will have to be selected for every gene to be represented in the array. Selection of oligonuclotides or primer pairs on a genome scale can be automated and a number of free and commercial programs are available that select oligos based on complete genome sequences and optimize them for a range of criteria (i.e., similar T_m, lack of secondary structure, and so on) (*[32–34]*; *also see* **Note 5**). A very important consideration is avoidance of potential cross-hybridization between microarray probes. Some programs incorporate steps that reject potentially cross-hybridizing probes.

One hitherto largely ignored parameter in amplicon-array design is that of probe size. The typical approach of many groups has been to maximize amplicon length given the ORF size. The resulting array is generally a collection of amplicon probes of vastly different lengths, which is exacerbated by the large number of small ORFs in bacterial genomes. This can cause significant problems upon subsequent acquisition and analysis of the data because of the direct relationship between amplicon length and average signal yield (**Fig. 2**). Laser scanning parameters that are optimal for short probes may be inadequate for longer probes owing to signal saturation, which may compromise the

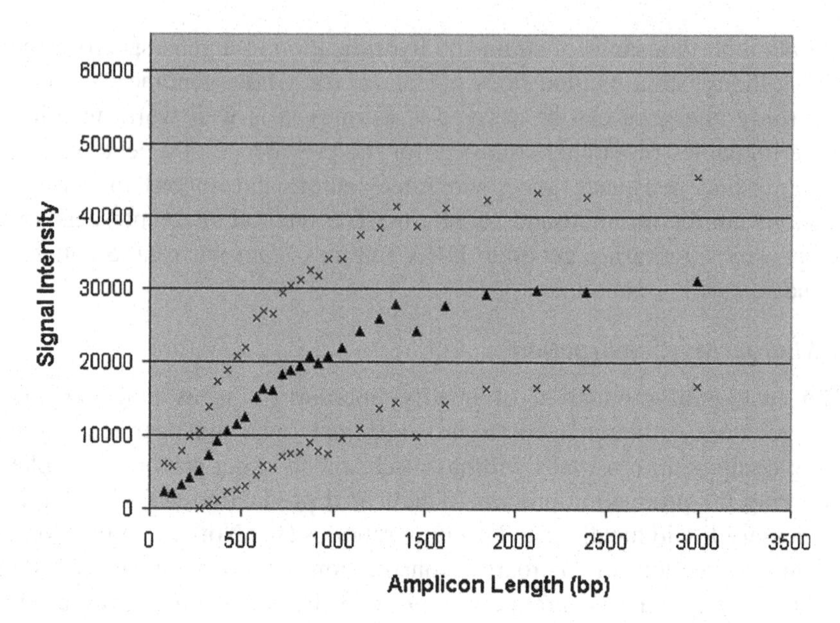

Fig. 2. Relationship between amplicon length and microarray signal. The average signal intensity obtained from the control channel using a *Campylobacter jejuni* amplicon array was tracked more than 150 hybridizations. Despite normalization of amplicon concentrations prior to arraying, decreasing signal yield is observed with decreasing amplicon length, affecting spot dynamic range and yielding erroneous log-ratios ***(42)***. Signal saturation for 16-bit scanning systems occurs at 65,535 units (▲ = median; **x** = ±1 standard deviation).

dynamic range of the array (*see* **Note 6**). With the benefit of hindsight, we currently advocate standardizing amplicon length within a relatively narrow range to minimize the considerable length effects on signal intensity.This results in more homogeneous signal yields and dynamic range across the array. Although, ideally, the average amplicon length would be selected so that it is similar to that of the shortest ORFs of interest, a complicating factor is the abundance of very short ORFs (i.e., 200 bp or less) in bacterial genomes. We have observed that amplicons from short ORFs can generate anomalous hybridization data ***(35)*** and, thus, a potential compromise could be to design amplicons in the 300 to 400 bp range for longer genes and as close to full-length as possible for shorter ORFs. Homogeneous amplicon lengths also simplify amplicon concentration normalization prior to spotting (*see* **Subheading 3.1.2., step 7**).

1.2.2. Microarray Design/Architecture

Given current densities achieved with spotted microarrays and the average gene content of a "typical" bacterial genome (2000–4000 genes), several key features are highly advisable from an array-design standpoint (**Fig. 3**):

1. A large number of positive and negative controls: these should be dispersed across the entire array to detect any potential spatial biases in foreground and background signal.
2. Multiple spot replicates: spot replicates not only increase the robustness of subsequent statistical analyses but are invaluable in case of individual spot failure. Many arrays are designed with two or three spot replicates that are printed side-by-side. If space allows, we advocate complete replicate arrays to be printed per slide so that some spot replicates are in a different physical location.
3. The inclusion of a "universal reference sample:" sheared whole-genomic DNA from the source strain used to make the array can act as an excellent positive control,

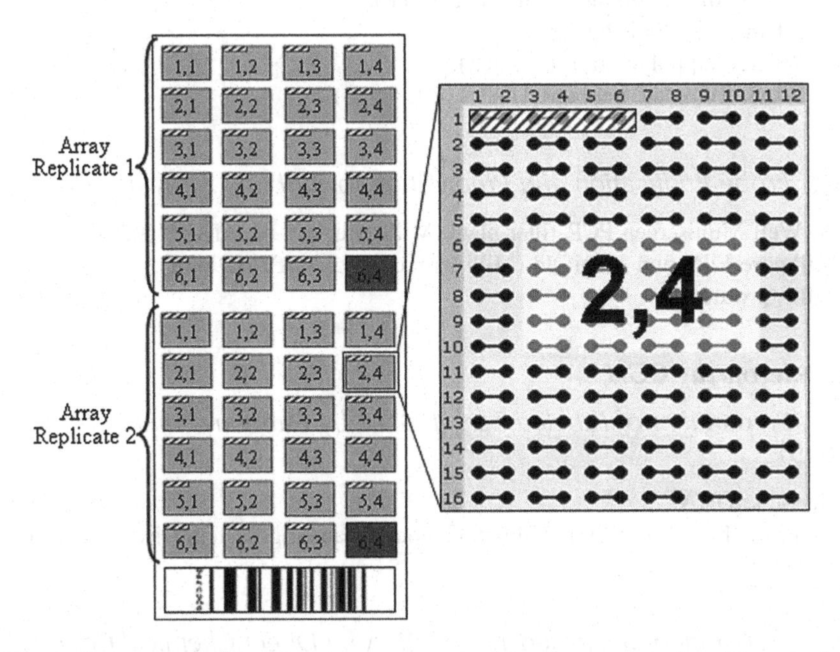

Fig. 3. An example of key features in a successful microarray design. These include: control spots of serially diluted control DNA on each subarray (cross-hatched boxes); two full array replicates; adjacent replicate spots (black circles). A complete subarray on each array replicate (dark gray boxes) is dedicated to control spots that can be used to quickly establish balanced scanning conditions without requiring scanning of the complete array surface.

and can be used to balance the signal from both strains being compared in the CGH experiment. This can greatly simplify image acquisition and analysis.

4. Normalized concentrations of amplicons for all the probes on the array (*see* **Note 7**).

2. Materials

2.1. Microarray Construction

2.1.1. Probe Amplification

1. Forward gene-specific primers (*see* **Note 8**).
2. Reverse gene-specific primers (*see* **Note 8**).
3. PCR master mix (per 96-well PCR plate).

   a. 6.9 mL Milli-Q water.
   b. 1.0 mL $MgCl_2$ (25 m*M*).
   c. 1.0 mL dNTP mixture (2 m*M* per dNTP).
   d. 1.0 mL 10X PCR buffer.
   e. 40 µL Taq polymerase (5 U/µL).
   f. 20 µL Template genomic DNA (100 ng/µL).

2.1.2. Probe Purification and Preparation for Robotic Arraying

1. 96-Well Multiscreen PCR filter plate (Millipore, cat. no. MANU3050).
2. Millipore Filtration Manifold (Millipore, cat. no. MAVM0960R).
3. Milli-Q water.

2.2. Microarray CGH

2.2.1. Preparation of Labeled Target DNA by Random Priming (see **Note 9**)

1. Genomic DNA.
2. BioPrime Plus Array CGH Indirect Genomic Labeling System (Invitrogen, cat. no. 18096-011).

2.2.2. Preparation of Labeled Target DNA by Direct Chemical Coupling

1. Genomic DNA (fragmented to average 1 Kb—*see* **Note 10**).
2. Mirus Label IT labeling kit (Mirus Bio Corp., Madison, WI; Cy-3, cat. no. MIR3600 and Cy-5, cat. no. MIR3700).
3. SigmaSpin Post-Reaction Clean-Up Columns (Sigma, cat. no. S-0185).
4. Qiaquick PCR Purification Kit (Qiagen, cat. no. 28106).

2.2.3. Microarray Hybridization (see **Note 11**)

1. Microarray (stored at room temperature in a dessicator until use).
2. DIG Easy hybridization solution (Roche, cat. no. 1-796-895).
3. Yeast tRNA (10 mg/mL).
4. Salmon sperm DNA (10 mg/mL).
5. High-humidity chamber.
6. Heat block (65 °C); hybridization oven (37 °C).

2.2.4. Washes

1. 20X SSC, 0.2-μm filtered (Sigma, cat. no. S-6639).
2. Washing buffers (prewarmed to 50 °C):

   a. Buffer 1: 1.0X SSC, 0.1% SDS.
   b. Buffer 2: 0.5X SSC.
   c. Buffer 3: 0.1X SSC.

3. Hybridization oven (50 °C).

3. Methods

3.1. Microarray Construction

3.1.1. Probe Amplification

1. Although the yield of each individual primer will vary, resuspend primers thoroughly in a common volume of Milli-Q water to an average concentration of 10 μ*M*.
2. Gene-specific amplicons are generated in 100-μL reactions in 96-well reaction plates: dispense 96 μL of master mix, 2 μL of forward primer, and 2 μL of reverse primer.
3. Perform reactions in a thermal cycler. Cycling conditions will vary but the cycling should include a "touchdown component" (*see* **Note 12**) to compensate for differences in the melting temperature among the primers in the reaction plate.
4. Visualize amplicons by gel electrophoresis and assess for the amplification of expected product sizes. Reactions yielding wrong sized or multiple amplicons may require redesign and synthesis of new primers (*see* **Note 13**). If low failure rates are achieved (<5 per 96-well plate), flag the failed amplicons and carry the plate through purification and arraying steps and simply ignore spots from failed probes during subsequent analysis. Generate additional amplicons resulting from resynthesized primers in additional plates (these can be purified subsequently). In case of higher failure rates, consolidate correct amplicons into fresh plates prior to probe purification to avoid large numbers of "failed spots" in the final microarray.

3.1.2. Probe Purification and Preparation for Robotic Arraying

Filtration using 96-well Multiscreen filter plates (Millipore) is an efficient method to remove unincorporated nucleotides and primers.

1. Wash off surfactants in the filter membranes by drawing 200 μL of Milli-Q water through the filter plates on the vacuum manifold filtration apparatus at the pressure recommended by the manufacturer.
2. Load reaction products onto the filter plates and filter through until the membranes are dry.
3. Add 50 μL of Milli-Q water to each well and filter through until the membranes are dry.
4. Repeat **step 3**.
5. Remove the plate from the apparatus and add 100 μL of Milli-Q water to each well. Seal the plate (precut sealing film or foil is available from many vendors), then place on a shaker and shake vigorously for 5–10 min to complete resuspension of the DNA.
6. Carefully remove the purified products using a multichannel pipet and transfer to new 96-well plates.
7. If access to a liquid handling robot and 96-well spectrophotometer is available, measure amplicon concentrations and transfer a common amount of each amplicon to fresh plates. Dry-down amplicons.
8. Prior to printing, resuspend amplicons in the appropriate printing buffer at a concentration of approx 0.2 μg/μL (concentration may vary depending on slide surface chemistry).

3.2. Microarray CGH

3.2.1. Preparation of Labeled Target DNA by Random Priming

1. Use 2 μg of genomic DNA and label target DNA using the BioPrime labeling kit according to the manufacturer's instructions.

3.2.2. Preparation of Labeled Target DNA by Direct Chemical Coupling

1. Bring up 5 μg of sheared genomic DNA to 40 μL with Milli-Q water, then add 5 μL of Buffer A, and 5 μL of dye (Cy-3 or Cy-5—*see* **Note 14**). Incubate in a light-tight container at 37 °C for 4–16 h.
2. To perform buffer exchange and remove the bulk of unincorporated dye, pass each labeling reaction through a SigmaSpin column according to the manufacturer's instructions.
3. Purify the labeled DNA from residual dye using Qiaquick column according to the manufacturer's instructions. (This additional purification step can reduce background signal. A second elution step can increase yields by an additional 10%.)

3.2.3. Microarray Hybridization

1. Calculate dye incorporation efficiency of probes by calculating the A_{260}, A_{550} (for Cy-3), and A_{650} (for Cy-5) (*see* **Note 15**). Tester and control probes should be matched so that samples with equivalent yields and incorporation efficiencies are cohybridized.
2. Pool matched tester/control targets and ethanol precipitate or dry-down in a vacuum centrifuge.
3. Prepare hybridization solution (per 100 µL : 90 µL of DIG Easy solution, 5 µL of yeast tRNA, and 5 µL of salmon sperm DNA). Warm the solution to 65 °C prior to its use to resuspend components that have precipitated out of solution.
4. Prepare high humidity chambers by quickly heating in a microwave to generate a small amount of water vapor. The humidity generated prevents the microarray from drying out (*see* **Note 16**). Place chambers at 37 °C until ready to use.
5. Carefully resuspend pooled labeled targets in appropriate volume of hybridization solution (*see* **Note 17**). If bubbles are introduced during mixing, remove them by quick centrifugation. Denature targets at 65 °C for a minimum of 5 min until ready to use.
6. To set up hybridization: place a clean cover slip onto a dust-free surface; pipet-labeled targets onto the cover slip without introducing bubbles; touch the microarray array-side down onto the liquid whereas allowing it to spread by capillary action; flip the microarray and place on a humidity chamber. After closing the lid to the chamber, proceed with the preparation of the next hybridization, working quickly to maintain humidity and temperature in the chamber. Carefully place chamber back into hybridization oven and hybridize the arrays for 16–24 h.

3.2.4. Washes (see **Note 18**)

1. Remove microarray from the humidity chamber (if using the chamber for more than one array, place it back into hybridization oven) and place it into a small container with enough prewarmed Buffer 1 to cover the entire array. Use gentle agitation to displace the cover slip.
2. Empty buffer and replace with fresh prewarmed Buffer 1. Incubate for 5 min at 50 °C with gentle agitation. Repeat.
3. Empty buffer and replace with prewarmed Buffer 2. Incubate for 5 min at 50 °C with gentle agitation. Repeat.
4. Empty buffer and replace with prewarmed Buffer 3. Incubate for 5 min at 50 °C with agitation. Allow the microarray to come to room temperature in Buffer 3. Arrays can be left in Buffer 3 until other microarrays have been processed.
5. Using forceps, carefully remove the microarray from the buffer. Shake excess buffer onto a lint-free paper towel until the array is nearly dry. It is important to avoid leaving large drops of buffer on the array surface to prevent the buffer from crystallizing onto the array surface (*see* **Note 19**).
6. Store arrays in a light tight slide holder until laser scanning. Repeat for additional arrays.

3.3. Array Detection and Analysis

As with all other microarray-based methods, the MCGH detection and analysis process involves: (1) the acquisition of images for both control and tester channels using (typically) a confocal laser-based scanning system; (2) analysis of the images to extract signal; and (3) analysis and interpretation of the signal data.

3.3.1. Image Acquisition

Image acquisition is an extremely important, and often neglected, piece of the microarray puzzle and improper scanning methods can seriously compromise subsequent data analysis.

An assumption made in MCGH studies is that the amounts of DNA from tester and control samples hybridized to the array are approximately equal for the majority of spots in the array. This stems from the expectation that most genes will be highly conserved; several lines of evidence including both genome sequence data and large-scale MCGH studies would indeed suggest that this assumption is solidly founded (e.g., **ref. *3***). Thus, in a typical MCGH microarray experiment one should expect the majority of spots to generate similar signal (*see* **Note 20**) and this prior knowledge provides an invaluable guiding principle during image acquisition. More specifically, laser-scanning parameters (laser power, photo-multiplier tube—PMT, detector gain) can be chosen which generate similar average signal levels in both control and tester channels (i.e., "balanced scans") for the highly conserved genes, which represent roughly 80% of the spots on the array. This minimizes any potential bias in the signal intensity data prior to subsequent analysis. Although data normalization techniques can remove some of this bias, Cui et al. ***(36)*** have shown that whereas data from unbalanced scans is very sensitive to the normalization method used (i.e., different normalization methods can yield very different results), data from balanced scans is quite robust when different normalization techniques are applied.

3.3.2. Signal Extraction and Log-Ratio Analysis

Analysis of microarray data requires the use of specialized software to quantify the signal from images acquired for each channel after laser scanning. Signal extraction is generally carried out in two steps: in the first step, spots are identified, the raw foreground and background signals are calculated for every spot on the array and spots with unreliable signal (e.g., bad spot morphology, very low signal—less than three to five times background) are filtered out from further analysis; in the second step, the data from all good quality spots is "normalized" by adjusting it for biases intrinsic to microarray technology. Although sophisticated mathematical approaches can be used to "clean-up dirty data," a far more sensible approach is to reduce any potential sources of bias when possible (i.e., through measures such as uniform probe lengths, matching of labeled targets according to concentration and dye incorporation efficiency, and balanced scanning practices) to minimize the extent of data adjustment necessary to remove any additional data bias.

Early methods of data normalization tended to be linear in nature, such that a similar adjustment was made to all spots. Newer approaches recognize that biases in the data can vary with respect to spot intensity (e.g., dye bias is intensity dependent) and with spatial position on the array (e.g., hybridization gradients, arrayer print-tip effects). Most array analysis programs now include the "print-tip loess" normalization method *(37)*, one of the most popular and robust intensity- and spatially dependent normalization methods. This method will generally produce excellent results from typical microarray data (*see* **Note 21**).

Two-channel microarray data is typically transformed by calculating the base 2 logarithm of the ratio of the channel 1 signal to channel 2 signal (i.e., $\log_2$ {channel 1/channel 2}), generally referred to as the "log-ratio" such that genes with differential signal have log-ratios that deviate significantly from 0. Excellent ways to visualize microarray data are the MA-plot (sometimes also referred to as the Ratio-Intensity or RI-plot), in which the log-ratio is plotted against the mean signal intensity of both channels *(37,38)* **(Fig. 4A,B)**, and the histogram of log-ratio frequencies **(Fig. 4C,D)**. Both methods enable the visualization of potential outliers in the data, which can be seen as spots that are removed from the main body of the log-ratio distribution.

3.3.3. MCGH Data Analysis and Interpretation

A number of programs, both freely and commercially available, have been developed to analyze microarray data (*see* **ref.** ***39***) for an excellent freely available microarray analysis suite developed by TIGR). However, in our experience, no single analysis platform has been developed specifically to deal with the idiosyncrasies of bacterial MCGH. We will instead focus our discussion on overall approach for analyzing and visualizing data.

3.3.3.1. Analysis of Replicates

Because of the technical variability inherent in microarray experiments and the biological variability in complex biological systems, statisticians have

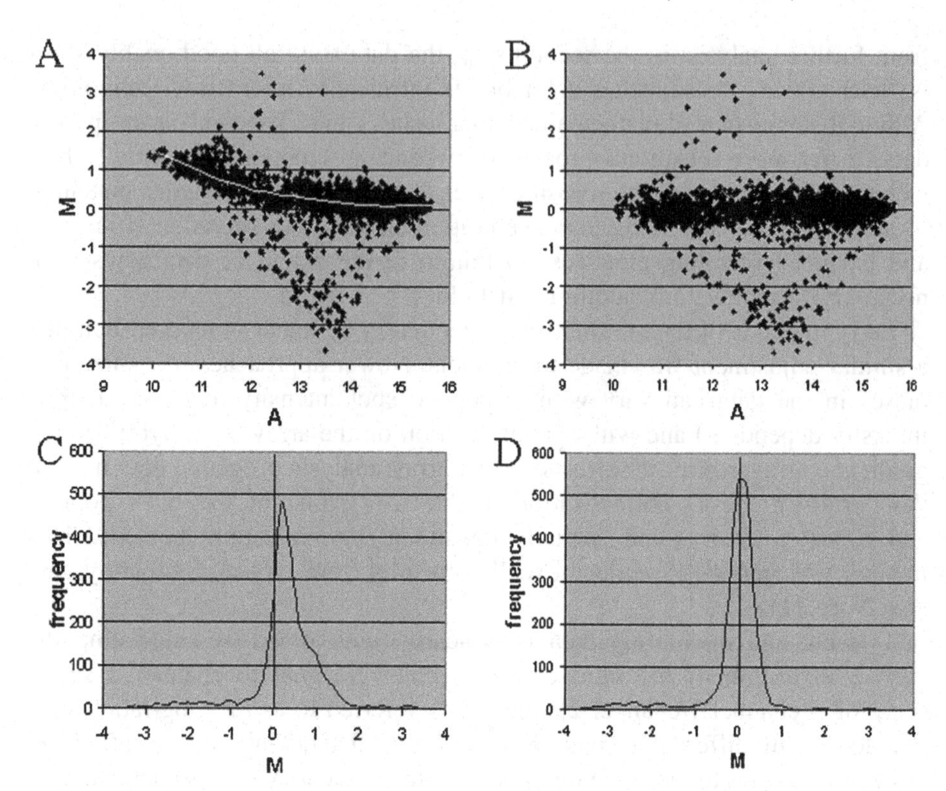

Fig. 4. Effects of normalization on log-ratio distribution. The log-ratio data from an MCGH experiment is shown using the "MA plot" (**A,B**) and "log-ratio histogram" (**C,D**) visualizations. Raw data is shown in **A** and **C**. Data after Loess normalization is shown in **B** and **D**. Note the increasing bias in the distribution in the low intensity range and the removal of the intensity-dependent bias in the normalized data.

strongly emphasized the importance of generating sufficient replicates for microarray-based transcript profiling experiments. To accurately measure gene expression, researchers typically perform several replicates (including technical, biological, and dye-swaps) and subject the expression ratios to rigorous statistical analysis to gain confidence based on the concordance of the results. A significant advantage of MCGH vs transcript profiling experiments is the fact that the biological sources of noise are essentially absent: genomic DNA is not subject to the short-term compositional fluctuations observed with mRNA. This fact alone overcomes the main analytical hurdle faced in microarray analysis because (with careful experimental design and technique) biological noise is generally a far greater source of variability than experimental noise. Replicates are important, and we encourage their inclusion in MCGH studies. However, the number required is much lower than with transcript profiling. Typically, we perform each experiment in duplicate or triplicate, with occasional dye-swaps to ensure that dye bias is not corrupting the data.

3.3.3.2. Outliers: Absent or Divergent Genes?

Outlier determination is possibly the single most significant component of microarray data analysis. MCGH outlier determination has generally depended on using a cut-off for assigning genes into conserved or divergent categories. In practice, this cut-off could be an empirically determined value based on analysis of a strain known to be missing (some) genes ***(6,40)***; or a cut-off derived from other published cut-off values (e.g., the twofold signal difference used in much of the early microarray literature). One problem with these approaches is that they apply the same cut-off to multiple arrays in a dataset despite the fact that technical variability and the number of divergent genes can greatly influence the shape of the log-ratio distribution. Although normalization discussed in **Subheading 5.2.** ensures that data within an individual microarray has been adjusted for bias, multiple arrays in a dataset might retain different degrees of bias with respect to one another such that not every array will display the same dynamic range and such that the same log-ratio value may not necessarily be "equivalent" in different arrays. To analyze multiple arrays with a fixed cut-off value, it is imperative that arrays be normalized against each other. A different approach altogether has been to develop methods that define dynamic cut-off values that are array-specific. Kim et al. have developed the program GACK, which determines thresholds specific to each array based on the deviation between the log-ratio distribution and an idealized normal distribution derived from the original log-ratio distribution, improving the sensitivity of divergent-gene detection ***(41)***. It is important, however, to acknowledge the limited reliability of sequence divergence estimates from hybridization data. We have found that although a trend between log-ratio and the level of sequence conservation is observable, the log-ratio distributions of genes with different levels of sequence conservation are partially overlapping, making accurate prediction of gene divergence levels extremely difficult ***(42)*** and, thus, advocate great caution with the interpretation of gene divergence from hybridization data.

Overall, we believe it is important to analyze data at two levels: a "high-sensitivity mode" where all potentially divergent genes are detected and a "high-accuracy mode" where those genes with the greatest likelihood of being highly divergent or absent are identified. At present, it is very important for groups to verify and carry out an empirical assessment of MCGH results (for example, by determining false-positive and false-negative rates based on MCGH characterization of other sequenced strains if available, or by PCR-based screening to confirm absent genes) to determine suitable parameters for analysis. The future is likely to bring advances in analytical algorithms and improvements in the accuracy of MCGH data analysis should be forthcoming.

3.3.3.3. Visualization and Higher-Order Analysis

From the point of view of CG, the main goals of visualization and high-order analysis are to compare and contrast the gene conservation profile of the strains being analyzed, and to reveal local and global trends in gene conservation.

Although MCGH cannot directly address gene order, or synteny, direct comparison of genome sequence data has shown that many closely related strains do indeed preserve synteny to a large degree. Thus, arranging MCGH data using the gene order from the reference strain used to build the array provides a useful starting framework for displaying gene conservation profiles. To compare the gene conservation profiles of two strains, simple ideogram plots can be used to display genomic regions with potential copy number differences, as seen in **Fig. 5**. To compare the gene conservation profiles of multiple strains, a number of programs that implement different data clustering algorithms have been developed for the analysis of gene expression data and work extremely well with MCGH data. Despite the increasing sophistication of the various clustering methods, even simple techniques such as hierarchical clustering *(43)* have been extremely effective at uncovering strain relationships based on whole-genome gene content, enabling genome-based typing or genomotyping *(23,44)*. In addition, because genome content is a valid form of phylogenetic signal *(45)*, MCGH data can be used in for genome-based phylogenetic analysis (or phylogenomics) (for examples, *see* **refs. *46*** and ***47***; *see also* **Fig. 6**).

To visualize global gene conservation trends for the species, the percentage of strains divergent at each gene can be plotted. Using this approach, numerous MCGH studies have shown the recurring theme of plasticity or hypervariable regions, spanning several adjacent genes that appear to be absent in many strains (for an example, *see* **ref. *35*** and **Fig. 7**). Interestingly, many of these

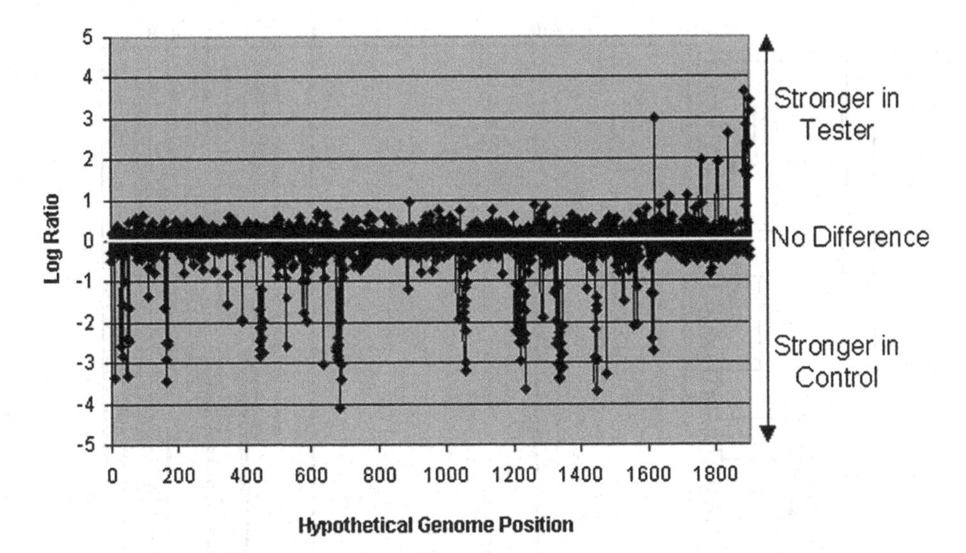

Fig. 5. Ideogram plot depicting the MCGH log-ratio data comparing a "Tester" *Campylobacter jejuni* strain against the "Control" strain. The data for each gene has been plotted using a hypothetical location based on the original genome location in the reference strain. Negative log-ratios, indicating stronger hybridization signal in the Control strain are observed throughout and represent possible divergent/absent genes. Positive log-ratios can be observed beyond genome position 1634, indicating stronger hybridization signal in the Tester strain. These genes are known to be absent in the Control strain but may be present in the Tester strain.

regions contain functionally related genes and tend to be over-represented with strain- or species-specific genes of unknown function.

3.4. Conclusions

Despite some shortcomings, which are technical and analytical in nature, MCGH has proven a very reliable technique in the study of bacterial CG and it currently remains the most accessible means of performing large-scale CG.

A major limitation is the limited "genome coverage" achieved by microarrays because only those genes included in the array can be tested for. Typically, microarrays have been based on the genome of a single strain, which undoubtedly places limitations on the resulting data because those genes represent only a fraction of the total gene pool for the species. However, an

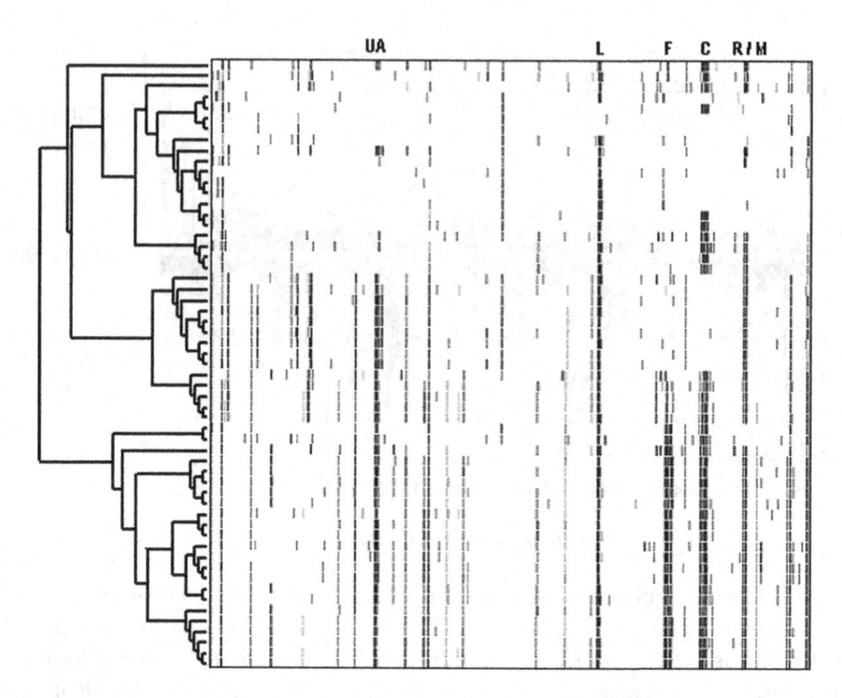

Fig. 6. Hierarchical clustering of MCGH profiles from 57 strains of *Campylobacter jejuni*. Clustering was performed using TMEV v3.0 ***(39)*** using the Euclidean distance metric. Significant differences in gene content (absent/divergent genes shown as black bars on white background) can be observed between clusters, although differences can also be observed within clusters (UA = *uxa/ald* locus; L = LOS locus; F = flagellar modification locus; C = capsular locus; R/M = restriction-modification locus).

increasing number of studies have begun to use "pan-genomic" arrays that comprise the nonredundant gene complement of two or more strains, though again the gene coverage achieved is limited by the availability of additional genome sequences.

It is interesting to note that among the most promising avenues for increasing the effective genome coverage of pan-genomic microarrays is the discovery of novel genes by MCGH-based screening of shotgun-arrays. Similarly, some of the most promising approaches for rapid whole-genome sequencing are to be found with ultra-high resolution MCGH using oligo-arrays. In the short term, much can be learned about the CG of a bacterial species using single genome-based microarrays. When coupled with novel MCGH-based methods for gene discovery and resequencing, MCGH can provide an integrated approach to bacterial CG.

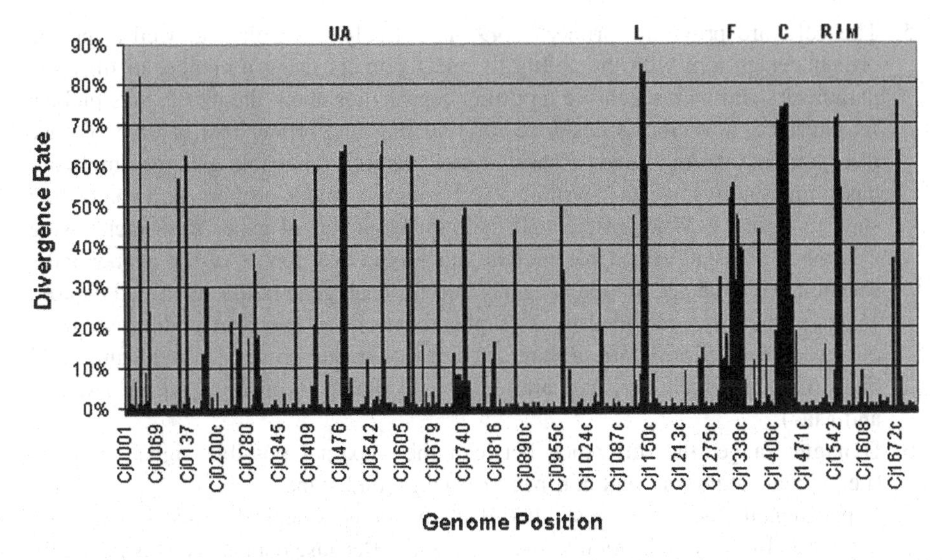

Fig. 7. Global gene conservation trends observed in 150 strains of *Campylobacter jejuni*. The divergence rate was calculated as the percentage of strains showing detectable divergence on MCGH for each gene on the array. Note the presence of several loci with significant levels of divergence across most *C. jejuni* strains surveyed. Among the most highly divergent loci are those involved in cell-surface modification (UA = *uxa/ald* locus; L = LOS locus; F = flagellar modification locus; C = capsular locus; R/M = restriction-modification locus).

4. Notes

1. In this chapter, we will conform to the nomenclature proposed by Duggan *(48)*, in which the DNA on the solid substrate is referred to as the "probe" and the DNA in solution is referred to as the "target."
2. *In situ* synthesis methods can yield microarrays with ~1000 spots/mm^2 (up to 8200, in the case of Affymetrix), compared to spotted microarrays, which yield <10–20 spots/mm^2.
3. Spotting is generally carried out through a commercial operation or through the increasing number of microarray production core facilities that have been established at a number of academic or governmental research institutions.
4. The specifics of the platform used, such as the slide surface chemistry, may require modifications to buffers and others. Some slide surface chemistries, for example Corning Ultra GAPS, work equally well with amplicons and long-oligos and require virtually no changes in hybridization or washing conditions. Protocols that have been optimized for your platform of choice are generally available from the production facility or slide vendor.

5. The software program "primer3" *(49)* is a freely available and well-respected primer design tool with the ability to select primers using a number of different parameters and with extensive reporting capabilities about the primer sets picked. Its interface, however, is designed for selecting one primer pair at time. To use it to generate the thousands of primer sets required for whole genome amplicon-based microarrays, we have written a Perl program (make_primers) as a controlling shell to primer3. Make_primers will submit large sets of genes to primer3 using a number of user-defined parameters and produces a report of the primer pairs selected and expected amplicons generated for each gene in the submitted dataset. Make_primers is controlled by a graphical interface, and is available free from our group (http://informatics/bio.nrc.ca/software/make_primers). It will run on all platforms for which Perl and primer3 are available (typically Linux/Unix, PC, and Mac).
6. Dynamic range: the difference between the maximal possible signal intensity (i.e., signal saturation) and nonspecific background signal.
7. Experimental design for most MCGH applications is such that two samples are compared by cohybridization to the microarray. Because data analysis is generally carried out by calculating the signal ratio of the two samples, "normalization" of amplicon concentrations yields prior to spotting is seldom considered because it has long been assumed that any concentration effects would be negated by the ratio calculation. Nevertheless, signal intensity does affect the dynamic range of a spot, affecting signal ratio values. Reamplification tags can be incorporated into gene-specific primers and used to perform a second round of PCR to reamplify the gene specific amplicons obtained in the first round of PCR *(50)*. Unlike the first round of amplification, which can generate very different yields for each amplicon, the reamplification reaction generally results in similar amplicon yields for all genes because common primers are used.
8. For high-throughput synthesis, oligos are synthesized in 96-well format with Forward and Reverse primers in matching plates. They are typically shipped in lyophilized form and should be reconstituted in Milli-Q water to the working concentration.
9. Most laboratories use random primer-based labeling *(51)* to generate labeled targets. Although we have also used it in the past, because of its simplicity and robustness, the chemical coupling protocol described in **Subheading 3.2.2.** is currently our method of choice. For random priming, we prefer to use the BioPrime indirect labeling system (Invitrogen). The indirect method is preferable over direct methods that directly incorporate dye-labeled nucleotides during synthesis to avoid biases as a result of the differential incorporation rate of Cy-3 and Cy-5 nucleotides by DNA polymerase.
10. Fragmentation can be achieved by digestion with restriction enzymes or sonication but we have found that the nebulization method of Bodenteich et al. *(52)* using a modified asthma nebulizer, and which is routinely used in shotgun library

construction, works efficiently and yields an excellent distribution of fragment sizes. Briefly, the nebulizer is modified by discarding the mouthpiece and sealing the large opening on the top cover using the cap from a 13-mL culture tube and sealing with Parafilm. Twenty-five micrograms of genomic DNA are suspended in 2 mL of 35% glycerol: 10 m*M* Tris:1 m*M* EDTA (pH 8.0) and loaded on the nebulizer chamber. The nebulizer is connected to a pressurized gas line at 15 PSI (i.e., compressed air, CO_2) using the tubing provided in the nebulizer kit. While on ice, the DNA mixture is nebulized for 1 min. The nebulizer is placed in a cushioned rotor bucket (use paper towel, Styrofoam) and the DNA mixture is collected at the bottom of the nebulizer by low-speed centrifugation (~1000*g* for 5 min). The DNA is recovered by ethanol precipitation.

11. Although many labs eventually opt for automated microarray hybridization stations to standardize hybridization conditions, the entry-level setup described can produce satisfactory results. A high-humidity hybridization chamber can be made using light-tight plastic microscope slide boxes containing a small amount of water and into which clean plain microscope slides have been placed to form a raised platform for the arrays. Each chamber can hold two or three arrays. Special microarray hybridization chambers are also available from several vendors including Corning and Bio-Rad.
12. A typical touchdown PCR profile will have 10 cycles over which the annealing temperature is decreased by 0.5 °C in each cycle, starting at the maximum T_m of any primer in the plate, followed by 20–25 cycles at an annealing temperature 3 °C below that of the lowest T_m of any primer in the plate.
13. Low yield reactions may require pooling of multiple reactions. In our hands, with automated primer design, relatively few reactions failures are owing to poorly designed primers. The majority of these failures are because of primer synthesis failure or errors during the preparation of the PCR reactions. In the vast majority of cases, repeating the PCR reaction of a failed sample, or resynthesizing the primers has resulted in the expected amplicon being produced.
14. Dyes are reconstituted in resuspension buffer provided in the kit. The dye should be aliquoted immediately into single use amounts into O-ring cryotubes. Dyes should be stored at −20 °C in a light-tight container. It is preferable to use freshly resuspended dyes whenever possible as coupling efficiency decreases with time.
15. To calculate dye incorporation efficiency, calculate the nucleotide to dye ratio:
    a. pmol nucleotides = [A260* volume (μL)* 50 ng/μL *1000 pg/ng] /324.5 pg/pmol.
    b. pmol Cy-3 = A_{550} * volume (μL)/0.15.
    c. pmol Cy-5 = A_{650} * volume (μL)/0.25.
    d. nucleotides/dye ratio = pmol nucleotides/pmol Cy-dye.
16. It is of crucial importance that the microarray be prevented from drying during either hybridization or washing steps to avoid unspecific background from probe precipitation or wash buffer crystallization.

17. The volume required depends on the size of the array, and the cover slip type and size and should be accurately determined by setting up mock hybridizations on standard microscope slides; both insufficient and excess volume can adversely affect the hybridization.
18. The stringency of washing conditions has to be empirically determined based on the length of probes, the GC content of the organism, and so on.
19. To eliminate any residual buffer on the array surfaces, microarrays can be spun dry. Dedicated bench-top mini-centrifuges for microarrays are available from several vendors.
20. Despite the fact that Cy-3 and Cy-5 differ in their chemical characteristics (i.e., absorption spectra, extinction coefficient, quantum yield, and so on) influencing signal yields, equivalent signal yields from both channels can be expected for the majority of spots if control and tester probes have been matched according to concentration and dye-incorporation efficiency.
21. Although more sophisticated methods exist and continue to be developed, in our view, if they become absolutely necessary it may be more judicious to simply repeat the experiment because, unlike some gene expression experiments on materials that may be in extremely limited quantity, genomic DNA for additional MCGH experiments should be easily obtainable.

References

1. Alm, R. A., Ling, L. S., Moir, D. T., et al. (1999) Genomic-sequence comparison of two unrelated isolates of the human gastric pathogen *Helicobacter pylori*. *Nature* **397,** 176–180.
2. Perna, N. T., Plunkett, G., Burland, V., et al. (2001) Genome sequence of enterohaemorrhagic *Escherichia coli* O157:H7. *Nature* **409,** 529–533.
3. Tettelin, H., Masignani, V., Cieslewicz, M. J., et al. (2005) Genome analysis of multiple pathogenic isolates of *Streptococcus agalactiae*: implications for the microbial "pan-genome". *Proc. Natl. Acad. Sci. USA* **102,** 13,950–13,955.
4. Maione, D., Margarit, I., Rinaudo, C. D., et al. (2005) Identification of a universal Group B *Streptococcus* vaccine by multiple genome screen. *Science* **309,** 148–150.
5. Behr, M. A., Wilson, M. A., Gill, W. P., et al. (1999) Comparative genomics of BCG vaccines by whole-genome DNA microarray. *Science* **284,** 1520–1523.
6. Salama, N., Guillemin, K., McDaniel, T. K., Sherlock, G., Tompkins, L., and Falkow, S. (2000) A whole-genome microarray reveals genetic diversity among *Helicobacter pylori* strains. *Proc. Natl. Acad. Sci. USA* **97,** 14,668–14,673.
7. Schena, M., Shalon, D., Davis, R. W., and Brown, P. O. (1995) Quantitative monitoring of gene expression patterns with a complementary DNA microarray. *Science* **270,** 467–470.
8. Schena, M., Shalon, D., Heller, R., Chai, A., Brown, P. O., and Davis, R. W. (1996) Parallel human genome analysis: microarray-based expression monitoring of 1000 genes. *Proc. Natl. Acad. Sci. USA* **93,** 10,614–10,619.

9. Kato-Maeda, M., Rhee, J. T., Gingeras, T. R., et al. (2001) Comparing genomes within the species *Mycobacterium tuberculosis*. *Genome Res.* **11,** 547–554.
10. Tsolaki, A. G., Hirsh, A. E., DeRiemer, K., et al. (2004) Functional and evolutionary genomics of Mycobacterium tuberculosis: insights from genomic deletions in 100 strains. *Proc. Natl. Acad. Sci. USA* **101,** 4865–4870.
11. Hakenbeck, R., Balmelle, N., Weber, B., Gardes, C., Keck, W., and de Saizieu, A. (2001) Mosaic genes and mosaic chromosomes: intra- and interspecies genomic variation of*Streptococcus pneumoniae*. *Infect. Immun.* **69,** 2477–2486.
12. Wick, L. M., Qi, W., Lacher, D. W., and Whittam, T. S. (2005) Evolution of genomic content in the stepwise emergence of Escherichia coli O157:H7. *J. Bacteriol.* **187,** 1783–1791.
13. Yao, Y., Sturdevant, D. E., Villaruz, A., Xu, L., Gao, Q., and Otto, M. (2005) Factors characterizing *Staphylococcus epidermidis* invasiveness determined by comparative genomics. *Infect. Immun.* **73,** 1856–1860.
14. Charbonnier, Y., Gettler, B., Francois, P., et al. (2005) A generic approach for the design of whole-genome oligoarrays, validated for genomotyping, deletion mapping and gene expression analysis on Staphylococcus aureus. *BMC Genomics* **6,** 95.
15. Wu, L., Thompson, D. K., Li, G., Hurt, R. A., Tiedje, J. M., and Zhou, J. (2001) Development and evaluation of functional gene arrays for detection of selected genes in the environment. *Appl. Environ. Microbiol.* **67,** 5780–5790.
16. Wilson, W. J., Strout, C. L., DeSantis, T. Z., Stilwell, J. L., Carrano, A. V., and Andersen, G. L. (2002) Sequence-specific identification of 18 pathogenic microorganisms using microarray technology. *Mol. Cell. Probes* **16,** 119–127.
17. Tiquia, S. M., Wu, L., Chong, S. C., et al. (2004) Evaluation of 50-mer oligonucleotide arrays for detecting microbial populations in environmental samples. *Biotechniques* **36,** 664–670.
18. Daran-Lapujade, P., Daran, J. M., Kotter, P., Petit, T., Piper, M. D., and Pronk, J. T. (2003) Comparative genotyping of the *Saccharomyces cerevisiae* laboratory strains S288C and CEN.PK113-7D using oligonucleotide microarrays. *FEMS Yeast Res.* **4,** 259–269.
19. Rajashekara, G., Glasner, J. D., Glover, D. A., and Splitter, G. A. (2004) Comparative whole-genome hybridization reveals genomic islands in *Brucella* species. *J. Bacteriol.* **186,** 5040–5051.
20. Wong, C. W., Albert, T. J., Vega, V. B., et al. (2004) Tracking the evolution of the SARS coronavirus using high-throughput, high-density resequencing arrays. *Genome Res.* **14,** 398–405.
21. Davignon, L., Walter, E. A., Mueller, K. M., Barrozo, C. P., Stenger, D. A., and Lin, B. (2005) Use of resequencing oligonucleotide microarrays for identification of Streptococcus pyogenes and associated antibiotic resistance determinants. *J. Clin. Microbiol.* **43,** 5690–5695.

22. Hayward, R. E., Derisi, J. L., Alfadhli, S., Kaslow, D. C., Brown, P. O., and Rathod, P. K. (2000) Shotgun DNA microarrays and stage-specific gene expression in Plasmodium falciparum malaria. *Mol. Microbiol.* **35,** 6–14.
23. Dorrell, N., Mangan, J. A., Laing, K. G., et al. (2001) Whole genome comparison of *Campylobacter jejuni* human isolates using a low-cost microarray reveals extensive genetic diversity. *Genome Res.* **11,** 1706–1715.
24. Call, D. R., Borucki, M. K., and Besser, T. E. (2003) Mixed-genome microarrays reveal multiple serotype and lineage-specific differences among strains of *Listeria monocytogenes*. *J. Clin. Microbiol.* **41,** 632–639.
25. Borucki, M. K., Kim, S. H., Call, D. R., Smole, S. C., and Pagotto, F. (2004) Selective discrimination of *Listeria monocytogenes* epidemic strains by a mixed-genome DNA microarray compared to discrimination by pulsed-field gel electrophoresis, ribotyping, and multilocus sequence typing. *J. Clin. Microbiol.* **42,** 5270–5276.
26. Samrakandi, M. M., Zhang, C., Zhang, M., et al. (2004) Genome diversity among regional populations of *Francisella tularensis* subspecies *tularensis* and *Francisella tularensis* subspecies *holarctica* isolated from the US. *FEMS Microbiol. Lett.* **237,** 9–17.
27. Poly, F., Threadgill, D., and Stintzi, A. (2004) Identification of *Campylobacter jejuni* ATCC 43431-specific genes by whole microbial genome comparisons. *J. Bacteriol.* **186,** 4781–4795.
28. Poly, F., Threadgill, D., and Stintzi, A. (2005) Genomic diversity in *Campylobacter jejuni*: identification of C. jejuni 81–176-specific genes. *J. Clin. Microbiol.* **43,** 2330–2338.
29. Porwollik, S., Santiviago, C. A., Cheng, P., Florea, L., and McClelland, M. (2005) Differences in gene content between *Salmonella enterica* serovar *Enteritidis* isolates and comparison to closely related serovars *Gallinarum* and *Dublin*. *J. Bacteriol.* **187,** 6545–6555.
30. Witney, A. A., Marsden, G. L., Holden, M. T., et al. (2005) Design, validation, and application of a seven-strain *Staphylococcus aureus* PCR product microarray for comparative genomics. *Appl. Environ. Microbiol.* **71,** 7504–7514.
31. Zhang, L., Srinivasan, U., Marrs, C. F., Ghosh, D., Gilsdorf, J. R., and Foxman, B. (2004) Library on a slide for bacterial comparative genomics. *BMC Microbiol.* **4,** 12.
32. Raddatz, G., Dehio, M., Meyer, T. F., and Dehio, C. (2001) PrimeArray: genome-scale primer design for DNA-microarray construction. *Bioinformatics* **17,** 98–99.
33. Xu, D., Li, G., Wu, L., Zhou, J., and Xu, Y. (2002) PRIMEGENS: robust and efficient design of gene-specific probes for microarray analysis. *Bioinformatics* **18,** 1432–1437.
34. Haas, S. A., Hild, M., Wright, A. P., Hain, T., Talibi, D., and Vingron, M. (2003) Genome-scale design of PCR primers and long oligomers for DNA microarrays. *Nucleic Acids Res.* **31,** 5576–5581.

35. Taboada, E. N., Acedillo, R. R., Carrillo, C. D., et al. (2004) Large-scale comparative genomics meta-analysis of *Campylobacter jejuni* isolates reveals low level of genome plasticity. *J. Clin. Microbiol.* **42,** 4566–4576.
36. Cui, X., Kerr, M. K., and Churchill, G. A. (2006) Transformations for cDNA Microarray Data. *Statistical Applications in Genetics and Molecular Biology* **2,** http://www.bepress.com/sagmb/vol2/iss1/art4.
37. Yang, Y. H., Dudoit, S., Luu, P., et al. (2002) Normalization for cDNA microarray data: a robust composite method addressing single and multiple slide systematic variation. *Nucleic Acids Res.* **30,** e15.
38. Smyth, G. K., Yang, Y. H., and Speed, T. (2003) Statistical issues in cDNA microarray data analysis. *Methods Mol. Biol.* **224,** 111–136.
39. Saeed, A. I., Sharov, V., White, J., et al. (2003) TM4: a free, open-source system for microarray data management and analysis. *Biotechniques* **34,** 374–378.
40. Dziejman, M., Balon, E., Boyd, D., Fraser, C. M., Heidelberg, J. F., and Mekalanos, J. J. (2002) Comparative genomic analysis of *Vibrio cholerae*: genes that correlate with cholera endemic and pandemic disease. *Proc. Natl. Acad. Sci. USA* **99,** 1556–1561.
41. Kim, C. C., Joyce, E. A., Chan, K., and Falkow, S. (2002) Improved analytical methods for microarray-based genome-composition analysis. *Genome Biol.* **3,** RESEARCH0065.
42. Taboada, E. N., Acedillo, R. R., Luebbert, C. C., Findlay, W. A., and Nash, J. H. (2005) A new approach for the analysis of bacterial microarray-based Comparative Genomic Hybridization: insights from an empirical study. *BMC Genomics* **6,** 78.
43. Eisen, M. B., Spellman, P. T., Brown, P. O., and Botstein, D. (1998) Cluster analysis and display of genome-wide expression patterns. *Proc. Natl. Acad. Sci. USA* **95,** 14,863–14,868.
44. Leonard, E. E., Takata, T., Blaser, M. J., Falkow, S., Tompkins, L. S., and Gaynor, E. C. (2003) Use of an open-reading frame-specific *Campylobacter jejuni* DNA microarray as a new genotyping tool for studying epidemiologically related isolates. *J. Infect. Dis.* **187,** 691–694.
45. Charlebois, R. L., Beiko, R. G., and Ragan, M. A. (2003) Microbial phylogenomics: branching out. *Nature* **421,** 217.
46. Champion, O. L., Gaunt, M. W., Gundogdu, O., et al. (2005) Comparative phylogenomics of the food-borne pathogen *Campylobacter jejuni* reveals genetic markers predictive of infection source. *Proc. Natl. Acad. Sci. USA* **102,** 16,043–16,048.
47. Read, T. D., Peterson, S. N., Tourasse, N., et al. (2003) The genome sequence of *Bacillus anthracis* Ames and comparison to closely related bacteria. *Nature* **423,** 81–86.
48. Duggan, D. J., Bittner, M., Chen, Y., Meltzer, P., and Trent, J. M. (1999) Expression profiling using cDNA microarrays. *Nat. Genet.* **21,** 10–14.
49. Rozen, S. and Skaletsky, H. (2000) Primer3 on the WWW for general users and for biologist programmers. *Methods Mol. Biol.* **132,** 365–386.

50. Lindroos, H. L., Mira, A., Repsilber, D., et al. (2005) Characterization of the genome composition of *Bartonella koehlerae* by microarray comparative genomic hybridization profiling. *J. Bacteriol.* **187,** 6155–6165.
51. Feinberg, A. P. and Vogelstein, B. (1983) A technique for radiolabeling DNA restriction endonuclease fragments to high specific activity. *Anal. Biochem.* **132,** 6–13.
52. Bodenteich, A. S., Chissoe, Y., Wang, F., and Roe, B. A. (1994) Shotgun cloning as the strategy of choice to generate templates for high throughput dideoxynucleotide sequencing, in *Automated DNA Sequencing and Analysis Techniques*, (Adams, M. D., Fields, C., and Venter, C., eds.), London, UK, Academic Press, pp. 42–50.
53. Fitzgerald, J. R., Sturdevant, D. E., Mackie, S. M., Gill, S. R., and Musser, J. M. (2001) Evolutionary genomics of *Staphylococcus aureus*: insights into the origin of methicillin-resistant strains and the toxic shock syndrome epidemic. *Proc. Natl. Acad. Sci. USA* **98,** 8821–8826.
54. Cassat, J. E., Dunman, P. M., McAleese, F., Murphy, E., Projan, S. J., and Smeltzer, M. S. (2005) Comparative genomics of *Staphylococcus aureus* musculoskeletal isolates. *J. Bacteriol.* **187,** 576–592.
55. Pearson, B. M., Pin, C., Wright, J., I'Anson, K., Humphrey, T., and Wells, J. M. (2003) Comparative genome analysis of *Campylobacter jejuni* using whole genome DNA microarrays. *FEBS Lett.* **554,** 224–230.
56. Leonard, E. E., Tompkins, L. S., Falkow, S., and Nachamkin, I. (2004) Comparison of *Campylobacter jejuni* isolates implicated in Guillain-Barré syndrome and strains that cause enteritis by a DNA microarray. *Infect Immun.* **72,** 1199–1203.
57. Smoot, J. C., Barbian, K. D., Van Gompel, J. J., et al. (2002) Genome sequence and comparative microarray analysis of serotype M18 group A *Streptococcus* strains associated with acute rheumatic fever outbreaks. *Proc. Natl. Acad. Sci. USA* **99,** 4668–4673.
58. Porwollik, S., Wong, R. M., and McClelland, M. (2002) Evolutionary genomics of *Salmonella*: gene acquisitions revealed by microarray analysis. *Proc. Natl. Acad. Sci. USA* **99,** 8956–8961.
59. Chan, K., Baker, S., Kim, C. C., Detweiler, C. S., Dougan, G., and Falkow, S. (2003) Genomic comparison of *Salmonella enterica* serovars and *Salmonella bongori* by use of an S. enterica serovar *typhimurium* DNA microarray. *J. Bacteriol.* **185,** 553–563.
60. Porwollik, S., Boyd, E. F., Choy, C., et al. (2004) Characterization of*Salmonella enterica* subspecies I genovars by use of microarrays. *J. Bacteriol.* **186,** 5883–5898.
61. Anjum, M. F., Marooney, C., Fookes, M., et al. (2005) Identification of core and variable components of the *Salmonella enterica* subspecies I genome by microarray. *Infect. Immun.* **73,** 7894–7905.
62. Doumith, M., Cazalet, C., Simoes, N., et al. (2004) New aspects regarding evolution and virulence of Listeria monocytogenes revealed by comparative genomics and DNA arrays. *Infect. Immun.* **72,** 1072–1083.

63. Dobrindt, U., Agerer, F., Michaelis, K., et al. (2003) Analysis of genome plasticity in pathogenic and commensal *Escherichia coli* isolates by use of DNA arrays. *J. Bacteriol.* **185,** 1831–1840.
64. Anjum, M. F., Lucchini, S., Thompson, A., Hinton, J. C., and Woodward, M. J. (2003) Comparative genomic indexing reveals the phylogenomics of *Escherichia coli* pathogens. *Infect. Immun.* **71,** 4674–4683.
65. Fukiya, S., Mizoguchi, H., Tobe, T., and Mori, H. (2004) Extensive genomic diversity in pathogenic *Escherichia coli* and *Shigella* Strains revealed by comparative genomic hybridization microarray. *J. Bacteriol.* **186,** 3911–3921.
66. Broekhuijsen, M., Larsson, P., Johansson, A., et al. (2003) Genome-wide DNA microarray analysis of *Francisella tularensis* strains demonstrates extensive genetic conservation within the species but identifies regions that are unique to the highly virulent *F. tularensis* subsp. *tularensis*. *J. Clin. Microbiol.* **41,** 2924–2931.
67. Wolfgang, M. C., Kulasekara, B. R., Liang, X., et al. (2003) Conservation of genome content and virulence determinants among clinical and environmental isolates of *Pseudomonas aeruginosa*. *Proc. Natl. Acad. Sci. USA* **100,** 8484–8489.
68. Hinchliffe, S. J., Isherwood, K. E., Stabler, R. A., et al. (2003) Application of DNA microarrays to study the evolutionary genomics of *Yersinia pestis* and *Yersinia pseudotuberculosis*. *Genome Res.* **13,** 2018–2029.
69. Zhou, D., Han, Y., Dai, E., et al. (2004) Defining the genome content of live plague vaccines by use of whole-genome DNA microarray. *Vaccine* **22,** 3367–3374.
70. Brunelle, B. W., Nicholson, T. L., and Stephens, R. S. (2004) Microarray-based genomic surveying of gene polymorphisms in *Chlamydia trachomatis*. *Genome Biol.* **5,** R42.
71. Cummings, C. A., Brinig, M. M., Lepp, P. W., van de Pas, S., and Relman, D. A. (2004) *Bordetella* species are distinguished by patterns of substantial gene loss and host adaptation. *J. Bacteriol.* **186,** 1484–1492.
72. Stabler, R. A., Marsden, G. L., Witney, A. A., et al. (2005) Identification of pathogen-specific genes through microarray analysis of pathogenic and commensal *Neisseria* species. *Microbiology* **151,** 2907–2922.
73. Paustian, M. L., Kapur, V., and Bannantine, J. P. (2005) Comparative genomic hybridizations reveal genetic regions within the *Mycobacterium avium* complex that are divergent from *Mycobacterium avium* subsp. *paratuberculosis* isolates. *J. Bacteriol.* **187,** 2406–2415.
74. Mongodin, E. F., Hance, I. R., Deboy, R. T., et al. (2005) Gene transfer and genome plasticity in Thermotoga maritima, a model hyperthermophilic species. *J. Bacteriol.* **187,** 4935–4944.

16

DNA Copy Number Data Analysis Using the CGHAnalyzer Software Suite

Joel Greshock

Summary

Recently developed microarray-based copy number measurement assays have drastically improved the accuracy and resolution to which DNA copy number alterations can be detected. As with any microarray assay, those designed to measure genome copy number produce large data sets for each sample. Furthermore, many successful studies of genome copy number require the concurrent comparison of many samples. Identifying software packages that provide the proper analytic tools is essential to effectively mine these valuable data.

CGHAnalyzer is a freely available, open source software suite designed specifically for the purpose of analyzing multiple-experiment microarray-based genome copy number data. This package can load data from a variety of formats, query large data sets for minimal common regions of gain and loss, integrate other genomic features into analyses (e.g., known/predicted genes), conduct higher order analyses such as hierarchical/k-means clustering and perform statistical tests to identify regions that are differentially altered between classes of samples. Each of these utilities represents common hurdles in approaching microarray copy number data sets. This passage provides an overview of practical, step by step analysis of array-based comparative genomic hybridization data using CGHAnalyzer and demonstrates specific techniques to improve the efficiency and accuracy of analysis.

Key Words: Microarray comparative genomic hybridization; CGH; CGHAnalyzer; software; analysis; clustering.

From: *Methods in Molecular Biology, vol. 396: Comparative Genomics, Volume 2*
Edited by: N. H. Bergman © Humana Press Inc., Totowa, NJ

1. Introduction

Genome copy number alterations have been implicated as important etiological factors in a wide range of human diseases, such congenital disorders *(1)* and cancer diagnoses *(2)*. The movement of copy number alteration detection assays to the microarray platform has expanded the resolution to which genome aberrations can be detected. The enormous potential of microarray-based genome copy number profiling is evidenced by its effective identification potential disease genes (e.g., **ref.** *3*). This expanded resolution of has necessitated the development of analytical methods and accompanying software for visualization. CGHAnalyzer is a freely available, open source software suite that is designed specifically for the visualization and analysis of multiexperiment microarray-based copy number data. This flexible tool allows users to load data from a variety of customized formats, visualize data in several standard views, and conduct high level analyses of large data sets.

A diversity of private and commercial microarray DNA copy number measurement assays are available. These include bacterial artificial chromosome (BAC) array-based comparative genomic hybridization (aCGH) *(4)* dual channel oligonucleotide arrays *(5)*, and single channel "SNP Chips" (e.g., Affymetrix Inc., Sunnyvale, CA) *(6)*. Although each is mechanically different, the resultant data produced by all copy number measurement assays are the same. Each consists of a "copy number ratio" for every mapped sequence that compose each array. Therefore, although the underlying assays are very different, approaches to their analysis are the same. For example, the copy number estimations of a specific gene in a tumor-derived cancer cell line run on any of the previously mentioned copy number assays would be made by analyzing the copy number ratio of the probes that map to the vicinity of that gene. CGHAnalyzer is capable of loading and analyzing data from all of these platforms.

Although many variations of the aCGH data analysis pipeline exist, most employ all or part of a generic, commonly used protocol. First, data are subject to a normalization step. In the case of aCGH data, normalization refers to an adjustment of all copy number ratios to diploid for each individual experiment (1 in the linear scale, 0 in log scale). This procedure is based upon the assumption that the mean/median copy number for all probes across the entire genome is diploid. Although this is not always a valid assumption *(7)*, it is a conventional procedure used in most published aCGH studies. Normalization is followed by a systematic estimation of the genomic locations of copy number breakpoints in a particular tumor. Several methods exist to accomplish this important step in aCGH analyses. Most simply, this can be done by applying a ratio threshold for every probe. For example, clones with test/reference ratios

>1.5 represent copy number gains, and those <0.5 are estimated as copy number losses. Alternatively, more sophisticated algorithms designate a probe as being gained or lost only when supported by evidence from those probes flanking it. Once breakpoints are properly estimated, aCGH data sets can be subjected to various types of analyses. Most commonly, a data set is queried for the most altered regions to distinguish potential gene targets. This procedure usually entails identifying minimal common regions of gain and loss across all samples. Clustering procedures are frequently used to determine which samples are most similar based upon their copy number profiles. Clustering can be used for sample class discovery and determining regions of coamplification. Class discrimination methods are also very useful in aCGH analyses. Methods such as adjusted t-test procedures *(8)* are used to identify features which may be useful in discriminating sample groups (e.g., **ref. *9***). Data query procedures and several standard clustering and group discrimination methods are implemented as part of the CGHAnalyzer software package.

2. Materials

1. The Java Runtime Environment (JRE), v1.4.1 or more recent, is required to run the CGHAnalyzer software. It is recommended that this be installed prior to downloading the CGHAnalyzer software. The JRE can be freely downloaded and installed from http://java.sun.com/.
2. The CGHAnalyzer software package can be downloaded from http://acgh.afcri.upenn.edu. Downloads require a simple, one time registration, so that users will be notified of software bugs and updates by email. The system requirements are minimal. CGHAnalyzer can run on any operating system. Approximately 80 Mb of free disk space is required and a minimum of 512 Mb RAM is recommended.
3. Data must be formatted prior to loading. Commercial tools such as Microsoft Excel are excellent for this purpose. More advanced users can easily format data using data-enabled programming languages such as Perl (freely available for Mac OSX/Unix at http://www.cpan.org; Windows at http://www.activestate.com).
4. Access to array-based comparative genomic hybridization data is required for using the CGHAnalyzer software.

3. Methods

3.1. Installing and Launching the CGHAnalyzer Software

CGHAnalyzer requires very few steps for its installation. First, the JRE must be installed. This is normally done as part of an automated procedure requiring little input from the user (*see* **Note 1**). CGHAnalyzer is downloaded as a zip file, which initially needs to be unpacked in a working directory (e.g., C:\Program Files on a Windows operating system). Once the software unzipped, little configuration remains. CGHAnalyzer is launched by double-clicking on the "CGHAnalyzer_dist.bat" file located in the base directory of the unzipped file. A window appearing indicates a successful launch (*see* **Note 2**).

3.2. Loading Data

The first hurdle in using the CGHAnalyzer suite is the data loading step. There are two essential components required to properly load data into CGHAnalyzer. First is the acquisition of array-based genome copy number data. This normally exists as tabular output from an image analysis package such as GenePix (Molecular Devices, Inc., Sunnyvale) (*see* **Note 3**). Although loading can be done from a number of formats, each file must represent a data table from one sample, and every table row contain data for a single probe. The file name will correspond to the sample name once data are loaded. At a minimum, there should be a column representing the copy number ratio for the corresponding probe. Additionally, these data must be normalized prior to their loading (*see* **Note 4**). The second component of loading data is a probe map file, which is required to relate each row's data to the proper location in the genome. This must contain matching probe names (case sensitive) to those in the data file/s as well as the chromosome number (e.g., *chr2* corresponds to chromosome 2) and base pair coordinates for each probe. Examples of these can be seen in **Table 1** (*see* **Note 5**).

3.2.1. Loading Data From a Single Platform

1. The "File" menu item "Load Data" is the entry point for all data loading. The resultant window offers several options for loading. First, the data format must be selected. Unless loading directly from GenePix files, the item titled "Generic Array-based copy number data files (.txt)" normally applies. The file extension must be ".txt" for this generic format.
2. Using the file navigation menu on the left side of the window, locate the directory where the copy number data files are stored. These should appear in the corresponding window to the right. The desired files should be selected (each corresponding to a single sample) and moved to the box to the right by using the adjacent "Add" button (*see* **Note 6**). The limit on sample sizes is only constrained by the computer memory available, though some of the visualization tools may become muddled when exceeding ~100 samples in a single view.
3. The probe mapping file must have the ".ann" file extension. Similar to the data files, the map file should be located and moved to the right window of the probe map window set.
4. Loading data can then be completed by clicking the "Load" button at the bottom of the window. A load is successful if all samples appear under one of the chromosome views.

Table 1
Examples of Data Formats Required for CGHAnalyzer Data Loading

Table 1A

Probe	Ratio
RP11-214A5	0.63
RP11-349D15	1.07
RP11-365J10	1.64

Table 1B

Probe	Chromosome	Start	Stop
RP11-214A5	chr3	116445139	116445342
RP11-349D15	chrX	93471836	93650217
RP11-365J10	chr2	27202263	27364595

[a] A. single file must contain data from one example. The two essential components are a columns of proper probe names and copy number ratios. B. A probe mapping file must have probe names that correspond to those in the data file/s, as well as their chromosome locations.

3.2.2. Loading Data From Multiple Platforms Simultaneously

1. Data from multiple platforms (i.e., different assays with disjoint probe sets) can be loaded simultaneously for analyses. The procedure is similar to one previously described. The lone difference is the need to load multiple map files. Data from multiple platforms will be aligned for their direct comparison.

3.3. Data Visualization

CGHAnalyzer is composed of several separate visualization modules. Each provides a customizable view of the data and distinct tools for analysis. All of the modules are interactive.

3.3.1. Chromosome Views Tool

The Chromosome View tool is the main window for viewing and analyzing data. Each column represents one sample, and the horizontal bars in each column depict the data for every probe in that sample on the selected chromosome. All loaded samples are displayed in this window simultaneously. The orientation of each probe along the chromosome is related to the aligned ideogram image. All basic analytic parameters must be selected on this screen immediately after loading data to ensure effective analyses.

1. By using the "Set Thresholds" option under the "CloneView" menu item, a user must first determine the test/reference ratio thresholds to be used in classifying each clone as being lost, gained, or unchanged. These parameters must be selected carefully as they have repercussions in downstream analyses.
2. A visualization color scheme must also be selected. The "Display Type" tab under the "CGH Display" menu allows a user to combine gain and loss estimates onto one side of the ideogram, or separate the estimated losses on the left and gains on the right. Other features under this menu item allow a user to customize the order and width of each displayed sample.
3. Under the "Display" tab, customizable display features such as color and log/linear value transformations can be set.

After setting the previously listed parameters, the estimated copy number alterations can be seen in the Chromosome Views window for each loaded sample. The left hand menu allows one to navigate the data for each chromosome. A key feature in this window is the right-click utility (*see* **Note 7**). Right-clicking on any probe provides links to raw data, the region's sequence represented in Ensembl and UCSC genome browser, and the complimentary CGH Browser tool (*see* **Subheading 3.2.2.**).

3.3.2. CGH Browser Tool

The CGH Browser tool provides the most basic means of visualizing raw data in CGHAnalyzer. This display, which plots the raw clone ratios with respect to their chromosomal position, is launched directly from right-clicking on data features in the Chromosome Views window (**Fig. 1**). Raw data occurs in a data

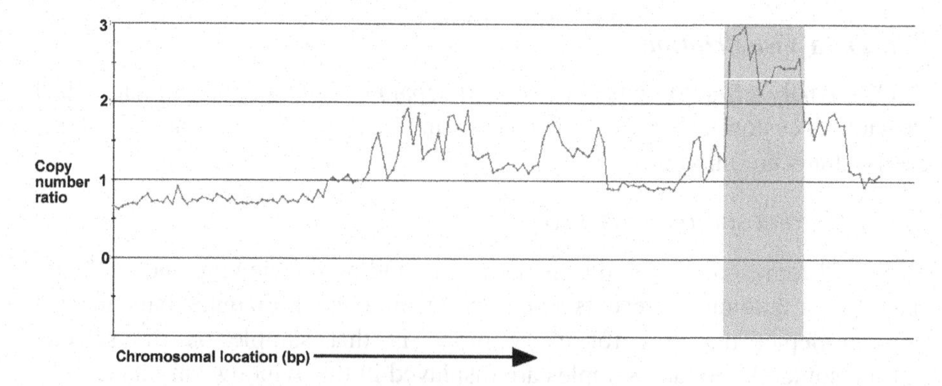

Fig. 1. Comparative genome hybridization Browser image view of chromosome 8 in a colon tumor published in **ref. *10***. The shaded area represents a chromosomal amplification on 8q.

table directly below the scatterplot. Users can navigate all samples from this window, displaying one chromosome at a time or all chromosomes at once. The "View" tab provides several important options for viewing data. For clarity, it is often desirable to display clones in sorted order, though clones can be shifted to their relative positions along the chromosome by selecting "Position" under the "X Axis Order" menu item. This may be important for gauging the actual spacing of clones along a chromosome. Also, clones designated as being unchanged based upon the previously set ratio-thresholds can be removed from the display by choosing the "Smooth Unconfirmed" option.

3.3.3. CircleViewer Tool

The CircleViewer tool supplies the crudest means of data display in CGHAnalyzer. This tool, which can be reached from the "Experiment Views" tab in the main menu, allows users to view all data for a single sample at once. This is done in a circular format, where the ring of spots represents probes on the microarray. Clones are in chromosomal order, starting at 9 o'clock extending clockwise. Each ring represents a separate chromosome where chromosome 1 is the outermost ring, and chromosome Y is the innermost. The shade of each spot reflects the ratio, where bright green spots are chromosomal gains, and reds are losses (**Fig. 2**). The utility of the Circle Viewer tool is to detect the overall aneuploidy in a single sample and provide right-click driven hot links to other modules and web-views.

3.4. Data Analysis

A wide range of analytical tools are available in CGHAnalyzer. These provide the utility to do the descriptive measures associated with most published aCGH studies as well as high level analyses such as clustering. Next are several examples that are often seen in published work.

3.4.1. Most Commonly Gained/Lost Regions

The identification of the most commonly gained/lost regions in an aCGH data set is a critical exercise and represents one of the most powerful utilities of CGHAnalyzer. The following steps must be followed to obtain the most accurate results.

1. Genome-wide copy number estimations must be made by using the "Flanking Regions" tab under the "CGH Display" menu. This protocol extends the gains and losses determined by the ratio thresholds to neighboring clones of a different copy number, thereby estimating genome-wide copy number profiles for each sample (*see* **Note 8**).

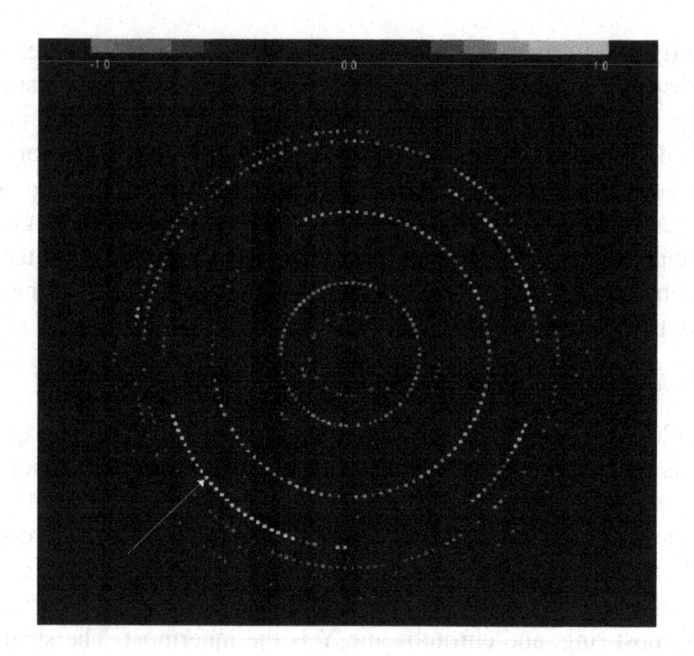

Fig. 2. Circle Viewer display of a colon tumor published in **ref. *10***. Bright greens represent chromosomal gains and reds represent losses. The arrows highlight an amplification of 8q23.3-24.13 seen in this tumor.

2. The "CGH Analysis" menu contains all the relevant query tools. For example, the "Region Amplifications" tab identifies the most commonly gained regions estimated by the Flanking Regions protocol mentioned above. Applying this to 48 colon cancer cell lines ***(10)*** indicates regions encompassing 20q13.13-13.2 have the most common amplifications when a gain ratio threshold of 1.25 is used (71%, 34/48 samples; **Fig. 3**) (*see* **Note 9**). Though chromosome 20 is commonly amplified, this region represents the minimal common region of gain and provides a more reasonable region to look for gene targets. Similar analyses can be done for chromosomal losses. The query results can be viewed in a tabular form from the left hand drop down menu.
3. All results can be exported in a tabular format for publication or further evaluation.

3.4.2. Clustering

1. Selecting the type of clustering analysis that best fits the specific analytic goals is the first step in performing clustering analyses. CGHAnalyzer supports several standard types of clustering, including hierarchical, K-means, and self organizing maps.
2. Loaded samples not wanted in the clustering must first be removed from the data set.

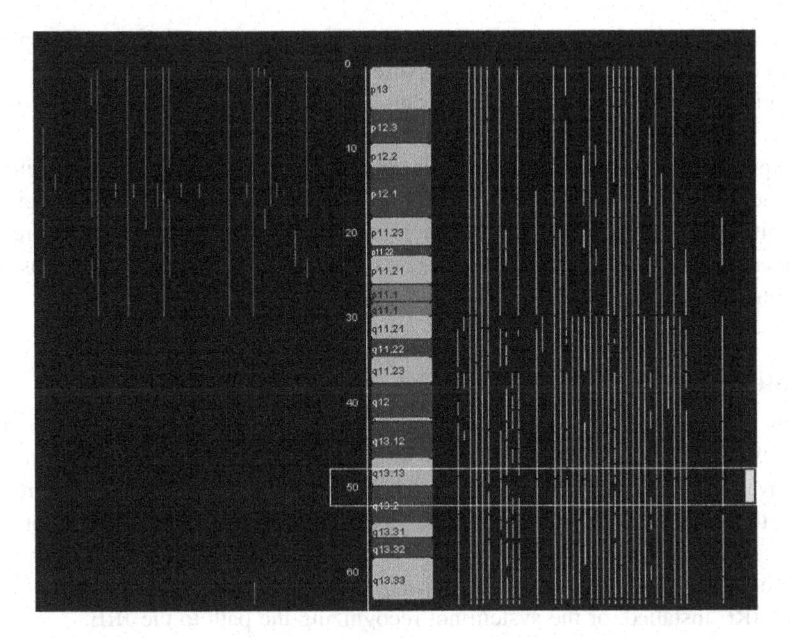

Fig. 3. Chromosomes view of chromosome 20 in 48 colon cancer cell lines ***(10)***. The most frequently gained regions, 20q13.13-13.2, are highlighted on the right.

3. Every clustering procedure requires the selection of genetic units to be used. This can be either probes, genes, or genomic regions (*see* **Note 10**). This parameter must be selected upon the initiation of any clustering protocol.
4. CGHAnalyzer can cluster on subsets of the genome. All clustering menus prompt users to select which chromosome/s on which to work. For instance, selecting only chromosome 1 results in the clustering of features located only on chromosome 1.
5. All clustering output can be accessed as a new menu item added to the "Analysis" tab in the left menu of the main window. Each display is interactive, providing associated statistics with the results. For instance, a hierarchical clustering procedure produces similarity plots for both samples and genomic regions as well as a heatmap (*see* **Note 11**). Each window can be exported as an image.

3.4.3. T-Test

1. The T-test analysis utility, a powerful method for discovering features that differentiate groups of samples, can be launched from the main window of CGHAnalyzer.
2. As with clustering procedures, users must select the features that will serve as units to differentiate samples. Selecting "Genome Regions" indicates that probe-independent genome regions will the subjects of the analyses. Also, one or more chromosomes can be queried for analysis.

3. The following window prompts users to group samples into classes. Samples can also be excluded from analyses (*see* **Note 12**).
4. The user-selected *p*-value parameters are used to determine the method of *p*-value calculation for each feature between classes (*see* **Note 13**).
5. Output for the t-tests procedure are listed under the "Analysis" tab in the left menu of the main window. The critical component of these analyses, the genes/region that significantly differentiate the sample groups given the input paramters, are listed under the "Gene Statistics" tab. These can be exported to a spreadsheet using the right-click utility.

4. Notes

1. To determine if the correct version of the JRE is being used, simply open a DOS window (or a terminal window on a Macintosh) and type the command "java –version." The first return line should state the version of the JRE installed. If the JRE version is earlier than 1.4.1, install the latest version at http://java.sun.com/.
2. The failure of CGHAnalyzer to properly launch is normally caused by not having the JRE installed, or the system not recognizing the path to the JRE.
3. Public repositories such as NCBI's Gene Expression Omnibus *(11)* have a growing number of aCGH data sets. Minimal reformatting is required to load these data into CGHAnalyzer.
4. Normalization is an essential step in aCGH analysis. Most simply, this can be done by dividing all ratios by the median/mean ratio. This can be done easily by using Microsoft Excel or a statistical tool such as the R programming language (freely available at http://cran.r-project.org/). Other options include intensity-dependent normalization, such as Lowess. A freely available tool that offers a variety of customized normalization options for GenePix files is available at http://acgh.afcri.upenn.edu.
5. Data from most copy number detection assays can be arranged to fit this format. For instance, data from the single channel Affymetrix SNP Chip (Affymetrix, Inc., Sunnyvale, CA) does not have true copy number ratios; however other software packages such as dChip *(12)* can calculate analogous ratios for each probe.
6. Reverse dye experiments ("dye-swaps") are commonly used to validate experiment results of aCGH studies. CGHAnalyzer can load and simultaneously display experiments with dye-swap results by formatting an annotation file. This can be done by properly formatting an annotation file with the name extension ".dye" (**Table 2**).
7. Right-click driven utilities are an essential part of CGHAnalyzer. If right click capabilities are not functional, this normally indicates that an out of data version of the JRE is being used (e.g., v1.3). Macintosh users should simultaneously hold the control key while clicking to access the right-click functions.

Table 2
A Dye Swap File Has Three Columns

Exp ID.	Sample	Dye
12435657	COLO 205	Cy3
12435658	COLO 205	Cy5
12435950	HCT 116	Cy3
12435949	HCT 116	Cy5

[a] A. unique experiment identifier can be related to a sample name and a test dye. Each sample name must have at least one experiment in both Cy3 and Cy5.

8. More sophisticated alternatives exist for genome-wide estimates of copy number. Several published examples that apply a diverse set of existing computational methods to this problem include the use of Hidden Markov Models ***(13)***, adaptive weights smoothing ***(14)***, and circular binary segmentation ***(15)***. All procedures of copy number breakpoint estimation involve a tradeoff in sensitivity and specificity. The use of one over another depends entirely on the specific analytic goals of the investigator.
9. As an alternative to working with genomic regions, genes can be the subjects of queries as well. CGHAnalyzer will automatically map known genes stored public databases to the genomic regions. The query results will then indicate the most gained or lost genes.
10. Selecting genes, probes, or genomic regions is a very important part of this procedure. This determines what the basic clustering units for the samples are. The use of genomic regions is the most powerful. The resultant clustered regions can always be later mapped to genes and probes after the clustering procedure is complete.
11. Heatmaps produced by clustering procedures can be quite large. This interactive image is often useful to analyze the members of a specific cluster to determine which regions/genes are driving the similarities between samples. Data from subsets of the heatmap can be exported by selecting portions of the tree and using the right-click function.
12. If these analyses are going to be done multiple times on the same sample set, class designations can be stored in a file and later reloaded. This can save a substantial amount of effort for large sample sizes.
13. Selecting the *p*-value parameters is critical. *p*-values based upon the permutation can offer more stringent results (fewer false-positives) with smaller critical values and large numbers of permutations. Of note, large numbers of permutations can cause lengthy run times.

References

1. Milunsky, J. M. and Huang, X. L. (2003) Unmasking Kabuki syndrome: chromosome 8p22-8p23.1 duplication revealed by comparative genomic hybridization and BAC-FISH. *Clin. Genet.* **64,** 509–516.
2. Look, A. T., Hayes, F. A., Shuster, J. J., et al. (1991) Clinical relevance of tumor cell ploidy and N-myc gene amplification in childhood neuroblastoma: a Pediatric Oncology Group study. *J. Clin. Oncol.* **9,** 581–591.
3. Albertson, D. G., Ylstra, B., Segraves, R., et al. (2000) Quantitative mapping of amplicon structure by array CGH identifies CYP24 as a candidate oncogene. *Nat. Genet.* **25,** 144–146.
4. Snijders, A. M., Nowak, N., Segraves, R., et al. (2001) Assembly of microarrays for genome-wide measurement of DNA copy number. *Nat. Genet.* **29,** 263–264.
5. Lucito, R., Healy, J., Alexander, J., et al. (2003) Representational oligonucleotide microarray analysis: a high-resolution method to detect genome copy number variation. *Genome Res.* **13,** 2291–2305.
6. Bignell, G. R., Huang, J., Greshock, J., et al. (2004) High-resolution analysis of DNA copy number using oligonucleotide microarrays. *Genome Res.* **14,** 287–295.
7. Davidson, J. M., Gorringe, K. L., Chin, S. F., et al. (2000) Molecular cytogenetic analysis of breast cancer cell lines. *Br. J. Cancer* **83,** 1309–1317.
8. Dudoit, S., Shaffer, J. P., and Boldrick, J. C. (2003) Multiple hypothesis testing in microarray experiments. *Statistical Science* **18,** 71–103.
9. Weiss, M. M., Kuipers, E. J., Postma, C., et al. (2004) Genomic alterations in primary gastric adenocarcinomas correlate with clinicopathological characteristics and survival. *Cell Oncol.* **26,** 307–317.
10. Douglas, E. J., Fiegler, H., Rowan, A., et al. (2004) Array comparative genomic hybridization analysis of colorectal cancer cell lines and primary carcinomas. *Cancer Res.* **64,** 4817–4825.
11. Edgar, R., Domrachev, M., and Lash, A. E. (2002) Gene Expression Omnibus: NCBI gene expression and hybridization array data repository. *Nucleic Acids Res.* **30,** 207–210.
12. Lin, M., Wei, L. J., Sellers, W. R., Lieberfarb, M., Wong, W. H., and Li, C. (2004) dChipSNP: significance curve and clustering of SNP-array-based loss-of-heterozygosity data. *Bioinformatics* **20,** 1233–1240.
13. Willenbrock, H. and Fridlyand, J. (2005) A comparison study: applying segmentation to array CGH data for downstream analyses. *Bioinformatics* **21,** 4084–4091.
14. Hupe, P., Stransky, N., Thiery, J. P., Radvanyi, F., and Barillot, E. (2004) Analysis of array CGH data: from signal ratio to gain and loss of DNA regions. *Bioinformatics* **20,** 3413–3422.
15. Olshen, A. B., Venkatraman, E. S., Lucito, R., and Wigler, M. (2004) Circular binary segmentation for the analysis of array-based DNA copy number data. *Biostatistics* **5,** 557–572.

17

Microarray-Based Approach for Genome-Wide Survey of Nucleotide Polymorphisms

Brian W. Brunelle and Tracy L. Nicholson

Summary

DNA microarrays can be used to detect polymorphic loci in addition to identifying genes or regions that are absent within a genome. A survey such as this offers greater insight into the level of diversification within a species or population, which is useful in organisms that have near-identical genomic content but differ in phenotype. The identification of such variable loci can then lead to the characterization of genes linked to unique biological attributes. Here, we describe a competitive hybridization assay using DNA microarrays as a comparative genomics tool to identify nucleotide polymorphisms among closely related strains of *Chlamydia trachomatis*.

Key Words: Microarray; *Chlamydia trachomatis*; competitive hybridization; polymorphism; nucleotide substitution; genomic survey.

1. Introduction

The identification of nucleotide polymorphisms between closely related organisms is valuable as these polymorphisms can serve as a basis for evolutionary and population analyses. Such nucleotide substitutions may also result in the replacement of amino acids in encoded proteins and, therefore, may underlie important phenotypic differences, such as virulence or tissue tropism. However, because sequencing numerous genomes is currently neither a rapid nor fiscally practical approach, DNA microarrays provide a useful alternative as they can survey and identify changes across an entire genome.

From: *Methods in Molecular Biology, vol. 396: Comparative Genomics, Volume 2*
Edited by: N. H. Bergman © Humana Press Inc., Totowa, NJ

DNA microarrays are competition-based assays in which a reference DNA sample and a test DNA sample are fluorescently labeled before hybridizing with immobilized DNA. Because the reference DNA is identical to the immobilized DNA, the ratio of the fluorescence at each locus is dependent on the nucleotide sequence of the test isolate. An equivalent signal ratio indicates uniform hybridization and signifies a very high nucleotide identity; the more variable the sequence of the test DNA at that locus, the larger the ratio becomes. This is a qualitative difference that loses its correlation when two sequences at a region fall below ~75% identity *(1)*.

The microarray assay described here involves (1) the detection of loci on the array that may be biased, (2) demonstrating that there is a correlation between the signal ratio and percent nucleotide identity of two sequenced genomes, and (3) testing samples to identify those regions that are variable. This procedure was conducted on 14 human serological variants of *Chlamydia trachomatis*, which are nearly identical on the genomic level *(2)* but are specific for one of three distinct tissue tropisms and pathologies *(3)*. Analyses revealed genes with nucleotide polymorphisms, some of which may be associated with the organism's fitness and success within a particular host niche.

2. Materials

2.1. Bacterial Strains, Growth Conditions

1. HeLa229 cells.
2. T150 tissue culture flask.
3. RPMI Media (Invitrogen, Carlsbad, CA).
4. Fetal bovine serum (Hyclone, Logan, UT).
5. Vancomycin (Sigma, St. Louis, MO).
6. Renografin (E.R. Squibb and Sons, Princeton, NJ).
7. Sucrose phosphate glutamate buffer, pH 7.4: 220 m*M* sucrose, 3.83 m*M* KH_2PO_4, 8.60 m*M* Na_2HPO_4, 4.26 m*M* glutamic acid.

2.2. Preparation of Genomic DNA

1. TE buffer, pH 7.6: 0.5 *M* Tris-HCl, 0.5 *M* ethylenediamine tetraacetic acid.
2. 10% (w/v) Sodium dodecyl sulfate (SDS).
3. 5 *M* Sodium chloride (NaCl).
4. 20 mg/mL Proteinase K.
5. Cetyltrimethylammonium bromide (CTAB)/NaCl solution: 10% CTAB (Sigma) in 0.7 *M* NaCl.
6. 24:1 (v/v) Chloroform/isoamyl alcohol.
7. 25:24:1 (v/v/v) Phenol/chloroform/isoamyl alcohol.
8. 70% Ethanol (EtOH).

9. Microcentrifuge (Eppendorf, cat. no. 5415D, or equivalent).
10. Isopropanol.

2.3. Slide Preparation

1. Premium plain glass slides (Fisher Scientific, Pittsburgh, PA).
2. Slide rack and staining dishes (Fisher Scientific, Shandon Lipshaw, cat. no. 121).
3. Poly-L-lysine (Sigma).
4. Sodium hydroxide (NaOH).
5. 95% EtOH.
6. Desiccator cabinet (Fisher Scientific).
7. Phosphate buffered saline.

2.4. Microarray Design and Fabrication

1. 96-Well PCR plates (MJ Research, South San Francisco, CA).
2. 96-Well thermocycler (Tetrad, MJ Research).
3. PCR reagents: Taq polymerase, 10X PCR buffer, 10 m*M* dNTPs, custom oligonucleotide primers.
4. Agarose (molecular biology grade).
5. EtOH/sodium acetate solution (3 *M* Na-acetate in 95% EtOH).
6. Table top centrifuge with plate adapters (Eppendorf, cat. no. 5810 or equivalent).
7. 70% EtOH.
8. 96-Well SpeedVac (Savant SC250, Thermo Electron Corporation, Milford, MA).
9. Aluminum sealing tape for Microtiter plates (Fisher Scientific).
10. 384-Well Genetix print plates (USA Scientific, Ocala, FL).
11. 20X Sodium chloride-sodium citrate (SSC) buffer, pH 7.0: 3 *M* Na_3 citrate, 3 *M* NaCl.
12. 3X SSC buffer, pH 7.0 (prepared from 20X stock).
13. Microarrayer (Biorobotics Microgrid or equivalent, Genomic Solutions, Ann Arbor, MI).

2.5. Postprocessing Slides

1. Slide rack and staining dishes (Fisher Scientific, Shandon Lipshaw, cat. no. 121).
2. Platform shaker.
3. Humid chamber (Sigma, cat. no. H6644).
4. Table top centrifuge with plate adapters (Eppendorf, cat. no. 5810 or equivalent).
5. 1 *M* Sodium borate (Na-borate), pH 8.0, filtered.
6. 1-Methyl-2-pyrrolidinone (Sigma).
7. Succinic anhydride (Sigma).
8. 95% EtOH.
9. Ultraviolet (UV) crosslinker (Fisher Scientific).

2.6. Labeling

1. RadPrime DNA Labeling System (Invitrogen).
2. 10X dNTP Mix (1.2 m*M* dATP, dCTP, dGTP, and 0.6 m*M* dTTP).
3. Cy3-dUTP (25 nmol) and Cy5-dUTP (25 nmol) (Amersham, Piscataway, NJ).
4. TE buffer, pH 7.4 (filtered).
5. Microcon 30 (Amicon) filters (Fisher Scientific).

2.7. Hybridizations

1. 10 mg/mL tRNA mix (Sigma).
2. Cover slips (Fisher Scientific).
3. 20X SSC, pH 7.0.
4. 2% (w/v) SDS.
5. Wash Solution 1: 1X SSC + 0.05% (w/v) SDS (1X SSC prepared from 20X stock).
6. Wash Solution 2: 0.06X SSC (prepared from 20X stock).
7. Elmer's rubber cement.
8. TE buffer, pH 7.4.
9. Glass staining dish and rack (Weaton, Fisher Scientific).
10. Gene Machines HybChamber (Genomic Solutions).
11. GENEPIX Scanner 4000A (Axon Instruments, Foster City, CA).
12. GenePix Pro 4.0 software (Axon Instruments).

3. Methods

3.1. Bacterial Strains, Growth Conditions

The genome of *C. trachomatis* strain D/UW-3 is fully sequenced *(4)* and was used to create the microarray *(5)*. Human-specific strains representing the major serological variants of *C. trachomatis* were selected for analysis to identify differences among those strains that infect one of three specific tissue niches (ocular, urogenital, or lymphatic). Also included in the study was the murine-specific mouse pneumonitis (MoPn) strain as it is closely related to the human-specific strains and its genome has been fully sequenced *(6)*.

1. Propagate *C. trachomatis* strains (A/Har1, B/TW-5, Ba/Apache-2, C/TW-3, D/UW-3, E/Bour, F/IC-Cal 3, G/UW-57, H/UW-4, I/UW-12, J/UW-36, K/UW-31, L1/440, L2/434, L3/404, MoPn) in HeLa229 cell monolayers in T-150 flasks containing RPMI medium supplemented with 10% (w/v) fetal bovine serum and 50 μg/mL vancomycin.
2. Isolate chlamydial elementary bodies (EB) by sonic treatments of cell suspensions and purify by ultracentrifugation over 30 and 30/44% (w/v) discontinuous Renografin gradients.
3. Freeze 100 μL aliquots at −80 °C in sucrose phosphate glutamate buffer until ready to use.

3.2. Preparation of Genomic DNA

1. Thaw chlamydial EB samples on ice.
2. Centrifuge at >13,000g for 5 min to pellet chlamydial EB samples and discard the supernatant.
3. Wash pellet three times with TE, pH 7.6.
4. Resuspend pellet in 1020 μL TE, pH 7.6 and vortex to mix samples thoroughly.
5. Split sample into two Eppendorf tubes (each containing 510 μL resuspended EB in TE, pH 7.6) and add to each tube: 60 μL 10% (w/v) SDS, 30 μL 20 mg/mL proteinase K.
6. Mix thoroughly and incubate for 4 h at 50 °C.
7. Add 100 μL 5 *M* NaCl, mix thoroughly, and incubate at 50 °C for 2 min.
8. Add 80 μL CTAB/NaCl solution, and incubate at 65 °C for 2 min.
9. Add 780 μL phenol/chloroform/isoamyl alcohol solution, mix thoroughly, and centrifuge at >13,000*g* for 5 min.
10. Remove aqueous viscous supernatant to a fresh tube, add an equal volume of chloroform/isoamyl alcohol, mix thoroughly, and centrifuge at >13,000*g* for 5 min.
11. Transfer supernatant to new Eppendorf tube and add 0.6 vol isopropanol to precipitate DNA.
12. Centrifuge at >13,000*g* for 10 min.
13. Discard supernatant, Air-dry DNA pellet and resuspend in 100 μL nuclease-free dH_2O.

3.3. Slide Preparation

1. Prepare NaOH Solution (*see* **Note 1**):

    a. Dissolve NaOH in dH_2O: 175 g/700 mL.
    b. Stir until completely dissolved.
    c. Add 1050 mL 95% EtOH.
    d. Stir until completely mixed.
    e. If solution remains cloudy, add water until clear.

2. Place slides in metal slide racks (30 slides/rack). Do not use defective slides.
3. Soak slides in the NaOH/EtOH/dH_2O solution for 2 h with gentle rotation.
4. Rinse slides extensively with dH_2O:

    a. Rinse each unit (slide/rack/container) vigorously with dH_2O for 5 min.
    b. Place slide racks in a large clean glass container, and tilt the container slightly for constant water flow.
    c. Wash under running water for 30 min.
    d. Do not let slides dry at any time.
    e. It is critical to remove all traces of NaOH/EtOH.

5. Prepare poly-L-lysine solution in plastic container.

   a. Mix together 100 mL phosphate buffered saline, 800 mL ddH_2O, 100 mL poly-L-lysine.
   b. Bring up the volume to about 1050 mL with ddH_2O to submerge up to three racks of slides.
   c. Mix and split the solution into three plastic containers.

6. Soak the slides in poly-L-lysine solution for 1 h with gentle rotation. Poly-L-lysine adheres to glass, therefore be sure to use a plastic container.
7. After the poly-L-lysine coating, rinse slides by gentle plunging up and down in two different changes of water.
8. Using a table top centrifuge, centrifuge at 72*g* for 5 min. Be sure to place paper towels below rack to absorb liquid.
9. Store slides in a desiccator for 3 wk prior to use. Slides that are older than 3 mo may result in faint printing and higher background.

3.4. Microarray Design and Fabrication

1. Choose primer pairs to amplify all open reading frames (the genome sequence of *C. trachomatis* strain D/UW-3 has 875 open reading frames *[4]*). Ensure that each primer is between 20 and 23 bases long and produces a PCR product between 150 and 800 bases. Primers should be produced in 96-well format.
2. Make 5 μM working stock plates of the forward and reverse primers.
3. Set up PCR reactions in a 96-well format using the following concentrations per 20 μL reaction: 2 U Taq Polymerase, 10 ng genomic DNA, 1X PCR buffer, 0.5 m*M* dNTPs, 0.4 μ*M* custom primer pair.
4. Amplify all open reading frames using the following PCR cycle conditions. Some annealing temperatures may need to be modified to optimize specific PCR products: 94 °C for 1 min; 30 cycles of 94 °C for 40 s, 52 °C for 40 s, 72 °C for 1.25 min; 72 °C for 5 min; 4 °C hold.
5. Run all 96-well reactions on a 1% (w/v) agarose gel and document results.
6. Add 150 μL EtOH/Na-acetate solution into each well.
7. Cover plates with aluminum sealing tape and store in −20 °C freezer overnight to precipitate.
8. Using a table top centrifuge, centrifuge at 3200*g* for at least 2 h at 4 °C.
9. Dump solution and pat dry on a napkin.
10. Fill each well with 100 μL 70% EtOH and centrifuge for another hour at 3200*g*.
11. Dump EtOH solution and dry plates using a 96-well SpeedVac or let plates air-dry.
12. Resuspend PCR products in 40 μL H_2O. Keep samples at 4 °C for at least 18 h, then transfer products to 384-well printing plates.
13. Dry plates down using a SpeedVac or let air-dry and resuspend in 12 μL 3X SSC.
14. Let plates resuspend at least overnight before printing.

15. Using a Microarrayer, spot DNA from all 384-well plates onto prepared poly-L-lysine treated slides. To provide duplicate assessments for each hybridization experiment, each PCR product should be printed twice per slide when possible.

3.5. Postprocessing Slides

1. Before starting:
   a. Warm a heat block to 90 °C and invert the insert.
   b. Place slide-staining chambers in front of the heating block.
   c. Fill slide staining chamber with H_2O.
   d. Get a dust free board needed for cross-linking step.
   e. Place the following items in a chemical hood: large glass beaker with ~1 L boiling H_2O (cover with foil while heating), stirring plate for preparation of blocking solution, and a shaker.
   f. For the preparation of blocking solution, have a clean 500 mL beaker ready.
   g. Have the following chemicals ready: 6 g succinic anhydride, 350 mL 1-Methyl-2-pyrrolidinone, 15 mL 1 *M* Na-borate, pH 8.0 (filtered).
2. Rehydrate the slides:
   a. Rehydrate slides by inverting (array side down) them over H_2O in a slide-staining chamber. Let the spots become glistening in appearance.
   b. Immediately flip slide (array side up) onto a heating block (with inverted insert) and watch the steam evaporate. When the array spots dry in a rapid wave-like pattern, remove slide from heating block. This usually takes about 5 s. Do only one slide at a time.
3. UV cross link slides:
   a. Place slides, array side up, on a flat and dust free board that fits into a UV crosslinker.
   b. Irradiate with 600 μJ UV light.
4. Block free lysines (wear lab coat when working with 1-Methyl-2-pyrrolidinone):
   a. Add 6.0 g succinic anhydride into 350 mL 1-Methyl-2-pyrrolidinone.
   b. As soon as the solids dissolve, quickly add 15 mL of 1 *M* Na-borate, pH 8.0, and pour the mixed solution into a slide washing tray.
   c. Quickly place slides into the succinic anhydride solution and vigorously plunge repeatedly for 60 s (do not pour solution over slides).
   d. Turn platform shaker on and rotate vigorously for 15 min.
   e. Remove the slide rack from the organic reaction mixture and place it immediately into the boiling water bath for 90 s.
   f. Transfer the slide rack to the 95% EtOH wash.
5. Immediately spin dry slides by centrifugation at 72*g* for 2 min.
6. Transfer slides to a dry slide box and store in a desiccator.

3.6. Labeling

Genomic DNA from each strain tested is randomly primed with Cy-5- or Cy-3-labeled nucleotides using the RadPrime DNA Labeling System. D/UW-3 genomic DNA was used as the reference DNA for all hybridizations and labeled with Cy5, whereas all other sample DNA was labeled with Cy3.

1. To an Eppendorf tube, add 0.2 μg genomic DNA and appropriate amount of water to bring volume up to 20 μL.
2. Add 20 μL of 2.5X RadPrime Buffer.
3. Boil sample for 5 min and snap cool on ice for 5 min.
4. Keep tubes on ice and add the following to each reaction in order: 5 μL 10X dNTP Mix, 3 μL of appropriate Cy3- or Cy5-dUTP, 2 μL Klenow Fragment (40 U/μL).
5. Incubate reactions for 10 min at 25 °C and then 90 min at 37 °C.
6. Add 5 μL Stop Buffer (0.5 *M* ethylenediamine tetraacetic acid).
7. Add the two labeled-DNA samples to be compared (appropriate Cy3 and Cy5 label reactions) together, transfer combined reactions to a Microcon 30 filter and add 400 μL TE buffer, pH 7.4.
8. Centrifuge samples at $>13,000g$ until ~25 μL remaining (~20 min). Then add 200 μL TE buffer, pH 7.4, to samples and spin until almost dry, <7 μL (~10 min). Recover labeled DNA samples by inverting Microcon filter into a new 1.5 mL collection tube and centrifuge at half max speed for 1 min.
9. Bring sample to 7 μL with TE buffer, pH 7.4, and transfer to a new 0.5-mL tube.

3.7. Hybridizations

To reduce the potential effects caused by variation in array quality, each hybridization should be performed at least twice. Because the chlamydial genome was printed twice per slide, there was a minimum of four data points for each locus.

1. Add the following to each sample: 0.67 μL tRNA 10 mg/mL, 1.90 μL 20X SSC, 1.35 μL 2% (w/v) SDS.
2. Heat denature samples by incubating at 95 °C for 2 min. Perform a quick spin in centrifuge to pellet samples and let samples cool at room temperature for 2 min before applying to microarray.
3. Apply each sample to microarray and seal with rubber cement.
    a. The array is invisible after postprocessing. Therefore, to ensure proper placement of a cover slip, make a template of your array using a clean glass slide, or by drawing on a piece of paper. Be sure to indicate the corners of the array and where the cover slip should lie.
    b. Lay each microarray slide on top of the template and, just inside the marked cover slip corners, apply a small microdot of rubber cement with a needle and syringe to each corner.

c. Apply the ~10 μL sample to the microarray slide. Place an appropriate size cover slip with bent precision forceps right on to of the four small dots of rubber cement. Try to free any air bubbles that might be present under the cover slip by gently tapping on the top of the cover slip with the forceps.
d. Completely seal the edges of the cover slip with the rubber cement. To avoid sticking of the cover slips to the top of the hybridization chambers, let the rubber cement dry before sealing the hybridization chambers.
e. Place the prepared and sealed microarray glass slide into a chamber compartment in a GeneMachines HyChamber and place ~100 μL of TE buffer, pH 7.4, in each chamber compartment.
f. Seal each GeneMachines HyChamber completely and place in appropriate temperature water bath for 12 h to overnight. (All chlamydial hybridizations took place under a glass cover slip in a 75 °C water bath overnight except for the MoPn vs D/UW-3 comparison, which transpired in a 65 °C water bath overnight [*see* **Note 2**].)

4. Take microarray glass slide out of the HyChamber and immediately submerge the slide into a flat container filled with Wash Solution 1.
5. Using a clean razorblade, remove the cover slip by lifting it at the corner and pulling up. Transfer the array to a slide rack submerged into a container filled with the Wash Solution 1 and plunge rack up and down for 2 min to wash slide.
6. Place the rack holding the array slides into a new container filled with Wash Solution 2 and plunge up and down for 2 min.
7. Place the rack holding the array slides into a new container filled with fresh Wash Solution 2 and again plunge up and down for 2 min.
8. To dry the slides, place rack holding array slides in table top centrifuge and spin for 5 min at 50*g*.
9. Scan each microarray glass slide using a GENEPIX Scanner 4000A and analyze the resulting 16-bit TIFF images using GenePix Pro 4.0 (or equivalent) software.

3.8. Calculation of Data

For each isolate, a normalization factor is calculated for the signal intensity of each wavelength and used to standardize any variation between replicate slides. The signal ratio of the normalized intensities between each wavelength is then determined for each locus. Because there are several data points for each locus of an isolate, the average ratio and its standard error is calculated for each region. In the chlamydial microarray assay, the signal ratio between the intensities from the reference DNA (635 nm) and the test DNA (532 nm) was used; it is the level of departure from a 1:1 ratio that indicates the degree of sequence polymorphism at a locus.

1. Normalize the signal intensity of each wavelength for each isolate to one chosen slide of that isolate (*see* **Note 3**).

   a. For each isolate, the normalization factor for a given wavelength for any slide Y is determined as the ratio of total signal intensity of a chosen slide X relative to the slide Y as shown next:

   $$NF_Y = \Sigma\, I_Y / \Sigma\, I_X$$

   Where NF_Y is the normalization factor for any slide of an isolate at a given wavelength, $\Sigma\, I_Y$ is the total intensity from any slide of an isolate, and $\Sigma\, I_X$ is the total intensity for the specific slide of the isolate chosen to be used as the basis for normalization.

   b. The normalized signal intensity of each locus on slide Y is then calculated by multiplying the signal intensity from each wavelength of that locus by its respective NF_Y for that slide.

2. For each slide of an isolate, calculate the ratio of the median signal intensity for each normalized locus on the array (*see* **Note 4**).
3. For each locus of an isolate, calculate the mean and standard error of the mean for the signal ratios determined in **step 2** (*see* **Note 5**).

3.9. Test for Intrinsic Slide Bias (Genome vs Self)

To detect loci on the array that may lead to false-positive data, the assay is performed using identical test and reference DNA, both of which are the same as the DNA used to construct the slide. With an expectation that all loci should have equal ratios of hybridization of the labeled DNA, those that are significantly different are eliminated from future analyses.

1. Follow the steps from **Subheading 3.8.** on replicates from hybridizations that used the same DNA for both the test and reference samples (*see* **Note 6**).
2. Determine which loci are statistical outliers (e.g., Z-test, 95% confidence interval).
3. Remove the data at these loci from all subsequent investigations (*see* **Note 7**).

3.10. Verification of the Relationship Between Signal Ratio and Nucleotide Identity (Known Genome vs Known Genome)

If two or more genomic sequences are known (and one of them was used to build the array), then these can be used to demonstrate the correlation of the signal ratio and sequence identity between the loci of the two genomes. However, this step can be eliminated if two genomic sequences are not available as it has been established for several bacteria ***(1,7–9)***.

1. Follow the steps from **Subheading 3.8.** for replicates from hybridizations that used test and reference DNA from samples that have fully sequenced genomes (*see* **Note 8**).
2. Determine percent nucleotide sequence identity between the two genomes for each locus represented on the microarray (*see* **Note 9**).
3. Graph the percent identity vs the log2 mean signal ratio for all the genes.

3.11. Identifying Polymorphic Loci in Uncharacterized Samples (Known Genome vs Unknown Genome)

By ranking the average ratio of the median from highest to lowest for each sample strain, assessments can be made about which regions are the most variable when compared to the reference strain. This also allows comparisons between strains regardless of total genomic identity.

1. Follow the steps from **Subheading 3.8.** for each sample.
2. Sort the average of the signal ratios for each locus from highest to lowest for each isolate, and rank them with one corresponding to the highest average.
3. Those ranked highest have the most sequence variability compared to the reference (*see* **Notes 10** and **11**).
4. Compare variable loci among the isolates and with observed attributes.

4. Notes

1. Prepare this solution fresh each time.
2. Higher temperatures were used to increase binding specificity; lowering the salt concentration or adding formamide may also help ***(10)***.
3. The value of the signal intensity used for each wavelength is the background subtracted from the median intensity.
4. The numerator should be the wavelength that corresponds to the reference strain; this will give ratios one and above.
5. If the standard error of the mean is large for a locus, look at the individual data points and their respective spots in the raw TIFF image file as there may have been a problem with the slide at that region. If there was something that artificially altered the data, that point should be eliminated from analysis.
6. For the chlamydial array, strain D/UW-3 was used as both the test and reference strain.
7. Although an unknown degree of bias may be introduced at these loci, a very high signal ratio (>5) in a test isolate may still be indicative of a gene deletion or high degree of polymorphism.
8. For the chlamydial array, strains D/UW-3 and MoPn were used.
9. Be sure to analyze only the sequences of the PCR products used to make the microarray.

10. A large mean signal ratio may be the result of a full or partial deletion; further analysis by other methods, such as PCR or Southern blot, would be necessary for confirmation.
11. Sequencing a few loci that were found to be either variable or highly similar for verification purposes would add confidence to all the findings in the study.

Acknowledgments

We would like to thank Dr. Eric M. Nicholson for his critical evaluation regarding the manuscript. We would also like to thank Dr. Gary Schoolnik and the members of his lab for their technical assistance with the microarray procedures described here. This work was performed in the laboratory of Dr. Richard S. Stephens, and we are grateful the support he provided. Mention of trade names or commercial products in this article is solely for the purpose of providing specific information and does not imply recommendation or endorsement by the US Department of Agriculture.

References

1. Brunelle, B. W., Nicholson, T. L., and Stephens, R. S. (2004) Microarray-based genomic surveying of gene polymorphisms in *Chlamydia trachomatis. Genome Biol.* **5,** R42.
2. Stephens, R. S. (1999) Genomic autobiographies of Chlamydiae, in *Chlamydia: Intracellular Biology, Pathogenesis, and Immunity*, (Stephens, R. S., ed.), American Society for Microbiology, Washington, DC, pp. 6–26.
3. Wang, S. P. and Grayston, J. T. (1988) Micro-immunofluorescence antibody responses to trachoma vaccines. *Int. Ophthalmol.* **12,** 73–80.
4. Stephens, R. S., Kalman, S., Lammel, C., et al. (1998) Genome sequence of an obligate intracellular pathogen of humans: *Chlamydia trachomatis. Science* **282,** 754–759.
5. Nicholson, T. L., Olinger, L., Chong, K., Schoolnik, G., and Stephens, R. S. (2003) Global stage-specific gene regulation during the developmental cycle of *Chlamydia trachomatis. J. Bacteriol.* **185,** 3179–3189.
6. Read, T. D., Brunham, R. C., Shen, C., et al. (2000) Genome sequences of *Chlamydia trachomatis* MoPn and *Chlamydia pneumoniae* AR39. *Nucleic Acids Res.* **28,** 1397–1406.
7. Dong, Y., Glasner, J. D., Blattner, F. R., and Triplett, E. W. (2001) Genomic interspecies microarray hybridization: rapid discovery of three thousand genes in the maize endophyte, *Klebsiella pneumoniae* 342, by microarray hybridization with *Escherichia coli* K-12 open reading frames. *Appl. Environ. Microbiol.* **67,** 1911–1121.

8. Kim, C. C., Joyce, E. A., Chan, K., and Falkow, S. (2002) Improved analytical methods for microarray-based genome-composition analysis. *Genome Biol.* **3,** RESEARCH0065.
9. Lindroos, H. L., Mira, A., Repsilber, D., et al. (2005) Characterization of the genome composition of *Bartonella koehlerae* by microarray comparative genomic hybridization profiling. *J. Bacteriol.* **187,** 6155–6165.
10. McConaughy, B. L., Laird, C. D., and McCarthy, B. J. (1969) Nucleic acid reassociation in formamide. *Biochemistry* **8,** 3289–3295.

18

High-Throughput Genotyping of Single Nucleotide Polymorphisms with High Sensitivity

Honghua Li, Hui-Yun Wang, Xiangfeng Cui, Minjie Luo, Guohong Hu, Danielle M. Greenawalt, Irina V. Tereshchenko, James Y. Li, Yi Chu, and Richeng Gao

Summary

The ability to analyze a large number of genetic markers consisting of single nucleotide polymorphisms (SNPs) may bring about significant advance in understanding human biology. Recent development of several high-throughput genotyping approaches has significantly facilitated large-scale SNP analysis. However, because of their relatively low sensitivity, application of these approaches, especially in studies involving a small amount of material, has been limited. In this chapter, detailed experimental procedures for a high-throughput and highly sensitive genotyping system are described. The system involves using computer program selected primers that are expected not to generate a significant amount of nonspecific products during PCR amplification. After PCR, a small aliquot of the PCR product is used as templates to generate single-stranded DNA (ssDNA). ssDNA sequences from different SNP loci are then resolved by hybridizing these sequences to the probes arrayed onto glass surface. The probes are designed in such a way that hybridizing to the ssDNA templates places their 3′-ends next to the polymorphic sites. Therefore, the probes can be labeled in an allele-specific way using fluorescently labeled dye terminators. The allelic states of the SNPs can then be determined by analyzing the amounts of different fluorescent colors incorporated to the corresponding probes. The genotyping system is highly accurate and capable of analyzing >1000 SNPs in individual haploid cells.

Key Words: High-throughput genotyping; single nucleotide polymorphism; multiplex PCR; microarray.

From: *Methods in Molecular Biology, vol. 396: Comparative Genomics, Volume 2*
Edited by: N. H. Bergman © Humana Press Inc., Totowa, NJ

1. Introduction

With the completion of the human genome project, millions of genetic markers consisting of single nucleotide polymorphisms (SNPs) have become available. Experimental systems capable of analyzing a large number of SNPs have become strongly demanded. In 1996, we reported a multiplex genotyping system for analyzing a large number of SNPs in a single assay by attaching universal sequences to the amplicons during PCR amplification and to convert the multiplex amplification to "singleplex." That allowed us to amplify 26 SNP-containing sequences in the same tube *(1)*. Very recently, several high-throughput genotyping methods have been developed by using the universal sequence strategy *(2–6)*. All these methods have a capacity of analyzing >1000 sequences in a single assay. However, because attaching universal sequences involves additional steps prior to or during PCR amplification, the detection sensitivity of these methods has been limited, and micrograms or submicrograms of genomic DNA are required for multiplexing >1000 SNPs in the same assay.

We have developed a high-throughput genotyping system without attaching universal sequences to the amplicons *(7)*. The system is based on using computer selected primers that unlikely generate nonspecific product during PCR amplification. After amplification, a small aliquot of the PCR product is used to generate single-stranded DNA (ssDNA). By hybridizing the ssDNA to the probes arrayed on a glass slide, sequences amplified from different SNP loci can be resolved. Allelic sequences from each SNP locus is then discriminated by the commonly used single-base extension assay *(8–12)* using corresponding dideoxyribonucleotide probes that can be allele – specifically labeled fluorescently with different cyanine dyes, Cy3 and Cy5. After scanning, the microarray image is digitized. Genotype of each locus in the sample is determined based on the signal intensities of the fluorescent colors in the respective spots.

2. Materials

2.1. Primer and Probe Selection

To select primers compatible with multiplex amplification, a computer program "Multiplex" is needed. The program is freely available for any noncommercial use from Dr. Honghua Li at holi@umdnj.edu. Users with commercial purpose may contact the Patent and Licensing Office, University of Medicine and Dentistry of New Jersey Robert Wood Johnson Medical School (732-235-9350) for sublicensing. (Note, the core technology including the program algorithm described in this chapter is covered by a pending PTC patent. Commercial use without licensing violates the international patent laws.)

2.2. Dissolving Oligonucleotides

1. TE buffer: 10 m*M* Tris-HCl, 1 m*M* EDTA. Adjust pH to 7.5 and autoclave at 125 °C and 19 psi for 1 h.
2. A dedicated laminar flow hood or a PCR workstation, free of DNA contamination.
3. A set of dedicated pipets, free of DNA contamination.

2.3. PCR Amplification and ssDNA Generation

1. 10× PCR buffer: 500 m*M* KCl, 1 *M* Tris-HCl, pH 8.3, 15 m*M* $MgCl_2$, 1 mg/mL gelatin. Autoclave at 125 °C and 19 psi for 1 h.
2. Solution containing dATP, dCTP, dGTP, and dTTP: 2 *M* each (Invitrogen, CA).
3. HotStart *Taq* DNA polymerase (Qiagen, CA),: usually 5 U/μL.
4. ddH_2O: autoclaved at 125 °C and 19 psi for 1 h (for dissolving all required reagents that cannot be autoclaved).
5. Solution containing all primers at a concentration of 100 n*M* each.
6. DNA solution: 1–2 ng/μL.

2.4. Microarray Preparation

1. EZ'nBrite microarray printing solution (GenScript, NJ), stored at −20 °C (*see* **Note 1**).
2. Glass slides for microarray printing (GenScript, NJ).
3. Oligonucleotide probes, 200 μ*M* each (*see* for detailed procedure for preparation in **Subheading 3.**).

2.5. Hybridization of ssDNA to Oligonucleotide Probes on a Microarray

1. Hybridization chamber (Corning).
2. LifterSlip cover slip (Eric Scientific Company)
3. 100× Denhardt's solution (Eppendorf).
4. 20× SSC: 3 *M* NaCl, 0.3 *M* sodium citrate, adjust the pH to 7.0; autoclave at 125 °C and 19 psi for 15 min.
5. 10% SDS: dissolve 10 g of SDS in 100 mL autoclaved ddH_20.

2.6. Probe Labeling

1. Sequenase (GE Healthcare Life Sciences) usually, 32 U/μL.
2. Sequenase buffer (supplied by the Sequenase vendor).
3. Dideoxyribonucleotides: fluorescent cyanine dye Cy3 or Cy5 labeled ddNTPs (PerkinElmer), usually 100 μ*M* each. Dissolve each solution with TE to 20 μ*M*. Which ones need to be purchased depending on the polymorphisms in the multiplex group.

3. Methods

3.1. Primer and Probe Selection

1. Extraction of SNP sequences: although automated procedures can be designed to extract SNP sequences from databases, it is difficult to design a universal computer program for this purpose because (1) users may have different database sources; and (2) the formats of most, if not all, databases are changed very frequently. Therefore, users need to develop their own method for this purpose (*see* **Note 2**).
2. Criteria for SNP selection. Because a large number of SNPs are available in the databases, and a major portion of the human genome is covered by a high SNP density, users often have choices to maximize the SNP quality. The following are the criteria that should be considered when selecting SNP:

   a. Heterozygosity. Heterozygosity is a measure of the fraction of heterozygous (or informative) individuals in a given human population. It should be pointed out that a large portion of SNPs may have different heterozygosities in different ethnic groups. Therefore, users should take the ethnicity of their subjects into consideration if the heterozygosity information is available.
   b. Availability of flanking sequences. If the available sequences are too short (<50 base), it may be difficult for primer selection. Usually 150 bases on each side of the polymorphic site are sufficient. The maximum length of the amplicons can be defined by users when running the program. However, users do not have to trim the input sequences. The program extracts the user defined length for each SNP automatically.
   c. Closely located SNPs. If a user chooses 150 (or some other number) bases on each side of the polymorphic site as the maximum length, he or she needs to make certain that there are no others known SNPs within the 300 bases. If additional SNPs are located within the 300 bases, they may be included into the selected primer sequences. In this case, they may compromise the genotyping accuracy.
   d. Repetitive sequences. Long stretches of repetitive sequences within the amplicon may affect primer design because primers with their 3′ bases located within these sequences may amplify these sequences located elsewhere in the genome and therefore, reduce the amplification specificity and the yield of the specific amplification. Usually SNPs with flanking sequences containing a stretch of >12 mononucleotide, >8 dinucleotides, >5 trinucloetides in the regions for primer selection should not be used. Users may write a small computer program to filter these sequences out.
   e. GC-content. Extreme GC-content may result in amplification difficulty. Usually GC content in a range from 25 to 75% is acceptable.
   f. Duplicated sequences. Because a large portion of the human genome contains duplicated or repetitive sequences, precaution should be taken to avoid selecting SNPs with these sequences in the flanking regions. This is because:

(1) if an "SNP" in a database was discovered by mistakenly treating the variation between duplicated sequences as allelic, the "SNP" may not be real; (2) if the sequences flanking an SNP site is duplicated, amplification of other copies of the duplicated sequences may cause genotyping error; and (3) when the duplicated sequence is amplified by the same set of primers and detected using the same probes, sequence variation located in the primer and probe regions may cause biased amplification or detection so that the final signal may be biased toward one of the alleles. Three computer programs may be used in a complementary way to learn whether a SNP-containing sequence is duplicated or is a repetitive sequence in the human genome: (1) BLAST on the server of National Center for Biological Information (NCBI, http://www.ncbi.nlm.nih.gov/BLAST), (2) BLAT at the website of University of California, Santa Cruz (http://www.genome.ucsc.edu/cgi-bin/hgBlat?db=hg8), and (3) RepeatMasker at the website of the Institute of Systems Biology (http://www.repeatmasker.org). SNPs located in the repetitive or duplicated sequences should be excluded to ensure the genotyping accuracy.

3. Primer selection. The computer program, MULTIPLEX, was written to select primers by taking a series of user defined parameters into consideration. These include:

   a. The range of primer lengths.
   b. The range of T_m of the primers.
   c. The maximal number (usually four) of allowed 3′ consecutive base matches between any two primers (including pair between different molecules of each primer).
   d. The maximal number (usually six) of allowed 3′ consecutive base matches with one mismatch between any two primers.
   e. The maximal number (usually seven) of allowed consecutive base matches between the 3′ bases of a primer and anywhere of other primers.
   f. The maximal number (usually nine) of allowed consecutive base matches with one mismatch between the 3′ bases of a primer and anywhere of other primers.
   g. The maximal allowed percentage (usually 75%) of matched bases between any primers.
   h. Amplicon length (150 bases or shorter) (*see* **Note 3**).

   The program comes with detailed instructions. It uses a Windows interface and is compatible with Windows XP and 2000.
4. Probe selection. Probe selection is different from primer selection. All probes have to have their 3′-ends next to the corresponding polymorphic sites. Probes can be designed in two directions, and used for genotyping in both directions for a high degree of accuracy. However, the criteria used for probe design may not be necessarily as stringent as those used for primers because (1) probes will not be used for amplification and therefore, may not have much effect even if a small

number of probe molecules hybridize to nonspecific templates; (2) since probes are immobilized to the glass surface, there should be no interaction among the probes. Under the hybridization conditions described next, the role of thumb for probe selection should be no stretch of consecutive matches longer than 14 bases between the bases from the 3′-end of the probes and any amplicon sequences. SNPs that cannot meet this requirement should be replaced.

3.2. Dissolving Oligonucleotides

1. Handle with care. Oligonucleotides synthesized as primers and probes need to be very carefully handled to avoid possible contamination (*see* **Note 4** for more details before handling).
2. Calculation of TE volume used to dissolve oligonucleotides to a desirable stock concentration (*see* **Note 5**). Both primers and probes need to be first dissolved to a stock concentration of 200 μM. Formula for converting OD260 units into μmole can be found at the website of Integrated DNA Technologies (http://www.idtdna.com/Analyzer/Applications/Instructions/Default.aspx?Analyer Definitions=true#MixedBaseDef).

$$\varepsilon_{260} = 2\,\mathrm{X}\,(\sum_{1}^{N-1}\varepsilon_{\mathrm{NearestNeighbor}}) - \sum_{2}^{N-1}\varepsilon_{\mathrm{Individual}}$$

where $\varepsilon_{\mathrm{NearestNeighbor}}$ is the extinction coefficient for a pair of neighboring bases, and $\varepsilon_{\mathrm{Individual}}$ is the extinction coefficient for an individual base, both of which can be found in **Table 1**; and N is the length of the oligonucleotide. (Note, 3′ and 5′ bases are not included in the second portion of the formula.) For example, an oligonucleotide "NU" has a sequence of GCA ACA GCT GCA TGC ATG CA, its OD_{260} units/μmole can be calculated as:

$$\begin{aligned} \mathrm{NU}\varepsilon_{260} &= 2\,\mathrm{X}\,(\mathrm{GC}\varepsilon_{260} + \mathrm{CA}\varepsilon_{260} + \mathrm{AA}\varepsilon_{260} + \ldots + \mathrm{TG}\varepsilon_{260} + \mathrm{GC}\varepsilon_{260} + \mathrm{CA}\varepsilon_{260}) \\ &\quad - (\mathrm{C}\varepsilon_{260} + \mathrm{A}\varepsilon_{260} + \mathrm{A}\varepsilon_{260} + \ldots + \mathrm{G}\varepsilon_{260} + \mathrm{C}\varepsilon_{260} + \mathrm{A}\varepsilon_{260}) \\ &= 2\,\mathrm{X}\,192.9 - 193.5 \\ &= 192.3\ (\mathrm{OD}_{260}\ \text{units}/\mu\text{mole}) \end{aligned}$$

If NU is measured to have 120 OD_{260} units, then

$$\text{Amount of NU in } \mu\text{mole} = 120\ \text{ODs}/192.3\ \mathrm{OD}_{260}\ \text{units}/\mu\text{mole} = 0.6240\ \mu\text{mole}$$

To dissolve it to a stock solution of 200 μM, the volume of TE can be calculated based on the equation of

Table 1
Extinction coefficient of single nucleotides and nearest neighbors

Nucleotide/ Dinucleotide	ε_{260}	Nucleotide/ Dinucleotide	ε_{260}
A	15.4	CG	9.0
C	7.4	CT	7.6
G	11.5	GA	12.6
T	8.7	GC	8.8
AA	13.7	GG	10.8
AC	10.6	GT	10.0
AG	12.5	TA	11.7
AT	11.4	TC	8.1
CA	10.6	TG	9.5
CC	7.3	TT	8.4

$$200\,\mu\text{mole}/1000\,\text{mL} = 0.6240\,\mu\text{mole}/\text{volume of TE}$$

Therefore,

$$\text{Volume of TE} = 0.6240\,\mu\text{mole} \cdot 1000\,\text{mL}/200\,\mu M$$
$$= 3.12\,\text{mL}$$

3. Preparation of oligonucleotide stock solution. Wear gloves at all time. Open the tubes containing oligonucleotides in a clean and dedicated laminar flow hood or PCR workstation. Use a dedicated and clean pipet to add a correct volume of the TE buffer as calculated in **step 2**. to a tube containing an oligonucleotide. Cover the tubes tightly and let it sit for 1 h in the hood. Vortex vigorously or shake the tubes in an up-and-down manner to ensure the oligonucleotide is dissolved uniformly. Incomplete vortexing may result in oligonucleotides dissolved in a gradient, which may cause experiment inconsistencies. Spin the tubes briefly in a microfuge, and store the stock solutions at −80 °C.
4. Prepare oligonucleotide mix. For each multiplex group, a large number of primers are needed. Therefore, it is necessary to prepare a solution containing all primers at a concentration convenient for later use. Because usually ~20 n*M* of each primer is used in multiplex PCR a solution containing 100 n*M* each primer will be convenient for use. The solution can be prepared by pooling an equal volume of all primer

stock solutions, when the total number of primers is 2000 (for 1000 SNPs). Less concentrated solution may also be prepared if the multiplex group exceeds 1000 SNPs, or in case the SNP number is less than 1000, autoclaved TE buffer may be added for a final concentration of 100 n*M* per primer.

3.3. PCR Amplification and ssDNA Generation

1. Preparation of PCR mix. PCR mix should be prepared in two separate locations. All components but DNA samples should be mixed under conditions that are free of DNA contamination, including wearing gloves all time, using clean and dedicated pipets, and manipulation in a DNA-free laminar flow hood or a PCR workstation. Mix the PCR components as follows:

3 µL	10× PCR buffer
3 µL	4-dNTP solution, 2 m*M* each
2 U/10 µL reaction	*Taq* DNA polymerase
6 µL	primer mix, 100 n*M* each
1 or 2 µL	DNA template

   Add ddH_2O to a total volume of 30 µL.
   Add the PCR mix to tubes containing DNA samples outside of the hood.
2. Multiplex PCR amplification. Use a PCR machine capable of ramping as slow as 0.01 °C/s. The following PCR machines have been used previously for this purpose: PTC100 Programmable Thermal Controller (MJ Research), T3 Thermocycler (Biometra), and PxE Thermal Cycler (Thermo Electron). Program the thermal profile as the following:
   a. Heat the sample to 94 °C for 15 min to activate the HotStart *Taq* DNA polymerase.
   b. Forty PCR cycles. Each cycle consists of 40 s at 94 °C for denaturation, 2 min at 55 °C, followed by 5 min of ramping from 55 to 70 °C for annealing and extension.
   c. After 40 cycles, 72 °C for 3 min to ensure completion of extension of all amplified strands.
3. Generation of ssDNA. ssDNA can be generated using the same conditions for multiplex PCR except: (1) 1–2 µL of the multiplex PCR product is used as templates, (2) one of the two probes in different directions is used as primers to enhance the specificity and yield; and (3) 45–50 cycles are used rather than 40. The correlation between primers and probes is illustrated in **Fig. 1** (*see* **Note 6**).

3.4. Microarray Preparation

1. Preparation of probes for microarray printing. Mix 1 vol of a probe (200 µ*M*) with 4 vol of microarray printing solution, EZ'nBrite, to a final concentration of 40 µ*M* in a well of a 384-well plate (*see* **Note 1**).

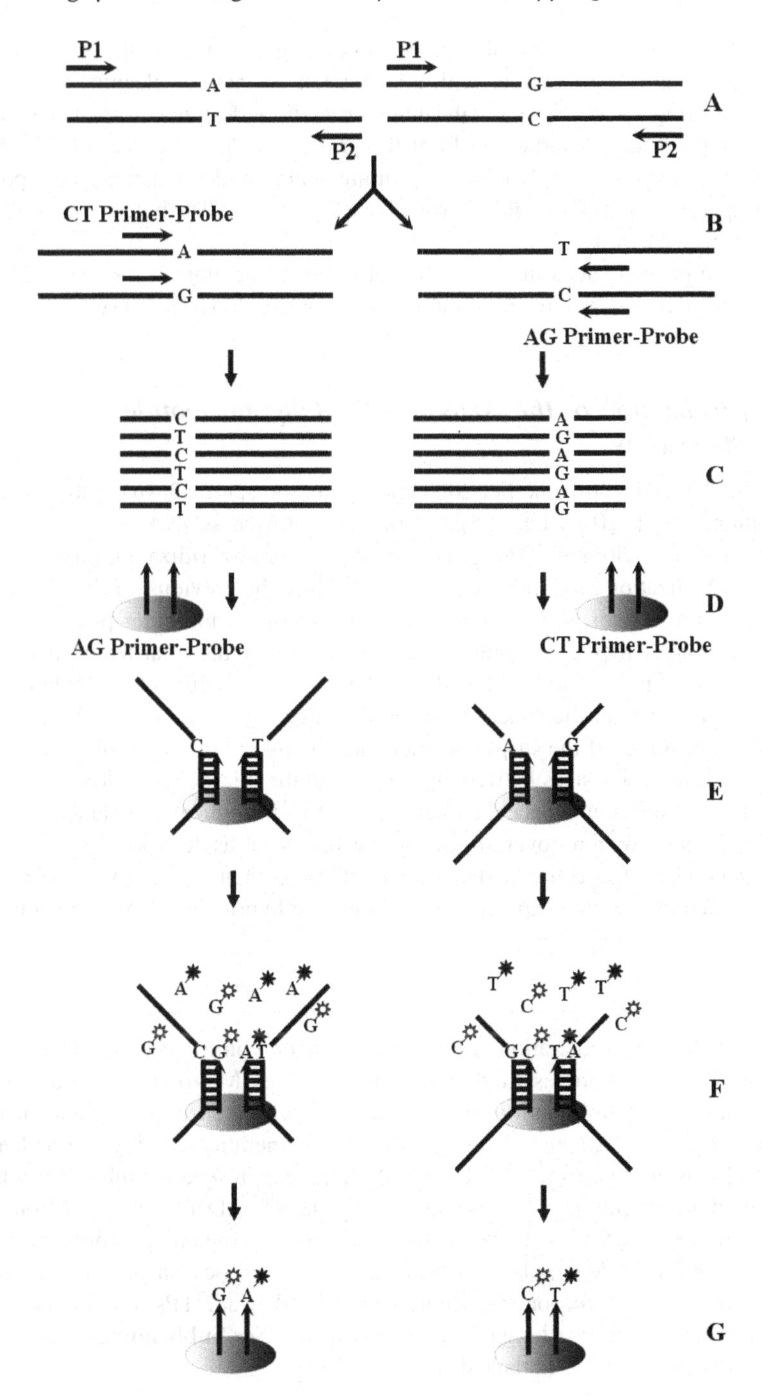
P1
P1
A
T
G
C
A
P2
P2
CT Primer-Probe
B
A
G
T
C
AG Primer-Probe
C
T
C
T
C
T
A
G
A
G
A
G
C
D
AG Primer-Probe
CT Primer-Probe
E
F
G

Fig. 1.

2. Microarray printing. Print the probes onto a glass slide with specially treated surface using a microarrayer with a sonication cleaner for cleaning the pins after each printing cycle. Program the microarray in a desirable format according to the manufacturer's instruction. Print the probes under a humidity of 50–55% and temperature of 22–25 °C. Check the printing quality under a microscope. Spots from high-quality printing should be round with granules distributed within the spots uniformly (*see* **Note 7**).
3. Store the printed slides in a clean box at room temperature for at least 12 h before use. Store the arrays in a vacuum desiccator for long-term use, up to at least 6 mo.

3.5. Hybridization of the ssDNA to the Oligonucleotide Probe Microarray

1. Prepare hybridization mix. For 2000 microarray spots, mix: 2.4 μL 100× Denhardt's solution, 1.5 μL 10% SDS, 7.5 μL 20× SSC, 18.6 μL ssDNA.
   Note, for an additional 1000 spots, add 15 μL more hybridization mix.
2. Add hybridization mix onto the microarray. Place the previously printed microarray slide into a hybridization chamber. Cut two strips of microtiter plate sealing film slightly longer than the length of the microarray (or the total length if more than one array is on the slide). The width of the strips is usually ~0.3 c*M*, but it should not be greater than the spaces between the edges of the slide and the microarray. Peel the backing off the strips and stick them along the two sides of the microarray, and leave narrow spaces between the strips and the microarray. (Note, wider spaces waste more solution.) Place the microarray into a hybridization chamber, and cover the array area with a cover slip using the film strip as support to leave a narrow space between the cover and the slide. If more than one array is on the slide, cover each area with a separate cover. Inject the hybridization mix to fill the space

Fig. 1. Schematic illustration of the multiplex genotyping procedure. Only one SNP is shown. Primers and probes are shown as arrowed lines. Microarray spots are indicated as ellipsoids. (**A**) Amplification of the polymorphic sequence. Two alleles using the same set of primers, P1 and P2, are shown. (**B**) Generation of single-stranded DNA (ssDNA) by using the primer-probes in both directions in separate tubes. Only the two allelic template strands in each reaction are shown; (**C**) ssDNA generated from **B**. (**D**) Addition of the ssDNA to the respective microarrays containing probes in different directions. (**E**) ssDNA templates hybridized to their probes on the microarrays. (**F**) Labeling probes by incorporating fluorescently labeled ddNTPs. (**G**) Labeled probes on the microarray after washing off all other reagents. *See* **Subheadings 3.3., 3.4., 3.5., and 3.6.** for more details. (Adapted from **ref. 7**.)

between the slide and the cover slip quickly to avoid trapping bubbles. Add clean H_2O to the two holes at the bottom ends of the chamber. Seal the chamber.

3. Hybridization. Incubate the hybridization chamber at 56 °C for 1 h in a water bath. Submerge the chambers in ice water for ~ 5 min before opening.
4. Wash the slide. Wash the slide with following the procedure:

   a. Remove the cover slip by rinsing the microarray in 1× SSC and 0.1% SDS at room temperature (note: attempt to peal off the cover may damage the array).
   b. Wash the slide at 56 °C for 10 min in a solution containing 1× SSC and 0.1% SDS, once.
   c. Rinse the slide at room temperature for 30 s in a solution containing 0.5× SSC, twice.
   d. Rinse the slide at room temperature for 30 s in a solution containing 0.2× SSC, twice.
   e. Dry the microarray slide by brief centrifugation. Place the microarray slide back in a hybridization chamber and cover the array with a new cover slip to prevent it from drying.

3.6. Probe Labeling (*see* Note 6)

1. Prepare labeling mix. For each 2000 spots, mix: 3.8 μL sequenase buffer, 10 U sequenase, 750 n*M* Cy3 and Cy5 labeled ddNTPs. Which ddNTPs should be used is based on the polymorphic sites in the multiplex group. For example, if only A · T to G · C transition SNPs are used and all probes are hybridized to the strands with A and G at the polymorphic sites, ddCTP and ddTTP should be used.
   Adjust the final volume to 25 μL with H_2O.
2. Labeling reaction. Perform the labeling reaction by using the procedure for hybridization except using the existing supporting strips and replace the hybridization mix by the labeling mix. Incubate the chamber at 70 °C for 10 min.
3. Wash and dry the slide using the procedure described in **Subheading 3.5., step 4**.

3.7. Microarray Scanning and Image Digitizing

1. Microarray scanning. Scan the microarray by following the scanner manufacturer's instructions.
2. Digitize microarray image. Digitize the resulting image using either GenePix (Axon Instruments) or ImaGene (BioDiscovery) software.
3. Genotype determination. Genotypes are determined after data normalization, background subtraction and low signal filtering. These can be done by using a computer program AccuTyping (freely available by email to holi@umdnj.edu). The program automatically carries out all previously listed processes with output files containing genotypes with comma separated values which can be opened by Microsoft Excel (*see* **Notes 8** and **9**).

4. Notes

1. Microarray printing solution. The microarray printing solution, EZ'nBrite, can be stored at −20 °C for 6 mo without losing it reactivity. However, its reactivity is reduced quickly when stored at room temperature (Gao and Li, unpublished data). Therefore, its exposure to room temperature should be minimized. To reuse the probes mixed with EZ'nBrite, the remaining portion should be place at 20 °C immediately after printing. Because EZ'Brite contains certain components that easily evaporate, plates containing probe-EZ'nBirte mix should not be left unnecessarily uncovered.
2. Using others' server for SNP sequence analysis. Manually submitting a large number of SNP-containing sequences to the servers for duplicate and repetitive sequence analyses is very tedious. Users may write scripts for automatic submission. However, automatic submission often increases the burden of the hosting servers. Publicly shared servers can be very slow. Therefore, if possible, users should take advantage of mirror sites. Both NCBI and USSC encourage mirror site installation and offer help.
3. Amplicon size. Higher multiplex efficiency and uniformity are observed when the amplicons are shorter than 150 bases. However, when the amplicons are too short, it becomes difficult for selecting desirable primers. Therefore, the lengths of the amplicons should be as short as possible only when an enough number of SNPs can be incorporated into the multiplex group.
4. Handling oligonucleotides to avoid contamination. Contamination can occur from many different sources. When an investigator talks over an open tube, his/her saliva droplets may fall into to the tube and a considerable number of cells may be carried in. If you use your hand with a glove to open a door, which is not a good habit, your glove may be contaminated with human DNA. Adding DNA samples to the PCR mix may contaminate the pipet. Multiplex PCR with high sensitivity is also sensitive to contamination. Therefore, investigator should take extra precaution to protect their stock solutions and PCR samples from contamination.
5. Oligonucleotide concentration in molarity is a better expression than weight/volume. Because oligonucleotides may have different lengths or molecular weights, the number of molecules at the same concentration in weight/volume may be significantly different. Using the estimated relation that is 33 μg/OD_{260} is not accurate because different bases have different absorbance. The correct calculation should be converted OD_{260} units into μmoles and the oligonucleotides should be dissolved in molarity.
6. Nucleotide sequence complementarity. Genotyping requires clear mind about the nucleotide sequence complementarity, which often confuses new or even experienced investigators. When confused, disappointing results may occur. For example, when a ssDNA is generated with one of the probes as primers, it should be hybridized to the microarray containing in the other direction probe. Hybridizing to the array consisting of the oligonucleotide used as primers will result in no

signal. For the same reason, the ddNTPs used for labeling need also to be figured out carefully. If the template contains base substitutions of A and G, ddCTP and ddTTP should be used.

7. Storage of microarrays. It was found that the microarrays can be stored under a vacuum for at least 6 mo. Better quality may be obtained after 1 wk of storage than the shorter stored ones, probably because of completion of the reaction for covalent attachment of the oligonucleotide probes to the glass surface.
8. Two direction genotyping. Because DNA has two strands, an SNP can be genotyped in two directions. The results can be compared to ensure a high accuracy. In this case, no additional cost is needed for oligonucleotide synthesize because a probe complementary to one strand can be used as a primer for generating ssDNA for another probe.
9. Multiplex genotyping involving a large number of SNPs. Computer programs should be written for many steps involved in the data analysis. In this way, data can be analyzed in a highly efficient way.

Acknowledgments

The work was supported by grants R01 HG02094 from the National Human Genome Research Institute, and R01 CA077363 and R33 CA96309 from the National Cancer Institute, National Institutes of Health to H.L.

References

1. Lin, Z., Cui, X., and Li, H. (1996) Multiplex genotype determination at a large number of gene loci. *Proc. Natl. Acad. Sci. USA* **93,** 2582–2587.
2. Yeakley, J. M., Fan, J. B., Doucet, D., et al. (2002) Profiling alternative splicing on fiber-optic arrays. *Nat. Biotechnol.* **20,** 353–358.
3. Hardenbol, P., Baner, J., Jain, M., et al. (2003) Multiplexed genotyping with sequence-tagged molecular inversion probes. *Nat. Biotechnol.* **21,** 673–678.
4. Kennedy, G. C., Matsuzaki, H., Dong, S., et al. (2003) Large-scale genotyping of complex DNA. *Nat. Biotechnol.* **21,** 1233–1237.
5. Fan, J. -B., Oliphant, A., Shen, R., et al. (2003) Highly parallel SNP genotyping. *Cold Spring Harb. Symp. Quant. Biol.* **68,** 69–78.
6. Matsuzaki, H., Loi, H., Dong, S., et al. (2004) Parallel genotyping of over 10,000 SNPs using a one-primer assay on a high-density oligonucleotide array. *Genome Res.* **14,** 414–425.
7. Wang, H. Y., Luo, M., Tereshchenko, I. V., et al. (2005) A genotyping system capable of simultaneously analyzing $>$ 1000 single nucleotide polymorphisms in a haploid genome. *Genome Res.* **15,** 276–283.
8. Shumaker, J. M., Metspalu, A., and Caskey, C. T. (1996) Mutation detection by solid phase primer extension. *Hum. Mutat.* **7,** 346–354.

9. Pastinen, T., Kurg, A., Metspalu, A., Peltonen, L., and Syvanen, A. C. (1997) Minisequencing: a specific tool for DNA analysis and diagnostics on oligonucleotide arrays. *Genome Res.* **7,** 606–614.
10. Syvanen, A. C. (1999) From gels to chips: "minisequencing" primer extension for analysis of point mutations and single nucleotide polymorphisms. *Hum. Mutat.* **13,** 1–10.
11. Lindblad-Toh, K., Tanenbaum, D. M., Daly, M. J., et al. (2000) Loss-of-heterozygosity analysis of small-cell lung carcinomas using single-nucleotide polymorphism arrays. *Nat. Biotechnol.* **18,** 1001–1005.
12. Pastinen, T., Raitio, M., Lindroos, K., Tainola, P., Peltonen, L., and Syvanen, A. C. (2000) A system for specific, high-throughput genotyping by allele-specific primer extension on microarrays. *Genome Res.* **10,** 1031–1042.

19

Single Nucleotide Polymorphism Mapping Array Assay

Xiaofeng Zhou and David T. W. Wong

Summary

Single nucleotide polymorphisms (SNPs) are the most frequent form of DNA variation present in the human genome, and millions of SNPs have been identified (http://www.ncbi.nlm.nih.gov/SNP/). Because of their abundance, even spacing, and stability across the genome, SNPs have significant advantages over other genetic markers (such as restriction fragment length polymorphisms and microsatellite markers) as a basis for high-resolution whole genome allelotyping. SNP scoring is easily automated and high-density oligonucleotide arrays have recently been generated to support large-scale high throughput SNP analysis. High-density SNP allele arrays have improved significantly and it is now possible to genotype hundreds of thousands SNP markers using a single SNP array. In this chapter, we will provide a detailed experimental protocol of Affymetrix GeneChip SNP Mapping Array-based whole genome SNP genotyping assay.

Key Words: Single nucleotide polymorphism; SNP; SNP array; SNP genotyping; loss of heterozygosity; copy number abnormality; Affymetrix GeneChip.

1. Introduction

The completion of the human genome project has allowed for the identification of millions of single nucleotide polymorphisms (SNP) loci (http://www.ncbi.nlm.nih.gov/SNP/), which makes them ideal markers for various genetic analyzes *(1)*. Because of their abundance, even spacing, and stability across the genome, SNPs have significant advantages over RFLPs and microsatellite markers as a basis for high-resolution whole genome allelotyping with accurate copy number measurements. High-density oligonucleotide

From: *Methods in Molecular Biology, vol. 396: Comparative Genomics, Volume 2*
Edited by: N. H. Bergman © Humana Press Inc., Totowa, NJ

arrays have recently been generated to support large-scale high throughput SNP analysis *(2)*. It is now possible to genotype hundreds of thousands SNP markers using a single GeneChip Human Mapping Array. The current available Affymetrix SNP array platforms include GeneChip Mapping 10K, GeneChip Mapping 50K Array Xba 240, GeneChip Human Mapping 50K Array Hind 240, GeneChip Mapping 250K Nsp Array and GeneChip Mapping 250K Sty Array. These platforms were initially designed for linkage analyses and case-control, family-based association studies *(3)*, but because of its unique design, high resolution, and user-friendliness, geneticists have quickly adapted them for many different genetic applications, such as loss of heterozygosity (LOH) and copy number abnormality (CNA) analyses *(4–6)*. In this chapter, we will provide a detailed experimental protocol of SNP array-based genomic mapping assay using GeneChip Mapping 50K Array Xba 240. The experimental procedure described here can be easily adapted to the other Affymetrix SNP array platforms.

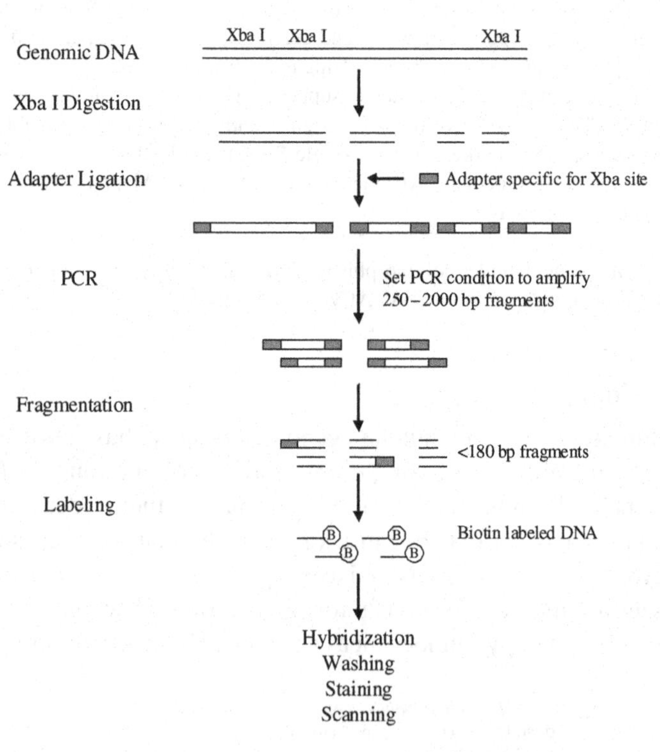

Fig. 1. Outline of the GeneChip mapping assay: *Xba*I.

The SNP array-based genotyping assay method devised for the Affymetrix GeneChip Mapping Assay is outlined in **Fig. 1**. About 250 ng of genomic DNA is digested with a specific restriction enzyme (Xba I for the GeneChip Mapping 50K Array Xba 240 array-based assay) and ligated to a specific adaptor recognizing the cohesive overhangs created by restriction enzyme digestion. All fragments resulting from restriction enzyme digestion, regardless of size, are substrates for adaptor ligation. A generic primer, which recognizes the adaptor sequence, is used to amplify the ligated DNA fragments using the PCR conditions that optimized to preferentially amplify fragments in the 250- to 1000-bp size range. The amplified DNA is then purified, fragmented to appropriate size (>180 bp), labeled and hybridized to GeneChip arrays. The arrays are washed and stained on a GeneChip fluidics station and scanned on a GeneChip Scanner 3000. The scanned array images are then analyzed using analytical software, such as GeneChip Genotyping Analysis Software (GTYPE) to reveal the allelic composition and produce the genotype calls for each SNP markers.

This procedure described here is adapted from the GeneChip Mapping 100K Assay Manual. The manual should also be consulted for any questions about the protocol (www.affymetrix.com).

2. Materials

2.1. Reagents Supplied by Affymetrix Affymetrix GeneChip Mapping 50K Xba Assay Kit P/N 900520

This kit includes sufficient reagent for 30 arrays. The items included in the kit are listed below:

1. Adaptor, Xba: 38 µL of 5 µ*M* annealed oligonucleotides (*see* below), specific for ligation to Xba restriction site.

   5' ATTATGAGCACGACAGACGCCTGATCT 3'

   3' AATACTCGTGCTGTCTGCGGACTAGAGATCT 5'
2. PCR Primer, 001: 1125 µL 10 µ*M* PCR primer (*see* below), to amplify ligated genomic DNA.

   5' ATTATGAGCACGACAGACGCCTGATCT 3'
3. Reference Genomic DNA, 103: 25 µL of 50 ng/µL human genomic DNA (single source).
4. GeneChip Fragmentation Reagent (20 µL): DNase I enzyme, formulated to fragment purified PCR amplicons.
5. 10X Fragmentation Buffer: 225 µL of 10X Buffer for fragmentation reaction
6. GeneChip DNA Labeling Reagent: 60 µL of 7.5 m*M* Biotin-labeled reagent for end-labeling fragmented PCR amplicons.

7. Terminal deoxynucleotidyl transferase (TdT): 105 μL of 30 U/μL TdT enzyme used to end-label fragmented PCR amplicons with the GeneChip DNA Labeling Reagent (7.5 m*M*).
8. 5X TdT Buffer: 420 μL of 5X Buffer for labeling reaction.
9. Oligo Control Reagent (60 μL).

2.2. Required Equipment and Software From Affymetrix, Inc.

1. GeneChip Fluidics Station 450 (or GeneChip Fluidics Station 400/250).
2. GeneChip Hybridization oven.
3. GeneChip Scanner 3000 with high-resolution scanning patch.
4. Affymetrix GeneChip Operating Software (GCOS).
5. Affymetrix GTYPE.
6. GeneChip Human Mapping 50K Array Xba 240.

2.3. Restriction Enzyme (XbaI) Digestion

1. Reduced ethylenediamine tetraacetic acid (EDTA) TE Buffer (TEKnova): 10 m*M* Tris-HCl, 0.1 m*M* EDTA, pH 8.0.
2. 20,000 U/mL *Xba*I (New England Biolab) also containing: 10X NE Buffer 2 and 100X: bovine serum albumin.
3. H_2O (BioWhittaker Molecular Applications/Cambrex): Molecular Biology Water.
4. 8-Tube strips, thin-wall (0.2 mL) and strip of 8 caps (MJ Research).
5. Thermal cycler.

2.4. Ligation

1. T4 DNA Ligase (New England Biolab), also containing T4 DNA Ligase Buffer.
2. Thermal cycler.

2.5. PCR Amplification

1. Platinum Pfx DNA Polymerase (Invitrogen Corporation): 2.5 U/μL, also containing 10X Pfx Amplification Buffer, 10X PCRx Enhancer, and 50 m*M* $MgSO_4$.
2. 2.5 m*M* Each of dNTP (Takara).
3. 2% TBE Gel (SeaKem Gold): BMA Reliant precast (2%).
4. All Purpose Hi-Lo DNA Marker (Bionexus).
5. Thermal cycler.

2.6. PCR Product Purification and Elution

1. Manifold – QIAvac multiwell unit (QIAGEN).
2. MinElute 96 UF PCR Purification Kit (QIAGEN).
3. Buffer EB (QIAGEN).
4. Vacuum Regulator to be used with QIAvac manifolds. The QIAGEN protocol requires ~800-mb vacuum.

2.7. Fragmentation and Labeling

1. 4% TBE Gel (4% NuSieve 3:1 Plus Agarose): BMA Reliant precast.
2. Thermal Cycler.

2.8. Hybridization

1. 5 *M* Tetramethyl ammonium chloride (Sigma).
2. MES hydrate SigmaUltra (Sigma).
3. MES sodium salt (Sigma).
4. DMSO (Sigma).
5. 0.5 *M* EDTA (Ambion).
6. Denhardt's solution (Sigma).
7. Herring sperm DNA (Promega).
8. Human Cot-1 (Invitrogen).
9. 20X SSPE (BioWhittaker Molecular Applications/Cambrex).
10. 10% Tween-20 (Pierce): Surfactamps.
11. 1 mg/mL Streptavidin, R-phycoerythrin conjugate (Molecular Probes).
12. 0.5 mg Biotinylated Anti-Streptividin (Vector).

3. Methods

This procedure is adapted from the Affymetrix 100K genotyping manual. The manual should also be consulted for any questions about the protocol.

3.1. XbaI Digestion

As this procedure processes DNA before PCR, precautions regarding contamination with template (human) DNA should be stringently observed (*see* **Note 1**).

1. Adjust the concentration of the target human genomic DNA to 50 ng/μL using reduced EDTA TE buffer (0.1 m*M* EDTA, 10 m*M* Tris-HCl, pH 8.0) in PCR staging room (*see* **Note 1**). *See* **Note 2** for the requirement of the target DNA.
2. Prepare digestion Master Mix (make 10% extra): this step should be performed in a pre-PCR clean room or a PCR clean hood, where no template DNA is ever present (*see* **Note 1**). Make the mix as listed in **Table 1**. Vortex and spin briefly and keep on ice.
3. Add 15 μL of the digestion Master Mix to 5 μL genomic DNA (50 ng/μL) in PCR tubes. Vortex and spin briefly.
4. Perform digestion in thermal cycler in the PCR staging room using following program: 37°C 2 h, 70°C 20 min, 4°C hold).
5. When the run is done, let it go down to 4°C and hold there for at least 2 min. Then, either proceed to ligation or store samples at −20°C.

Table 1
***Xba*I Digestion Mix**[a]

Water	10.5 μL
NE Buffer 2	2 μL
10X BSA[b]	2 μL
*Xba*I	0.5 μL
Total	15 μL

[a] Amounts for one reaction.
[b] BSA comes as 100X solution from New England Biolab. Diluted to 10X with H_2O before use.

3.2. Ligation

As this procedure processes DNA before PCR, precautions regarding contamination with template (human) DNA should be stringently observed (*see* **Note 1**).

1. Prepare the ligation Master Mix as listed in **Table 2** for number of samples that were digested plus 10% extra (to be performed in the pre-PCR clean room or a PCR clean hood, where no template DNA is ever present, *see* **Note 1**).
2. Aliquot 5 μL of the ligation Master Mix to each tube that digestion was performed in (final volume in the PCR tube is 25 μL). This step should be performed in the PCR staging room.
3. Perform ligation reaction in the thermal cycler in the PCR staging room using following program: 16°C 2 h, 70°C 20 min, 4°C hold.
4. When finished, proceed to PCR amplification or store sample at −20°C.

Table 2
Ligation Master Mix[a]

Water	0.625 μL
T4 DNA ligase buffer[b]	2.5 μL
Adaptor (*Xba* or *Hind*)	1.25 μL
T4 DNA ligase	0.625 μL
Total	5 μL

[a] Amounts for one reaction.
[b] Thaw T4 DNA ligase buffer on ice for at least 1–1.5 h. If digestion was done on the same day, you can start thawing the buffer when you set up the digestion.

3.3. PCR Amplification

Three PCR amplification reactions are needed for each ligation product (sample) to produce sufficient product for array hybridization.

3.3.1. Prepare PCR Master Mix (in Pre-PCR Clean Room)

1. Take the Pfx polymerase and PCR Enhancer out of the freezer and put on ice.
2. Vortex all of the reagents that were frozen and spin briefly to ensure they are well mixed.
3. Prepare PCR Master Mix as listed in **Table 3** for the number of samples that were digested three times, plus 5 % extra.

3.3.2. Set Up PCR Reaction (in the PCR Staging Room)

1. Vortex the master mix on the vortexer in the hood to make sure it is well mixed. Store on ice.
2. Add 75 μL water to each tube in which ligation was performed. Mix by pipetting up and down several times.
3. For each ligation reaction (sample), aliquot 90 μL of the PCR Master Mix to three PCR tubes.
4. Aliquot 10 μL of diluted ligation into the appropriate PCR tubes. For the PCR negative control, use 10 μL water.
5. Cover all tubes with strip caps, vortex, and spin briefly.

3.3.3. PCR Reaction (in the Mail Lab)

1. Place the tubes in the PCR thermal cycler in the Main Lab. Recommended model MJ Tetrad PTC-225 96-well block (Bio-Rad), or ABI GeneAmp PCR System 9700–96 well block (Applied Biosystems).

Table 3
PCR Master Mix[a]

Water	44 μL
PCR amplification buffer (10X)	10 μL
PCR enhancer (10X)	10 μL
MgSO4 (50 m*M*)	2 μL
dNTP (2.5 m*M* each)	12 μL
PCR primer (10 μ*M*)	10 μL
Pfx polymerase (2.5 U/μL)	2 μL
Total	90 μL

[a] Amounts for one reaction.

Table 4
PCR Programs for Amplification

MJ thermal cycler			ABI GeneAmp thermal cycler		
Temperature	Time	Cycles	Temperature	Time	Cycles
94°C	3 min	1X	94°C	3 min	1X
94°C	15 s		94°C	30 s	
60°C	30 s	30X	60°C	45 s	30X
68°C	60 s		68°C	60 s	
68°C	7 min	1X	68°C	7 min	1X
4°C	HOLD		4°C	HOLD	

2. Run following PCR program as listed in **Table 4**.
3. When the run is done, vortex the tubes, and spin briefly.
4. Check the PCR amplification product by run 3 μL of each PCR product on a 2% TBE agarose gel, at 120 V for approx 1 h. The PCR products should appear as smear on the gel, with average size between 250 and 2000 bp (**Fig. 2**).

3.4. Purification of the PCR Amplification Product

1. Connect a vacuum manifold to a suitable vacuum source able to maintain ~800 mbar, e.g., QIAvac 96 or QIAvac Multiwell Unit (QIAGEN). Place a waste tray inside the base of the manifold.

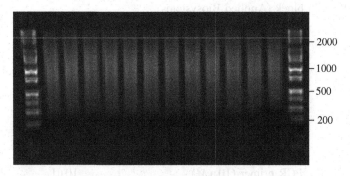

Fig. 2. Sample image of the PCR amplification product run on a 2% TBE agarose gel. To check the PCR amplification step, 3 μL of each PCR product should be run on a 2% TBE agarose gel, at 120 V for 1 h. The PCR products should appear as smear with average size between 250 and 2000 bp.

2. Place a MinElute 96 UF PCR Purification Plate on top of the manifold. Cover the entire plate with the aluminum cover. Cut off a strip for the wells you will use. Note: Mark the row with a marker so that we know it has been used.
3. Turn on the vacuum. Adjust the regulator to 800–900 mbar.
4. Consolidate the PCR product for each sample (three PCR reactions) into one well of the MinElute plate.
5. Maintain the ~800 mbar vacuum until the wells are completely dry.
6. Wash the PCR products by adding 50 μL molecular biology water and dry the wells completely (taking about 20 min). Repeat this step two more times for a total of three washes.
7. Switch off vacuum source and release the vacuum.
8. Carefully remove the MinElute plate from the vacuum manifold.
9. Gently tap the MinElute plate on a stack of clean absorbent paper to remove any liquid that might remain on the bottom of the plate.
10. Add 50 μL EB buffer to each well. Cover the plate with PCR plate cover film.
11. Moderately shake the MinElute plate on a plate shaker, e.g., Jitterbug (Boekel Scientific, cat. no. 130000), for 15 min.
12. Recover the purified PCR product by pipetting the eluate out of each well. For easier recovery of the eluates, the plate can be held at a slight angle. Make sure you take 50 μL off each well.
13. Take a 4-μL aliquot to measure the concentration of the purified PCR product using spectrophotometer.
14. Adjust the DNA concentration to 40 μg of per 45 μL solution using EB buffer (10 m*M* Tris-Cl, pH 8.5).
15. Transfer 45 μL of each purified DNA (40 μg) to new PCR tubes for the fragmentation reaction.

3.5. Fragmentation (to be performed in the Main Lab)

1. Take out the fragmentation buffer, labeling reagent, and TdT buffer and thaw on ice.
2. Preheat the PCR thermal cycler to 37°C.
3. Aliquot 45 μL of each sample into fresh PCR tubes. Put on ice.
4. Add 5 μL 10X Fragmentation Buffer to each sample in the tubes on ice and vortex at medium speed for 2 s (with tubes covered).
5. Get fragmentation reagent from the freezer and keep in enzyme block.
6. Prepare a dilution of the fragmentation reagent as listed in **Table 5**, based on the enzyme concentration on the stock tube.
7. Add 5 μL of diluted Fragmentation Reagent (0.04 U/μL) to each sample tube on ice. Pipet up and down several times to mix. Perform this step quickly.

Table 5
Fragmentation Reagent

Fragmentation reagent stock concentration	2 U/μL	3 U/μL
Fragmentation reagent	3 μL	2 μL
10X Fragmentation buffer	15 μL	15 μL
Water	132 μL	133 μL
Total	150 μL	150 μL

8. Cover the tubes with strip caps and seal tightly.
9. Vortex the tubes at medium speed for 2 s, and spin briefly.
10. Place the fragmentation plate in preheated thermal cycler (37°C) as quickly as possible.
11. Run the following program: 37°C 35 min, 95°C 15 min, 4°C hold.
12. After the run, spin the tubes briefly.
13. Check the fragmentation product by run 4 μL of each fragmentation product on 4 % TBE agarose gel, at 120 V for approx 1 h. The product size should <180 bp (**Fig. 3**). The recommended gel: 4 % TBE Gel, BMA Reliant precast.

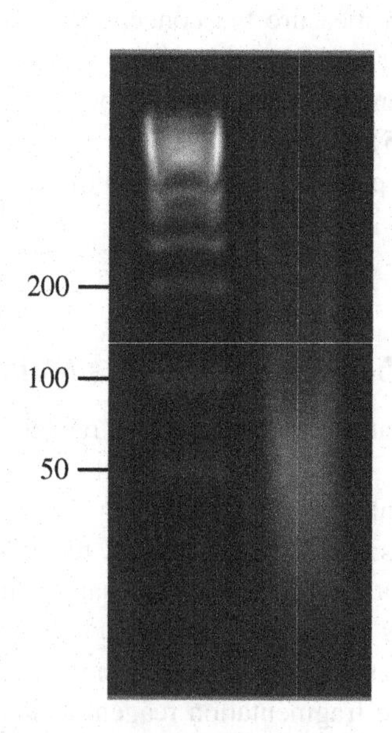

Fig. 3. Sample image of the fragmentation product run on a 4 % TBE agarose gel. To check the fragmentation product, 4 μL of each reaction should be run on 4 % TBE agarose gel, at 120 V for approx 1 h. The product size should <180 bp.

Table 6
Labeling Mix[a]

5X TdT buffer	14 µL
GeneChip DNA labeling reagent	2 µL
TdT (30 U/µL)	3.5 µL
Total	19.5 µL

[a] Amounts for one reaction.

3.6. Labeling

1. Prepare Labeling Mix as master mix on ice as listed in **Table 6** and vortex at medium speed for 2 s. Prepare the mix for the number of samples you have plus 1.
2. Aliquot 19.5 µL of Labeling Master Mix into each tube containing 50.5 µL of fragmented DNA.
3. Cover the tubes with caps.
4. Vortex and spin briefly.
5. Put on the thermal cycler and run the following program (37°C 2 h, 95°C 15 min, 4°C hold).
6. Briefly spin the tubes when the program is done. The samples can be stored at −20°C if not proceeding to hybridization immediately.

3.7. Array Hybridization

1. Prepare the Hybridization Cocktail Master Mix as listed in **Table 7**. Make 10% extra.
2. Aliquot 190 µL into Eppendorf tubes.

Table 7
Hybridization Cocktail Master Mix[a]

12X MES	12 µL
100% DMSO	13 µL
50X Denhardt's solution	13 µL
0.5 *M* EDTA	3 µL
10 mg/mL hsDNA	3 µL
Oligonucleotide control	2 µL
Human Cot-1 (1 mg/mL)	3 µL
Tween-20 (3%)	1 µL
5 *M* TMACL	140 µL
Total	190 µL

[a] Amount for one sample.

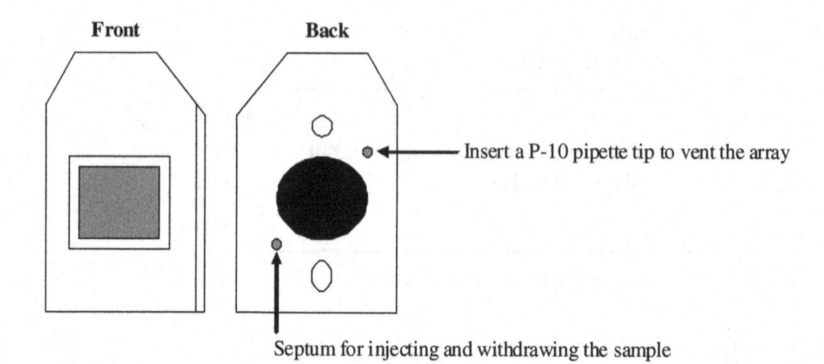

Fig. 4. Illustration of loading the sample into GeneChip array cartridge.

3. Add 70 μL of labeled DNA into each tube. These samples can be used immediately or stored at −20°C.
4. Heat the 260 μL of hybridization mix and labeled DNA at 95°C in a heat block for 10 min to denature.
5. While it is heating, label the appropriate chips. Make sure to use the correct type of array(s). In this chapter, we will use GeneChip Mapping 50K Array Xba 240.
6. Cool down on ice for 10 s, and spin briefly in a microfuge.
7. Place the tubes at 48°C for 2 min.
8. Insert a fresh, P10 pipet tip into the upper septum on the array to vent it. Inject 200 μL denatured hybridization into the array using an appropriate pipet. *See* **Fig. 4** for illustration.
9. Hybridize overnight at 48°C for 16–18 h at 60 rpm in the GeneChip Hybridization oven.
10. After hybridization, remove the hybridization cocktail from the probe array and set it aside in a microcentrifuge vial. Store at −80°C for long-term storage.
11. Fill the probe array completely with 250 μL of Array Holding Buffer (*see* **Subheading 3.9.3.** for making the Array Holding Buffer). If necessary, the probe array can be stored in the Array Holding Buffer at 4°C for up to 3 h before proceeding with washing and staining. Equilibrate the probe array to room temperature before washing and staining.

3.8. Array Washing and Staining

This section contains protocol of using the GeneChip Fluidics Station 450 to automate the washing and staining of GeneChip Mapping 50K Arrays. The Fluidics Station is controlled by GeneChip Operating Software (GCOS). Please refer to the Fluidics Station User's Guide and the GCOS Instruction for the detailed instructions of installation, operation, maintenance of the Fluidics Station.

Table 8
Wash A: Nonstringent Wash Buffer (6X SSPE, 0.01 % Tween-20)

20X SSPE	300 mL
10 % Tween-20	1.0 mL
Water	699 mL
Total	1000 mL

Filter the buffer through a 0.2-μm filter, and store at room temperature.

Before start, register the experiment in the GCOS as New Experiment following the instruction of GCOS. Enter the information into the appropriate boxes. Then save the experiment in the database.

1. Prepare the Washing Buffer A and Washing Buffer B as listed in **Tables 8** and **9**, respectively.
2. Prepare the following stock solutions that needed for making the buffers for staining.
   a. 0.5 mg/mL Anti-Streptavdin antibody: resuspend 0.5 mg anti-Streptavdin antibody in 1 mL of water. Store at 4°C.
   b. 12X MES Stock Buffer: make the 12X MES Stock Buffer (1.22 *M* MES, 0.89 *M* [Na+]) as listed in **Table 10**. Filter the buffer through a 0.2-μm filter. Do not autoclave. Store at 2 – 8°C, and shield from light. Discard the buffer if it turns to yellow.
3. Prepare the Stain Buffer (**Table 11**), Streptavidin phycoerythrin Solution Mix (**Table 12**), Antibody Solution Mix (**Table 13**), and 1X Array Holding Buffer (**Table 14**) for array staining.

Table 9
Wash B: Stringent Wash Buffer (0.6X SSPE, 0.01 % Tween-20)

20X SSPE	30 mL
10 % Tween-20	1.0 mL
Water	969 mL
Total	1000 mL

Filter the buffer through a 0.2-μm filter and store at room temperature.

Table 10
12X MES Stock Buffer

MES hydrate	70.4 g
MES sodium salt	193.3 g
Water	800 mL

Mix and adjust volume to 1000 mL. The pH should be between 6.5 and 6.7.

4. Priming the fluidics station: change the intake buffer reservoir A to Non-Stringent Wash Buffer, and intake buffer reservoir B to Stringent Wash Buffer. Run the Prime program. Follow the instructions on the LCD.
5. In the Fluidics Station dialog box on the workstation, select the correct experiment name from the drop-down Experiment list. The Probe Array Type appears automatically.
6. In the Protocol drop-down list, select Mapping100Kv1_450 (**Table 15**), to control the washing and staining of the probe array.
7. Choose Run in the Fluidics Station dialog box to begin the washing and staining. Follow the instructions in the LCD window on the fluidics station.
8. Insert the appropriate probe array into the designated module of the fluidics station while the cartridge lever is in the down or eject position.
   When finished, verify that the cartridge lever is returned to the up or engaged position.
9. When prompted to "Load Vials 1-2-3," place the three vials into the sample holders 1, 2, and 3 on the fluidics station.

   a. Place one vial containing 500 μL streptavidin phycoerythrin stain solution mix in sample holder 1.
   b. Place one vial containing 500 μL anti-streptavidin biotinylated antibody stain solution in sample holder 2.
   c. Place one vial containing 800 μL Array Holding Buffer in sample holder 3.

Table 11
Stain Buffer

Water	666.7 μL
SSPE (20X)	300 μL
Tween-20 (3 %)	3.3 μL
Denhardt's (50X)	20 μL
Total (volume needed for one array)	990 μL

Table 12
Streptavidin Phycoerythrin Solution Mix

Stain buffer	495 μL
1 mg/mL Streptavidin phycoerythrin	5.0 μL
Total	500 μL

d. Press down on the needle lever to snap needles into position and to start the run.

Once these steps are complete, the fluidics protocols begin. The Fluidics Station dialog box at the workstation terminal and the LCD window displays the status of the washing and staining steps.

10. When staining is finished, remove the microcentrifuge vials containing stain and replace with three empty microcentrifuge vials as prompted.
11. Remove the probe arrays from the fluidics station modules by first pressing down the cartridge lever to the eject position.
12. Check the probe array window for large bubbles or air pockets.

    a. If bubbles are present, the probe array should be filled with Array: Holding Buffer manually, using a pipet. Take out one-half of the solution and then manually fill the probe array with Array Holding Buffer.
    b. If the probe array has no large bubbles, it is ready to scan on the GeneChip Scanner 3000. Pull up on the cartridge lever to engage wash block and proceed to array scan.

13. Shut down the Fluidics Station by running the ShutDown protocol.

3.9. Array Scanning

This section contains protocol of using the GeneChip Scanner 3000 to scan the probe arrays. The scanner is also controlled by GCOS. Please consult appropriate manuals for the detailed instructions of installation, operation, maintenance of the GeneChip Scanner.

Table 13
Antibody Solution Mix

Stain buffer	495 μL
0.5 mg/mL Biotinylated antibody	5.0 μL
Total	500 μL

Table 14
1X Array Holding Buffer (100 mM MES, 1 M [Na+], 0.01 % Tween-20)[a]

12X MES Stock buffer	8.3 mL
5 *M* NaCl	18.5 mL
10 % Tween-20	0.1 mL
Water	73.1 mL
Total	100 mL

[a] Store at 2 – 8°C, and shield from light.

Make sure the laser is warmed up prior to scanning by turning the GeneChip Scanner 3000 laser on at least 10 min before use. If the array was stored at 4°C, allow to warm to room temperature before scanning. After completing the procedures described in this chapter, the scanned probe array image (.dat file) is ready for analysis.

1. Go to the scanner window in the GCOS. Select the experiment that registered at the beginning of the array washing.
2. Click Load/Eject button on GCOS to load the array to be scanned. Load the array by open the sample door of the scanner and insert the array into the holder.
3. Click Start. Then click the OK in the Start Scanner dialog box to start the scanning. The scanned array image will appear on the screen as scan progresses.

3.10. Data Analysis

The raw data (scanned array image, which is saved as .dat file) acquired using the GCOS will be analyzed using GTYPE to generate genotyping calls

Table 15
FS-450 Fluidics Protocol - Mapping100Kv1_450

Post Hyb Wash no. 1	6 cycles of 5 mixes/cycle with Wash Buffer A at 25°C
Post Hyb Wash no. 2	6 cycles of 5 mixes/cycle with Wash Buffer B at 45°C
Stain	Stain the probe array for 10 min in SAPE solution at 25°C
Poststain wash	6 cycles of 5 mixes/cycle with Wash Buffer A at 25°C
Second stain	Stain the probe array for 10 min in antibody solution at 25°C
Third stain	Stain the probe array for 10 min in SAPE solution at 25°C
Final wash	10 cycles of 6 mixes/cycle with Wash Buffer A at 30°C. The final holding temperature is 25°C
Filling array	Fill the array with Array Holding Buffer

for each SNP markers on the array. The outline of the data analysis workflow is shown in **Fig. 5**.

In this section, we will provide a brief introduction of the data analysis for GeneChip Mapping array using GTYPE. For advanced data analysis, please refer to Affymetrix GeneChip Genotyping Analysis Software User's Guide (www.affymetrix.com).

1. Open the Batch Analysis window. Then, select the appropriate array data (.cel files) for analysis by dragging the files into the Batch Analysis window.
2. Start the analysis by clicking Analyze and the selected files will be analyzed.
3. After Batch Analysis is complete a report will be displayed summarizing data from the samples. The report contains information that can be used to assess the performance of the sample, such as call rate. In general, a successful experiment should have a call rate of >90 %.
4. Following analysis, the .chp files are generated for each sample. The .chp files can be opened by double-clicking on the files in the GTYPE data file tree. Multiple .chp files can be opened simultaneously. Once opened the .chp files will display a Dynamic Model scatter plot in the upper portion of the window and a data table in the lower portion of the window as shown in **Fig. 6**. In the data table a genotype call for each SNP along with a confidence score is displayed.
5. The SNP call data can be exported for use in third party software products in either tab delimited file format, Merlin format, GeneHunter format, or Microsoft Excel format. To export data:

   (a) Open the .chp files you would like to export in GTYPE.
   (b) Select Export and Table in the menu bar.
   (c) Specify the location and format of the export data.

GCOS

1) Experiment registration
2) Washing and staining arrays
3) Scanning arrays
4) Automatically generate .CEL files

GTYPE

5) Batch analyze data
6) Generate report
7) Generate .CHP files
8) Export SNP genotyping data from .CHP files for additional analyses

Fig. 5. Data analysis flow for GeneChip Mapping 50K Arrays when using GeneChip Operating Software and GeneChip Genotyping Analysis Software.

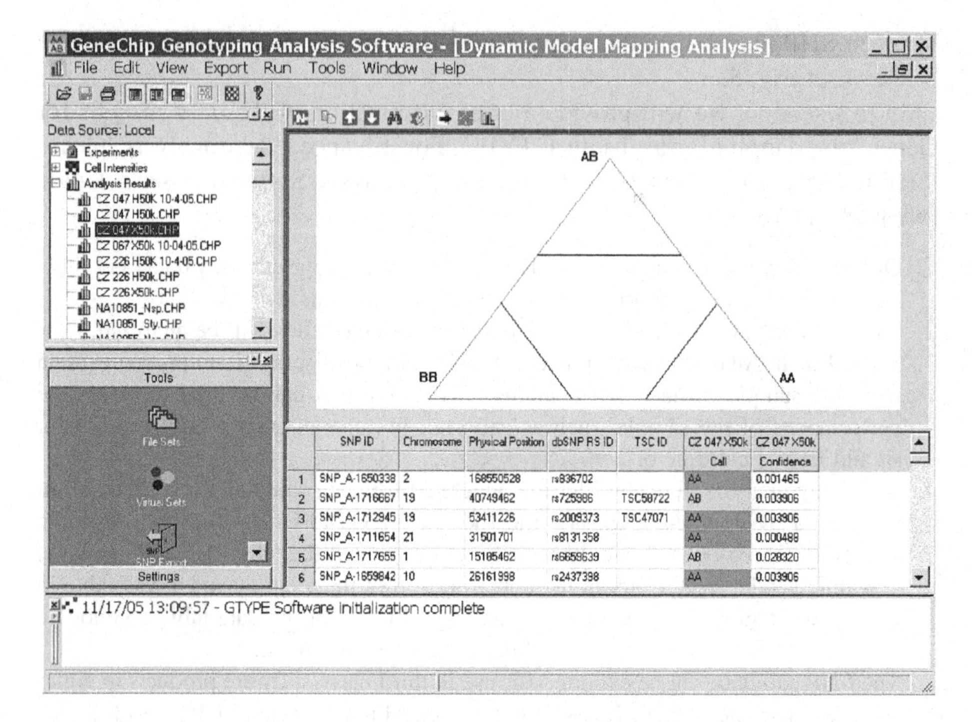

Fig. 6. Screen shoot of the analyzed data (.chp files) displayed using GTYPE.

4. Notes

1. One major reason of reduced genotyping accuracy is contamination. The most likely potential source of contamination for SNP mapping array assay is the previously amplified PCR product. It is recommended to set up the laboratories as pre-PCR Clean Room, PCR Staging Room, and Main Lab. The pre-PCR Clean Room and PCR Staging Room can be substitute with a PCR clean hood and a PCR Staging hood, such as the AirClean 600 PCR Workstation (AirClean Systems). To avoid the potential cross-contamination, each step of the SNP array mapping assay should performed in the dedicated room/area as illustrated in **Table 16**. It is also highly recommended to have dedicated equipments for each dedicated room/area, such as pipet, vortexer, desktop centrifuges, thermal cycler, and so on.
2. The quality of the input genomic DNA is critical for the success of the GeneChip Mapping Assay. DNA must be double-stranded (not single-stranded) and free of PCR inhibitors (such as high concentrations of heme and EDTA). DNA must not be contaminated with other human genomic DNA sources, or with genomic DNA from

Table 16
Laboratory Setup

	Assay steps	Template (genomic DNA)	PCR product
Pre-PCR clean room	Step 1: Reagent preparation	–	–
PCR staging room	Step 2: Enzyme digestion		
	Step 3: Ligation	+	–
	Step 4: PCR (set up)		
Main lab	Step 4: PCR (reaction)		
	Step 5: PCR clean up		
	Step 6: Fragmentation	+	+
	Step 7: Labeling		
Microarray core	Step 8: Hybridization		
	Step 9: Washing, staining, and scanning	+	+

other organisms. DNA must not be highly degraded. The recommended genomic DNA isolation methods are:

(a) SDS/ProK digestion, phenol-chloroform extraction, Microcon or Centricon (Millipore) ultrapurification and concentration.
(b) QIAGEN; QIAamp DNA Blood Maxi Kit.

3. The most common problem encountered for this SNP array-based mapping assay is the low SNP call rate. The followings are potential causes and respective recommendations.

(a) Over fragmentation of DNA sample due to incorrect dilution of Fragmentation Reagent (DNase I) stock. Use correct concentration of Fragmentation Reagent (DNase I). Check concentration (U/μL) on the label of the Fragmentation Reagent stock and prepare the fragmentation dilution as described in **Subheading 3.5.** Work quickly and on ice, transfer reaction tubes to preheated thermal cycler (37°C).
(b) Failed labeling reaction. Use a new vial of Terminal Dideoxy Transferase. Verify the labeling reagents and repeat labeling.
(c) Suboptimal quality of the input genomic DNA. Ensure DNA samples are of high quality (*see* **Note 2**). Genomic DNA that has been preamplified using certain methods (e.g., Phi29 DNA polymerase-based genome amplification) may give reasonable results *(7)*. Pilot test should be planned to evaluate preamplification of specific DNA samples with this assay.
(d) Mix up the samples. Repeat the assay using correct samples.

Acknowledgments

The authors would like to thank the Affymetrix, Inc. for allowing adapting their GeneChip Mapping 100K Assay protocol for this chapter. The authors also would like to thank Dr. Hui Ye at University of Illinois at Chicago and Dr. Zugen Chen at UCLA Microarray Core for their technical assistances. This work was supported in part by National Institutes of Health PHS grants R21 CA97771 and R01 DE015970-01 (to Wong), K22 DE014847, RO3 DE016569, RO3 CA114688, and a TRDRP grant 13KT-0028 (to Zhou).

References

1. Sachidanandam, R., Weissman, D., Schmidt, S. C., et al. (2001) A map of human genome sequence variation containing 1.42 million single nucleotide polymorphisms. *Nature* **409,** 928–933.
2. Matsuzaki, H., Loi, H., Dong, S., et al. (2004) Parallel genotyping of over 10,000 SNPs using a one-primer assay on a high-density oligonucleotide array. *Genome Res.* **14,** 414–425.
3. Sawcer, S. J., Maranian, M., Singlehurst, S., et al. (2004) Enhancing linkage analysis of complex disorders: an evaluation of high-density genotyping. *Hum. Mol. Genet.* **13,** 1943–1949.
4. Zhou, X., Li, C., Mok, S. C., Chen, Z., and Wong, D. T. W. (2004) Whole genome loss of heterozygosity profiling on oral squamous cell carcinoma by high-density single nucleotide polymorphic allele (SNP) array. *Cancer Genet. Cytogenet.* **151,** 82–84.
5. Zhou, X., Mok, S. C., Chen, Z., Li, Y., and Wong, D. T. (2004) Concurrent analysis of loss of heterozygosity (LOH) and copy number abnormality (CNA) for oral premalignancy progression using the Affymetrix 10K SNP mapping array. *Hum. Genet.* **115,** 327–330.
6. Lin, M., Wei, L. J., Sellers, W. R., Lieberfarb, M., Wong, W. H., and Li, C. (2004) dChipSNP: significance curve and clustering of SNP-array-based loss-of-heterozygosity data. *Bioinformatics* **20,** 1233–1240.
7. Zhou, X., Teman, S., Chen, Z., Ye, H., Mao, L., and Wong, D. T. (2005) Allelic imbalance analysis of oral tongue cancer by high-density single nucleotide polymorphism arrays using whole genome amplified DNA. *Hum. Genet.* **118,** 504–507.

20

Molecular Inversion Probe Assay

Farnaz Absalan and Mostafa Ronaghi

Summary

We have described molecular inversion probe technologies for large-scale genetic analyses. This technique provides a comprehensive and powerful tool for the analysis of genetic variation and enables affordable, large-scale studies that will help uncover the genetic basis of complex disease and explain the individual variation in response to therapeutics. Major applications of the molecular inversion probes (MIP) technologies include targeted genotyping from focused regions to whole-genome studies, and allele quantification of genomic rearrangements. The MIP technology (used in the HapMap project) provides an efficient, scalable, and affordable way to score polymorphisms in case/control populations for genetic studies. The MIP technology provides the highest commercially available multiplexing levels and assay conversion rates for targeted genotyping. This enables more informative, genome-wide studies with either the functional (direct detection) approach or the indirect detection approach.

Key Words: Single nucleotide polymorphism; genetic variation; whole genome association; linkage analysis.

1. Introduction

Molecular inversion probes (MIP) is a single-stranded DNA molecule containing several features enabling target identification, detection, and quantification. These probes have allowed high level of multiplexing and being applied to various applications. By far the most popular application is single nucleotide polymorphisms (SNP) genotyping *(1)*, which is being used for microbial detection, expression profiling, methylation analysis, comparative genomic hybridization, and interrogation of alternative splicing. This

From: *Methods in Molecular Biology, vol. 396: Comparative Genomics, Volume 2*
Edited by: N. H. Bergman © Humana Press Inc., Totowa, NJ

technology was initially developed at Stanford University and has been further developed and commercialized by ParAllele BioScience, currently part of Affymetrix (www.affymetrix.com). The underlying principle for this technique is the use of unique sequence tag (molecular tag) in target-specific oligonucleotides called MIP *(1)*. These probes carry two target-specific regions each about 20 nucleotides, one unique sequence tag about 20 nucleotides, and two sequences each of 20 nucleotides serving as general site for two PCR primers. In addition, two endonuclease sites have been engineered in the backbone of these probes.

Multiplexing of more than 20,000 genotypes has been performed with very high accuracy *(2)*. An assay conversion of ~90% (including *in silico* dropouts) and a trio concordance rate >99.6% and completeness level >98% have been obtained using a recently developed algorithm for data analysis *(3)*. In addition to custom assay panel, several standard panels are now available off-the-shelf waiving the assay development cost. These standard panels provide comprehensive coverage of high-value functional SNPs or contain the most informative SNPs to cover specific disease areas like inflammation. There are two targeted approaches to study design: (1) indirect detection using haplotype tag SNPs *(4)* and (2) direct detection using high-value, functional SNPs. In the indirect approach, tag SNPs act as highly informative surrogates for other SNPs in the genome. In the direct approach, functional SNPs are chosen based on their probability of being causative—such as amino-acid changing SNPs and those in regulatory regions. Targeted approaches to study design can increase the efficiency of an association study by a factor of 3–10 over random-SNP approaches by leveraging informative and functional SNPs. The MegAllele Genotyping Human 20K cSNP panel contains over 20,000 amino acid changing SNPs in a single assay *(2)*. This product has demonstrated high data quality **(Fig. 17)** enabling direct detection of disease alleles without relying on linkage disequilibrium. The advantage of the functional SNP approach is its affordability and the immediate relation of a genetic association found to changes in a protein or a biological pathway *(5)*.

2. Materials (6)

2.1. Anneal

1. Water.
2. Assay panel (*see* **Note 1**).
3. Buffer A.
4. Enzyme A.
5. Kit control DNA.

2.2. Gap Fill, dNTP, Ligate, Invert, First PCR

1. Gap fill mix.
2. Exo mix.
3. Cleavage enzyme.
4. Cleavage tube.
5. Amp mix.
6. Plate of dNTPs.
7. 134 μL Stratagene Taq polymerase.

2.3. Second PCR

1. HY A allele tube.
2. HY C allele tube.
3. HY G allele tube.
4. HY T allele tube.
5. Clontech TITANIUM Taq DNA polymerase.

2.4. Target Digest

1. HY digest mix.

2.5. Sample Hybridization

1. The following GeneChip® universal tag arrays as appropriate:
   (a) GeneChip universal 3K tag array.
   (b) GeneChip universal 5K tag array.
   (c) GeneChip universal 10K tag array.
   (d) GeneChip universal 25K tag array.
   (e) GeneChip hybridization oven 640 with carriers.
2. Hyb cocktail.

2.6. Stain and Wash

1. GeneChip fluidics station 450.
2. Buffer H.
3. Stain cocktail.
4. Wash solution A.
5. Wash solution B.

2.7. Scan Arrays

1. GeneChip scanner 3000 7G 4C.

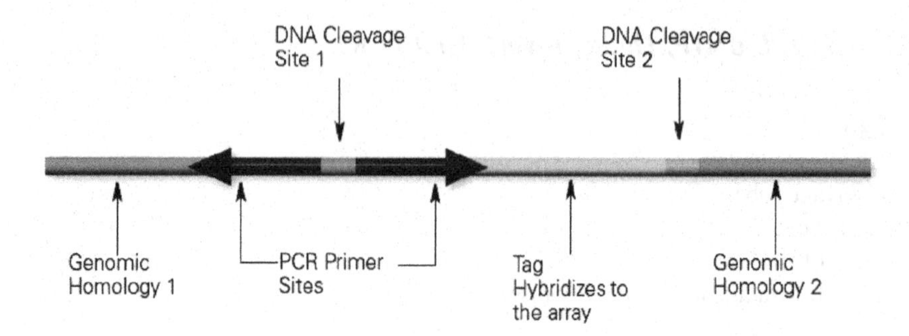

Fig. 1. Schematic presentation of different features of molecular inversion probe.

3. Methods (6)

3.1. Anneal

This stage should be done in the preamp lab and it would take approx 45 min. During this stage, the assay panel probes are annealed to genomic DNA target samples. Allow reagents to warm to room temperature on the bench top and keep them on ice until ready to use. Leave enzyme A at −20°C until ready to use. Genomic DNA samples (150 ng/μL), *3K/5K* 13.4 μL, *10K/20K* 26.7 μL, the assay panel probes, and anneal cocktail reagents (water from kit, *3K/5K* 21.6 μL, *10K/20K* 8.3 μL; buffer A, 5 μL; assay panel, 5 μL; enzyme A, 0.0625 μL) are mixed in an *anneal plate*. Samples are transferred from a sample plate to an anneal plate (*see* **Note 2**).

An anneal plate is a 96-well PCR plate with the barcode designation ANN<*barcode*>. All the plates are barcoded to be traceable. The plate is then placed on a thermal cycler and the program, Meg Anneal, is run (*see* **Note 3**)

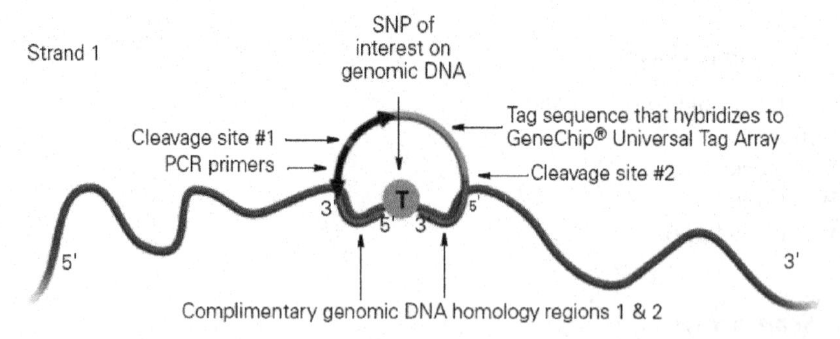

Fig. 2. Assay panel probe annealed to genomic DNA sample.

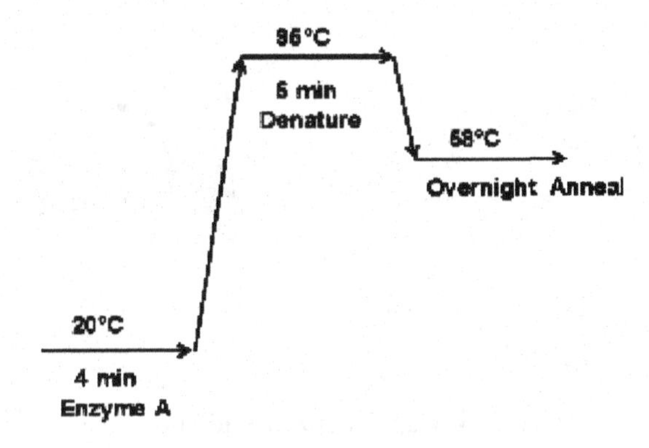

Fig. 3. Meg anneal thermal cycler program.

(Fig. 3). Because the samples must be left to anneal for ≥16 h, this stage is typically started at the end of the day, and the anneal program is allowed to run overnight.

3.2. Gap Fill, dNTP, Ligate, Invert, First PCR

This stage should be done in the preamp lab and it would take approx 2.5 h. During this stage, samples are transferred from the anneal plate to two assay plates. During the transfer, samples are split into four equal aliquots. As described next, several additions are then made to each sample, one addition at a time, at specific intervals **(Fig. 9)**. The thermal cycler program run for these additions is determined by the size of the assay panel you are using.

1. Gap fill mix addition: the first addition, Gap fill mix, is a cold addition. Two and a half microliters of Gap fill mix is added to each sample on ice. After the overnight incubation the thermal cycler program is stopped and 2.5 μL of Gap fill mix is added to each sample in the anneal plate. During the 10 min of incubation, gap fill enzymes find and bind to the single base gap in the assay panel probe (**Fig. 4**). The gap is centered where the SNP of interest is located in the genome.
2. dNTP addition and ligation: each sample is now split into four equal aliquots in the assay plate, followed by the addition of 4.0 μL of dNTPs. Nine microliters of one row of samples is transferred from the anneal plate to four rows of an assay plate, splitting each sample into four aliquots of equal size. For this cold addition, a different nucleotide is added to each row of assay plate on ice. The plate is placed back on the thermal cycler and the program is resumed. In the example shown in **Fig. 5**, during the 10 min of incubation only the aliquot containing dATP will undergo a reaction, wherein the gap fill enzymes will use dATP to fill the gap in the probe (*see* **Note 4**).

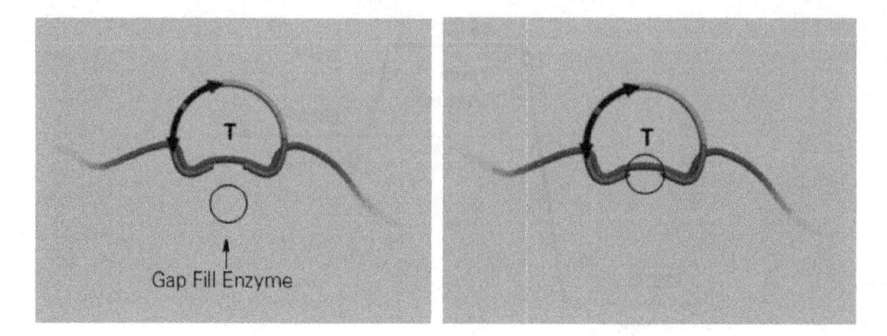

Fig. 4. Gap fill enzyme addition.

Once the gap in the probe is filled, the backbone of the assay probe is covalently sealed. The sealed probe is now referred to as a padlocked probe.

3. Exo mix addition: the third addition is the exo mix, a cold addition. In this step, 4 μL of exonucleases are added to each aliquot on ice. The plate is placed back on the thermal cycler and the program is resumed. The exonucleases digest the

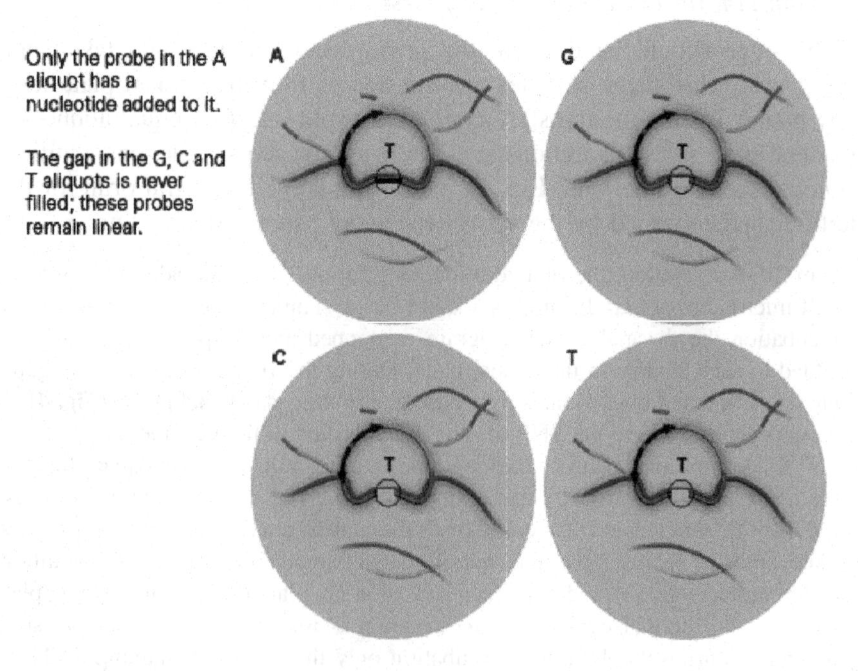

Only the probe in the A aliquot has a nucleotide added to it.
The gap in the G, C and T aliquots is never filled; these probes remain linear.
A
G
C
T

Fig. 5. dNTP addition. This example depicts one assay probe only. In reality, thousands of probes are undergoing the same process simultaneously.

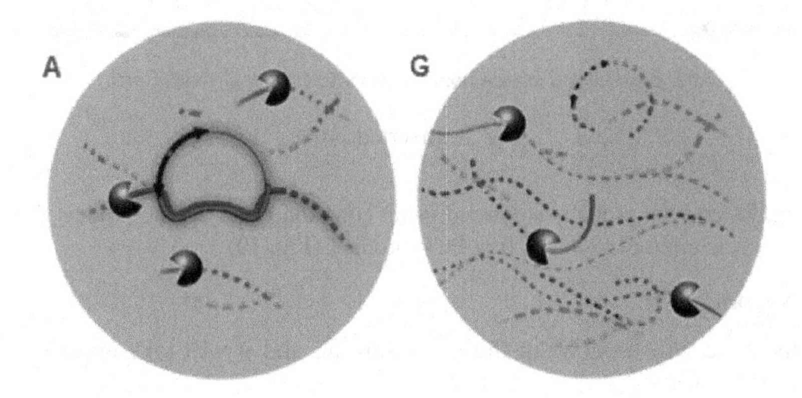

Fig. 6. Exo mix addition showing two aliquots only.

linear probes and single-stranded DNAs that are present **(Fig. 6)**. Linear probes are present in the aliquots where the gap was not filled by a dNTP. Single-stranded DNA is present in all aliquots (*see* **Note 5**).

4. Cleavage mix addition and inversion: the fourth addition is the cleavage mix, a hot addition. Twenty-five microliters of cleavage mix is added to each reaction while the assay plate remains on the thermal cycler. The padlocked probe is cleaved at cleavage site 1 in the assay probe backbone, thereby releasing it from the genomic DNA **(Fig. 7)**.

   Once the assay probe has been cleaved, it releases from the remaining genomic DNA and becomes a linear molecule. Because the orientation of the PCR primer sequences has changed from the original orientation, the probe is now referred to as an *inverted probe* **(Fig. 8)**.
5. Amp mix addition and first PCR: the fifth addition is the amp mix, a hot addition. To prepare the amp mix 67 μL of Taq polymerase is added to the amp mix tube. Twenty-five microliters of amp mix is then added to each reaction in the assay plate on the thermal cycler. Once the amp mix is added, a PCR reaction takes place

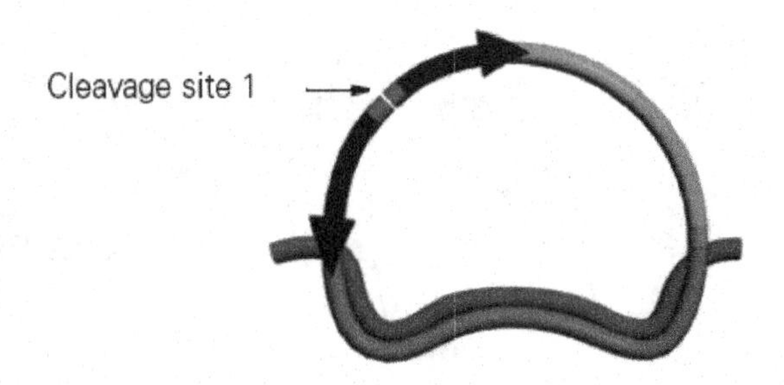

Fig. 7. Cleavage mix addition.

After cleavage and release from the genomic DNA, the probe *inverts.*

Fig. 8. Inverted probe.

(Fig. 9), and the probe is amplified using common primers. The product from this reaction is referred to as the first PCR product **(Fig. 10)**.

3.3. Second PCR

This stage will be done in the post-amp lab and it will take approx 25 min (hands-on time).

The thermal cycler time is 30 min. During this stage, label plates (96-well PCR plates with the barcode designation LBL<*barcode*>) are prepared. Four microliters of each sample is transferred from the assay plate to a label plate. After the transfer, a different allele-specific primer is added to each row as a label **(Fig. 11)**, and the second PCR is performed **(Fig. 12)**. The thermal cycler program used is determined by the size of the assay panel. For 3K and 5K assay panels, Meg Hypcr 3–5K is used. For 10K and 20K assay panels, Meg Hypcr 10–20K is used **(Fig. 13)**.

1. To prepare the allele tube mixes add 22 μL of Taq polymerase to each allele tube.
2. Aliquot 31 μL of each allele tube mix to the corresponding reaction.
3. The first quality control gel is run while the second PCR thermal cycler program is running (Meg Hypcr 3–5K or Meg Hypcr 10–20K). Samples are taken from the assay plates. To run the first quality control gel 7 μL of each sample is transferred from an assay plate to a 96-well PCR plate. Three microliters of loading buffer are added and mixed. The mix is loaded onto a gel. The experiments with low signal strength should be repeated from stage 1.

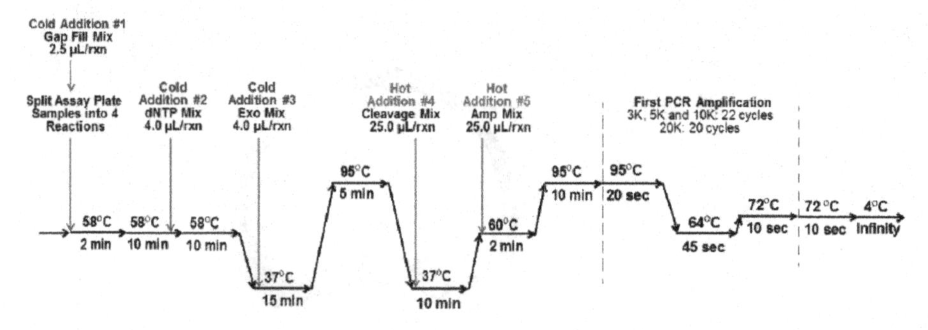

Fig. 9. Thermal cycler programs for the first PCR.

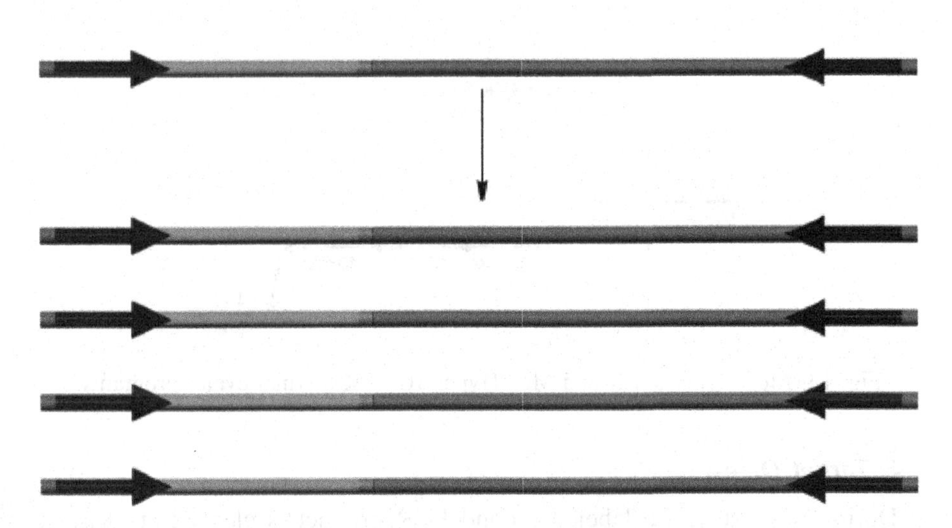

Fig. 10. Amplified probes from first PCR.

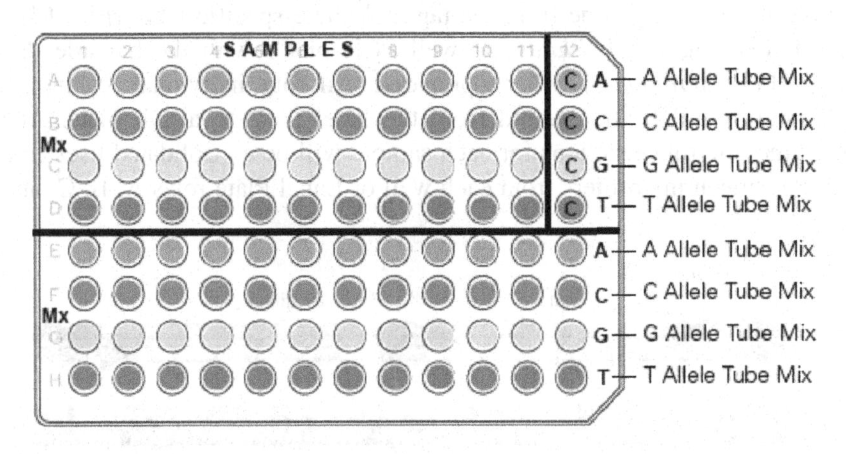

Fig. 11. Label plate labeling and sample placement.

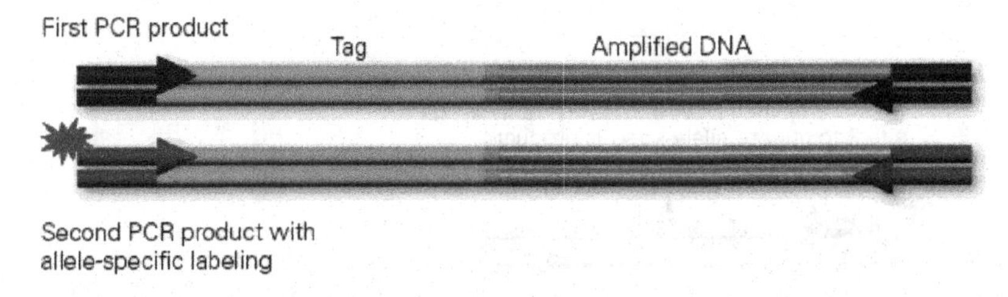

Fig. 12. Second PCR.

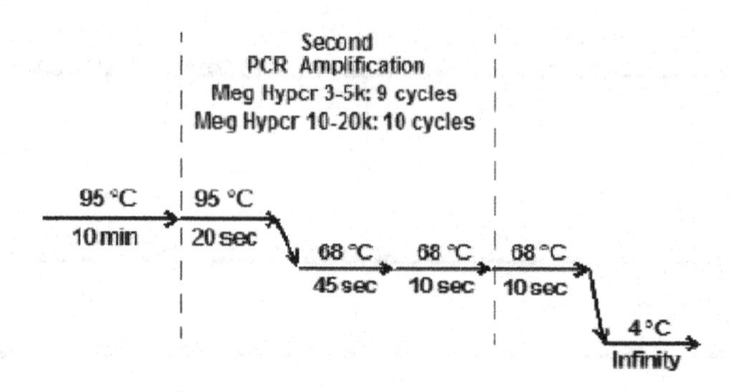

Fig. 13. Meg Hypcr 3–5K and Meg Hypcr 10–20K thermal cycler programs.

3.4. Target Digest

During this stage, the labeled second PCR product is cleaved at cleavage site 2 in the backbone. This process removes the amplified genomic DNA portion of the assay probe from the tag and allele-specific label **(Fig. 14)**. As part of this stage, *Hyb Plates* (96-well PCR plates with the barcode designation HYB<*barcode*>) will be prepared. All four individual reactions for each sample will be transferred from the label plate to a hyb plate. At the same time, the four reactions for each sample will be consolidated back to one well. Seventeen microliters from each well of Label Plate rows A, B, C, and D

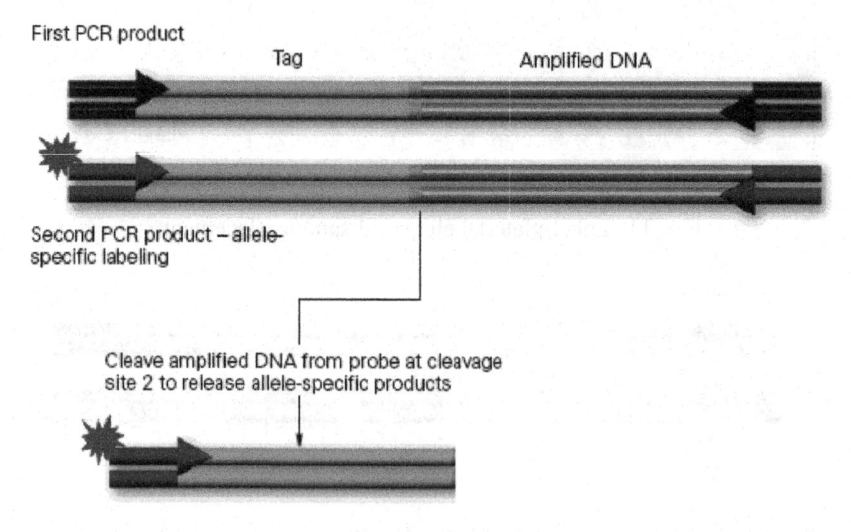

Fig. 14. Target digest.

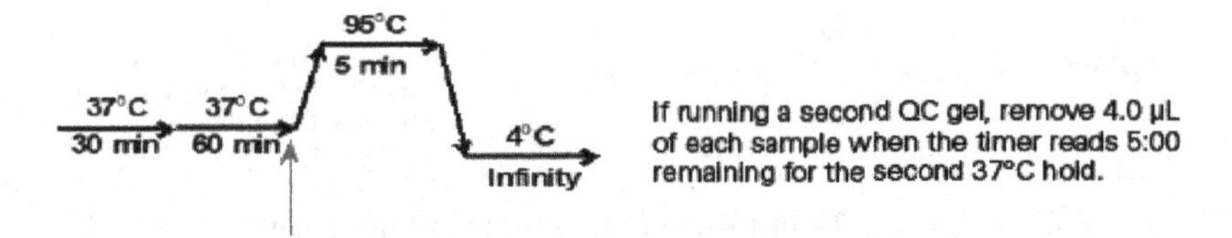

Fig. 15. Meg Hydigest-a thermal cycler program.

is transferred to hyb plate row A. The hyb plates are then placed on thermal cyclers and the Meg Hydigest-a program is run (**Fig. 15**).

While transferred from label to hyb plates, all four reactions for each sample will be consolidated into one well.

1. Six microliters of HY digest mix is added to each sample.
2. Each plate is placed on a thermal cycler and the lid is closed. The program, Meg Hydigest-a is started.
3. If running a second quality control gel, sample from the hyb plate will be removed when 5 min remaining for the second 37°C hold during the thermal cycler program. Four microliters of each reaction is removed for the gel. Three microliters of loading buffer and 3 μL of DI water or 1X TE Buffer is added to each reaction. The gels are load and run. The experiments with incomplete digestion should be repeated from the assay plate stage 5 (second PCR).

3.5. Sample Hybridization

During this stage, hyb cocktail will be added to each sample. The samples are then placed on a thermal cycler and denatured. To denature the samples each hyb plate is placed on a thermal cycler, and the Meg Denature program is run (**Fig. 16**).

After denaturation, each sample is loaded onto a GeneChip universal tag array. The arrays are placed into a hybridization oven that has been preheated to 39°C. Samples are left to hybridize for 12–16 h.

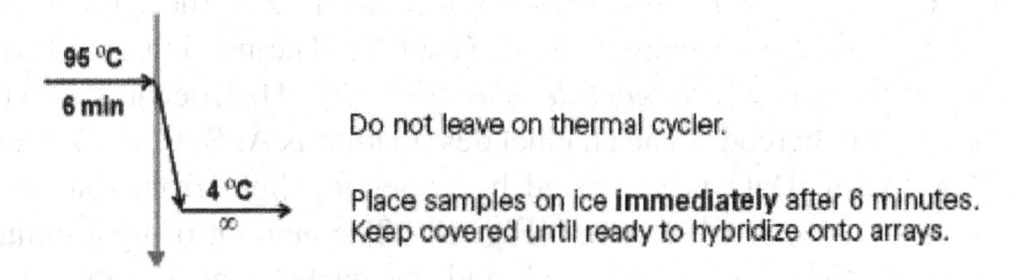

Fig. 16. Meg denature thermal cycler program.

Fig. 17. Hybridization of the probe to the c-tag.

1. Thirty microliters of hyb cocktail is added to each reaction.
2. Ninety microliters of each reaction from the hyb plate is loaded onto an array.
3. The array is placed and left in a hybridization oven for 12–16 h.

Each feature on an array contains multiple *complementary tags* (*c-tags*) (*see* **Note 6**). The tag in the assay probe hybridizes to the c-tags on the array surface (**Fig. 17**).

3.6. Stain and Wash

During this stage, the arrays are loaded onto a GeneChip fluidics station. Each array is washed and stained in preparation for scanning.

1. To prepare the storage cocktail 11.4 μL of buffer H is diluted using 179 μL of wash solution B.
2. For every array, 190 μL of stain cocktail is transferred to an amber 2.0-mL Eppendorf tube.
3. For every array, 190 μL of storage cocktail is transferred to a clear 2.0-mL Eppendorf tube.
4. Arrays are loaded onto the fluidics station.
5. The prompts on the fluidics station should be followed. Stain and storage cocktails are loaded in appropriate positions on the fluidic station.

3.7. Scan Arrays

During this stage, the arrays are loaded onto the GeneChip Scanner 3000 7G 4C (Scanner 3000 7G 4C). Each array is scanned individually, and the data collected is stored in four files referred to as .cel files. Each .cel file contains the data for one channel (A, C, G, or T). The naming convention for .cel files is *<filename><channel designation>*.cel. The filename is typically an appended array barcode. The channel designation is A, B, C, or D. Example: @510866004A.cel. Data is generated by collecting light from four different wavelengths—one for each channel (**Fig. 18**). The amount of light emitted by each feature on the array is collected, and the background is subtracted (*see* **Note 7**).

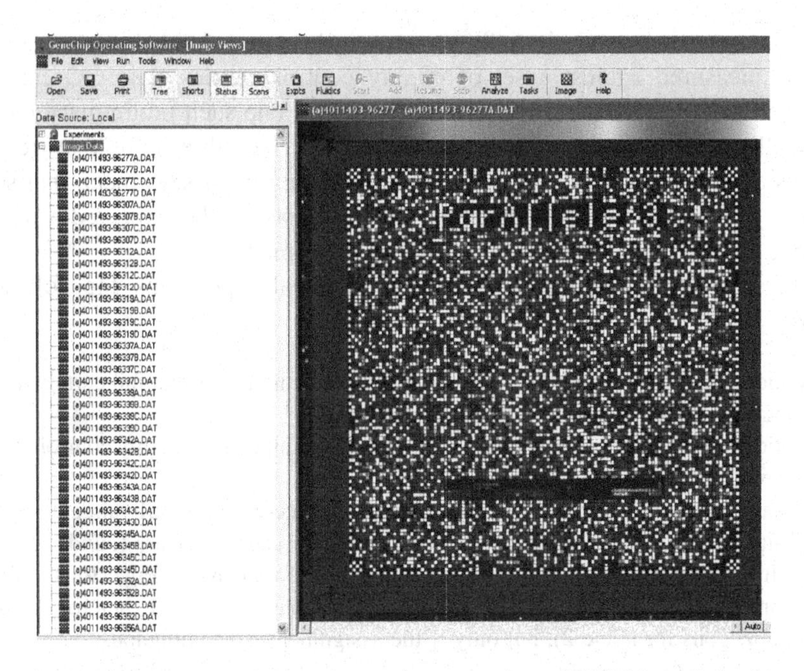

Fig. 18. Data generated from four colors scanning of 20K MegAllele chip.

3.8. Data Analysis (6)

GeneChip targeted genotyping analysis software and GeneChip operating software allow one to create projects and enter and track sample information during the whole experiment. After scanning the arrays one can import and process array data from GeneChip operating software. During processing, the data is compared against a set of QC metrics. Then you can cluster data and generate genotypes using GeneChip targeted genotyping analysis software.

Generating genotypes using GTGS is commonly referred to as clustering or cluster genotyping. Clustered data are data that are similar to each other. In graphical plots, clustered data appears bunched together. An essential feature of SNP genotyping is that measured signals for a SNP (one assay) do not distribute uniformly. Instead, the signals distribute in clumps that generally denote 1 of only 3 calls: AA, AB, or BB. Cluster genotyping is genotyping that assigns calls to data by:

1. Determining the cluster locations for each SNP across a set of samples.
2. Then assigning each data point to a cluster.

When experiments are initially processed, the data is loaded into a database and is measured against a set of quality control metrics. The QC call rate reported in the experiment QC summary table is one such preliminary metric. QC call rate is only an estimate. It is based on fixed thresholds that treat all assays equally. Differences in biochemistry among the assays mean that some assays generate better quality data. GTGS adapts to the expected variation among assays, resulting in more accurate genotyping.

The cluster genotyping operation automatically performs the following steps:

1. Validation of the individual assays for an experiment: assays that fail certain quality control metrics are removed from the experiment.
2. Validation of entire experiments: experiments that fail certain quality control metrics are removed from the data set.
3. Assay clustering: each assay is automatically inspected across all validated experiments, and the locations of up to three clusters for each assay are found.
4. Preliminary genotyping: each validated data point for an assay (**Fig. 19**) is assigned to one of up to three possible clusters (homozygous for allele 1, heterozygous, or homozygous for allele 2), but only if the assignment is unambiguous.

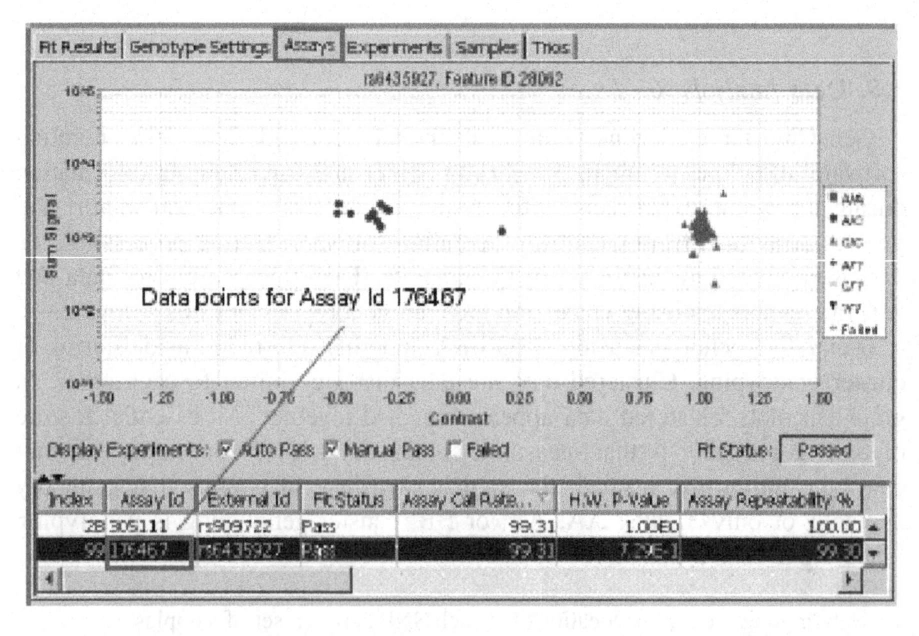

Fig. 19. Data points for one assay.

5. Validation of genotyped assays: an entire assay can be removed from the data set if it has:

   (a) A high proportion of experiments did not generate genotypes for an assay.
   (b) An unusual distribution of genotypes among the clusters.
   (c) Low repeatability in reporting a genotype for different experiments using the same sample.
   (d) Low Mendelian consistency.

4. Notes (6)

1. Most of the reagents are provided by Affymetrix in the Gene Chip SNP kit.
2. 12-Channel pipets are used to aliquot reagents to one row at a time and 24-channel pipets are used to aliquot reagents to two rows at a time.
3. Optimal incubation time is 16–24 h. Do not incubate samples for more than 30 h.
4. This example is based on a homozygous SNP locus. If the sample was heterozygous for this locus, two of the four aliquots would have a nucleotide added to the probe.
5. For homozygotes: after the exo mix addition, three of the four aliquots contain digested DNA only for each assay probe. In the fourth aliquot, the genomic DNA is mainly digested away except for the small region where it is double-stranded because of its association with the padlocked probe. For heterozygotes: after the exo mix addition, two of the four aliquots contain digested DNA only. In the other two aliquots, the genomic DNA is mainly digested away except for the small region where it is double-stranded because of its association with the padlocked probe.
6. If the sample is that of a homozygote, only one of the four types of allele-specific labeled probes will hybridize to the c-tags on the feature for the target SNP (for example, probes with the dATP label). If the sample is that of a heterozygote, two of the four types of allele-specific labeled probes hybridize to the c-tags on the feature for the target SNP (for example, probes with the dATP label and probes with the dCTP label).
7. If the sample is from a homozygote, only a single wavelength of light is emitted from the feature for the target SNP. If the sample is from a heterozygote, two different wavelengths of light are emitted.

References

1. Hardenbol, P., Baner, J., Jain, M., et al. (2003) Multiplexed genotyping with sequence-tagged molecular inversion probes. *Nat. Biotechnol.* **21,** 673–678.
2. Wang, Y., Moorhead, M., Karlin-Neumann, G., et al. (2005) Allele quantification using molecular inversion probes (MIP). *Nucleic Acids Res.* **33,** e183.
3. Moorhead, M., Hardenbol, P., Siddiqui, F., et al. (2006) Optimal genotype determination in highly multiplexed SNP data. *Eur. J. Hum. Genet.* **14,** 207–215.

4. Hardenbol, P., Yu, F., Belmont, J., Mackenzie, J., et al. (2005) Highly multiplexed molecular inversion probe genotyping: over 10,000 targeted SNPs genotyped in a single tube assay. *Genome Res.* **15,** 269–275.
5. Ireland, J., Carlton, V.E., Falkowski, M., et al. (2006) Large-scale characterization of public database SNPs causing non-synonymous changes in three ethnic groups. *Hum. Genet.* **119,** 75–83.
6. Affymetrix® GeneChip® scanner 3000 targeted genotyping system user guide.

21

novoSNP3

Variant Detection and Sequence Annotation in Resequencing Projects

Peter De Rijk and Jurgen Del-Favero

Summary

The same high-throughput techniques used to make genomic sequences generally available, are also useful in mapping the genetic differences between individuals. Resequencing of a genomic region in a set of individuals is considered the golden standard for the discovery of sequence variants. However, with the available high-throughput sequencing technology data analysis has become the rate-limiting step in data management and analysis of large resequencing projects. To solve this issue we developed a software package novoSNP that conscientiously discovers single nucleotide polymorphisms and insertion-deletion polymorphisms in sequence trace files in a fast, reliable and user friendly way. Furthermore, it can also be used to create databases containing annotated reference sequences, add and align trace data, keep track of validation status of variants, annotate variants, and produce reports on validated variants and genotypes. novoSNP is available from http://www.molgen.ua.ac.be/bioinfo/novosnp. There are versions for MS Windows as well as Linux.

Key Words: SNP detection; insertion/deletion detection; resequencing; DNA sequence analysis; genetic variant; high-throughput analysis; automation.

1. Introduction

Technological improvements shifted sequencing from low-throughput work intensive gel based systems to high-throughput capillary systems. This resulted in the completion of the human and numerous other eukaryotic genome sequences ***(1–5)***, which can now be fully explored in the post-genome era.

From: *Methods in Molecular Biology, vol. 396: Comparative Genomics, Volume 2*
Edited by: N. H. Bergman © Humana Press Inc., Totowa, NJ

A major challenge of the post-genome era is mapping genetic differences between individuals providing valuable information about quality of life of human beings. Discovery of these sequence variants by resequencing a genomic region in a set of individuals is considered the golden standard *(6)*. The direct result of this evolution is that data analysis has become the rate limiting step in large-scale genomic resequencing experiments of a gene or a candidate chromosomal region. To provide a solution for this problem, we developed novoSNP that allows automated analysis and management of large sequence data sets *(7)*. novoSNP aligns sets of trace files to a reference sequence, and automatically detects both single nucleotide polymorphisms (SNPs) and insertion/deletion variants (INDELs). Tools for visual inspection of the obtained variants and easy validation are provided. Reference sequences can be annotated with the position of genes and other regions of interest. novoSNP can automatically determine the impact of variants on these genes. A number of reports listing variants, genotypes or displaying specific traces can be generated.

2. Materials

1. Reference sequence: the reference sequence is preferably longer than the resequenced region, because variants will only be detected in regions where reference sequence is available. The position of variants will be measured from the start of the reference sequence. novoSNP supports sequences in FASTA, Genbank, and EMBL formats. Annotation information present in Genbank and EMBL files used as a reference sequence will be automatically imported and used in novoSNP.
2. Trace files: novoSNP accepts sequence trace files in several commonly used formats (abi, scf, alf, ctf, ztr). Trace files are preferably organized in directories (runs) containing the sequencing results of one forward and one matching reverse primer. Trace files should be base called (*see* **Note 1**). If trace files follow a naming convention that includes the sample name, novoSNP can extract this information and can use it in subsequent analysis and reporting steps (*see* **Note 2**).

3. Methods

3.1. Basic Project Setup and Variant Detection

1. Start novoSNP. Opening the novoSNP program will display the main novoSNP interface that consists of a menu, a tool bar, and three frames (**Fig. 1**). The top left tree frame shows the reference sequences and associated runs and traces in the project (**Fig. 1C**). The variant table below it lists all variants (**Fig. 1D**). The traces frame displays the actual trace data (**Fig. 1G**). The tool bar is divided in two parts: the first part presents the most commonly used actions and is always present (**Fig. 1B**). The second part is interchangeable, and can present either display, filter, or trace options.

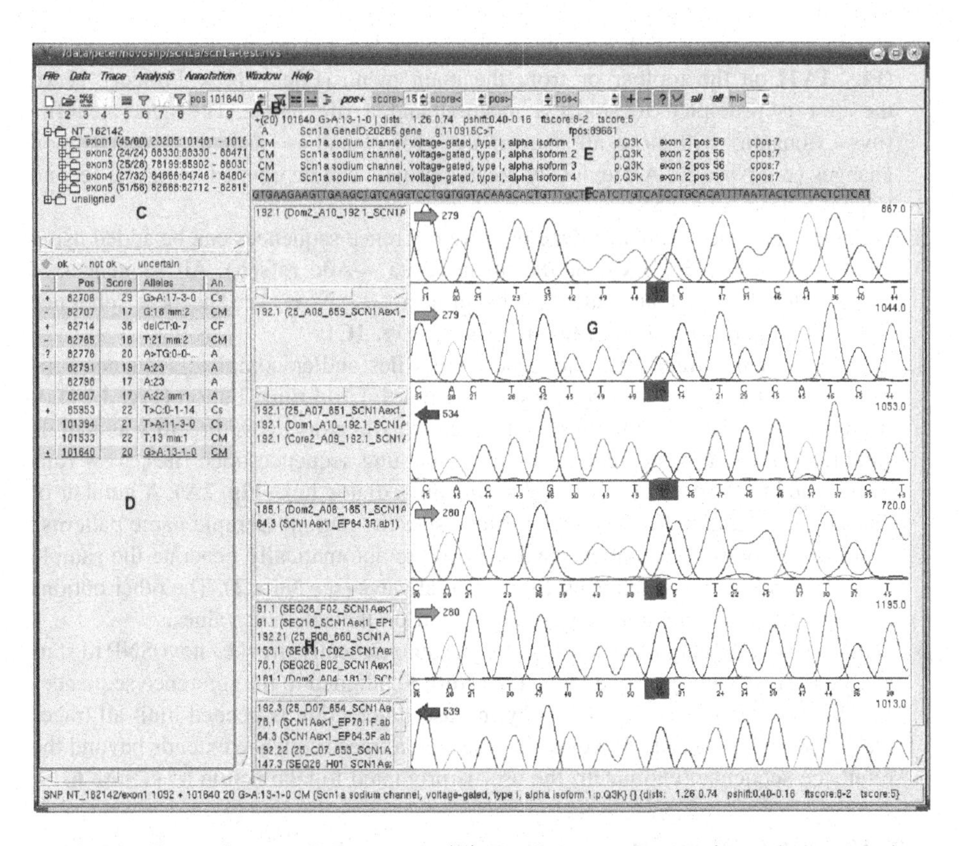

Fig. 1. The main novoSNP screen. (**A**) Toolbar common actions: 1: new project, 2: open project, 3: add reference sequence, 4: add traces, 5: alignment, 6: filter, 7: Options, 8: change toolbar, 9: position (in reference sequence). (**B**) Variable part of the tool bar. Currently, the filter tools are shown. This can be switched to the trace display settings using button 8. (**C**) Tree window showing the reference sequence and the associated runs. Between brackets are the number of traces actually aligned to this reference sequence vs. the total number of traces in the run. Traces not aligned to any reference sequences are added to the "unaligned" branch of the tree. The aligned position of the traces relative to the reference sequence is also shown (minimum and maximum values separated by a colon). (**D**) Variant table frame showing information on the obtained variants. The position, score, variant type, and annotation are shown. The three check boxes are used to indicate and change the status of the currently selected variant. (**E**) Annotation information of the currently displayed variant. (**F**) The selected variant is underlined and shown in red surrounded by the color annotated reference sequence.(**G**) Traces frame displaying the trace data anchored on a C>T SNP. The variant position is indicated by a gray or red (for traces that were important in finding the variant) background around the base. The direction of the trace is indicated by the arrow in the

2. Creating a new project. A new project is created by clicking the first (New) button **(Fig. 1A1)** on the toolbar or from the main menu (File – New Project). Next, the user is prompted to select a location and a file name. The generated file (nvs extension) will store all data of the project and is instantly updated upon changes (*see* **Note 3**). A previously created project can be opened using the second (Open) button **(Fig. 1A2)**.
3. Adding a reference sequence. One or more reference sequences can be added using the third button **(Fig. 1A3)** or the menu (Data – Add refseq). All sequences in the selected file will be added to the project (*see* **Note 4**). The new reference sequence(s) is (are) shown in the tree frame **(Fig. 1C)**.
4. Adding sequence trace files. Sequence trace files and/or complete sequence runs can be added to the project by clicking the fourth (Add runs) button **(Fig. 1A4)** or from the menu (Data – Add runs) and selecting a directory. This directory and all its subdirectories will be searched for files containing sequence trace files. New runs to be added to the project can be selected from a dialog box **(Fig. 2A)**. A number of options are available for this dialog. Make sure the option "Sample name patterns" is set correctly as these patterns will be used to automatically generate the sample name for each trace, which is useful in several ways (*see* **Note 2**). The other options related to variant detection can usually be left on their default values.
5. Data analysis. Adding new runs to a project automatically invokes novoSNP to start analysis. Traces are quality clipped (*see* **Note 5**), aligned to the reference sequences and all variants are scored. Normally, no user interaction is needed until all traces are added and analyzed. However, in case e.g., a trace sequence extends beyond the reference sequence (endmatch) the user is prompted to take action (*see* **Note 6**).

3.2. Variant Assignment

1. Scoring. novoSNP automatically assigns a variant score to each position in the alignment. The higher this score, the more likely a true variant is present at that position (*see* **Note 7**). Only positions with a score of three or more are added to the variant table shown in the bottom left frame of the main novoSNP interface window **(Fig. 1D)**.
2. Scoring filter options. Low scoring variants can be filtered out by changing the variant filter (*see* **Note 8**): the minimum score for inclusion in the SNP list is actually very low (*see* **Note 7**), allowing different trade-offs by setting different minimum scores. Setting a high minimum score (e.g., 20) will filter out most false-positives, but only clear-cut variants will be shown. Setting a lower score limit (e.g., 10)

Fig. 1. *(Continued)* top left of the trace display. **(H)** Sample names and sequence file names of the grouped trace files. Any of these can be selected to display the corresponding trace data.

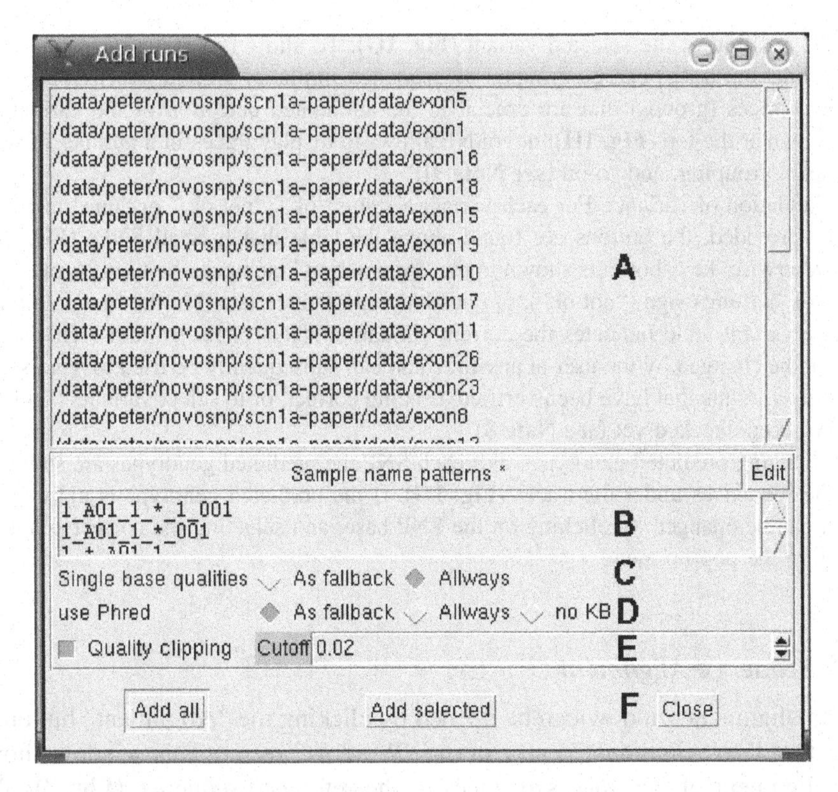

Fig. 2. The Add runs dialog. **(A)** Runs can be selected or deselected in the top listbox. **(B)** Sample name patterns parameter (*see* **Note 2**). **(C)** Single base qualities: when set to "As fallback" novoSNP will use quality data in the file for clipping if available; If no external quality data is available, novoSNP will calculate qualities based on the trace data itself. If set to "Always" novoSNP will always determine base qualities itself. This is the default option as it usually produces better results. **(D)** Use Phred: if Phred is available to novoSNP, it can be used to base call traces for those trace files without base calling information (As fallback). Optionally Phred can be used always, which is useful for traces that have been base called with older, less accurate base callers. **(E)** Quality clipping: turn quality clipping off or on, and set the cutoff (*see* **Note 5**). **(F)** The button labeled "Add selected" will add the selected runs to the project, and start processing them. The button "Add all" can be used to add all runs shown without first selecting any.

allows the detection of variants that are less clear-cut, but more false-positives will have to be checked.

3. Visual inspection of variants. Clicking on a potential variant from the variant list **(Fig. 1D)** will load the corresponding traces and will display them centered at

the position of the selected variant (**Fig. 1G**). To allow fast visual inspection and validation, traces can be grouped by sequence similarity around the SNP position, and traces (groups) that are crucial in the automated detection of the variant are shown at the top (**Fig. 1H**). novoSNP allows to display traces in a number of ways (size, grouping, and so on [*see* **Note 9**]).

4. Validation of variants. For each variant a status "ok," "not ok," or "uncertain" can be provided, the buttons are found above the table that lists all SNPs (**Fig. 1D**). Otherwise, key shortcuts shown in the "Annotation" menu can be used. A plus sign ("ok"), minus sign ("not ok"), or question mark ("uncertain") in the first column of the variant table indicates the current validation status of the variants. This status can be changed by the user at any time and can subsequently be used to, e.g., select only variants that have been verified as being correct, or to select variants that have not been checked yet (*see* **Note 8**).
5. Changing predicted genotypes. For each SNP the predicted genotypes are shown in the sequences under the traces (**Fig. 1G**). If the predicted genotype is not correct, it can be changed by clicking on the SNP base, and selecting the correct genotype from the pop-up menu.

3.3. Sequence Alignment

An alignment window can be opened by clicking the "Alignment" button or selecting the "Alignment" entry in the "Window" menu. This window shows the alignment of the trace sequences to the reference sequence (**Fig. 3**). The user can change the order of the traces in the alignment using the sort menu buttons at the top. Only regions covered by aligned traces are used in variant analysis, whereas clipped regions (*see* **Note 5**) in the beginning and end of the trace sequence, indicated by a gray background, are not included in the analysis.

The alignment window allows rapid inspection of the sequence coverage by the trace files and of (crucial) samples coverage of the region with good quality sequences.

If a run contains sequence traces that align to separate regions of the reference sequence, the alignment for only one of these regions will be shown. The other regions can be selected in the region menu at the top.

3.4. Automated Sequence Annotation

1. Annotation of the reference sequence. The annotation of the reference sequence is used to indicate the position of different (functional) elements (UTR, CDS,...) of one or more genes relative to the reference sequence(s). The easiest way to generate an annotated reference sequence is using a reference sequence containing

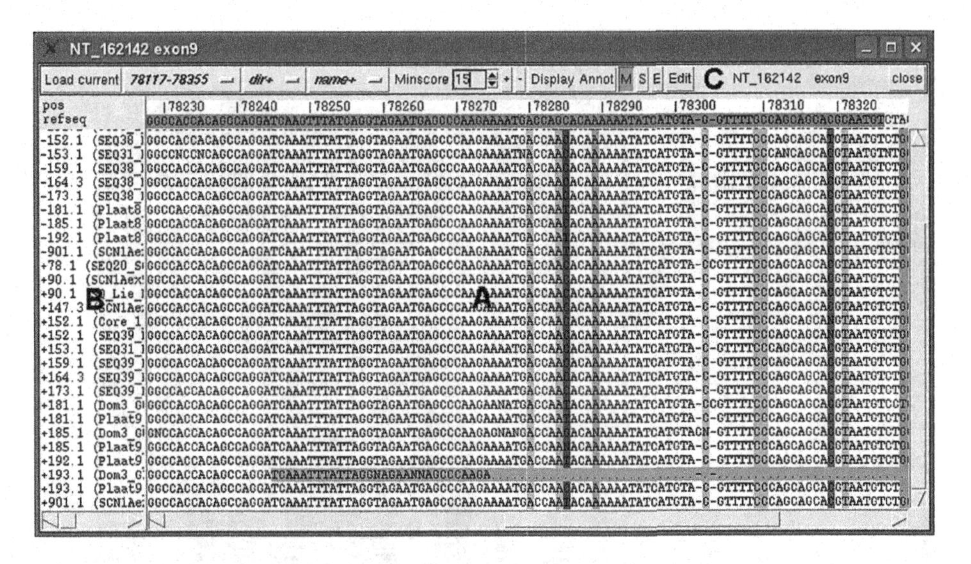

Fig. 3. The Alignment window. (**A**) Alignment. The reference sequence is locked, the aligned trace files can be scrolled using the scrollbar. Variants are indicated using colors ranging from light yellow for low scoring positions to dark red for very high scoring positions. (**B**) Sample names (and full file names). The direction of the trace is indicated by a plus or a minus sign before the name. (**C**) Toolbar with in order of appearance: (**1**) Load current: updates the alignment to the currently selected reference sequence and run. (**2**) Alignment region: displays the region of the reference sequence the currently shown alignment covers. If different traces align to separate regions of the reference sequence, other regions can be selected here. (**3**) Value used to order sequences in the alignment: some possible options are: direction, name, start or end, quality. (**4**) Secondary value used to order sequences. (**5**) Minimum variation score displayed in the alignment. (**6**) Use + and − to increase or decrease font size of the alignment. (**7**) Selection of the annotation to display on the reference sequence. (**8**) Clicking on a position or variant will move the main window to the selected position or variant (M). When the S or E button is selected, clicking in the alignment will change the clipping start and end positions. (**9**) The Edit button allows editing of the alignment. (**10**) The close button closes the alignment window.

annotation data (Genbank or EMBL format). New annotation may be added or existing annotation edited using the Annotation window (**Fig. 4**), which is invoked from the Window Annotation menu. There can be several annotation entries (e.g., different genes). Different entries may overlap the same region as in the case of different isoforms of one gene.

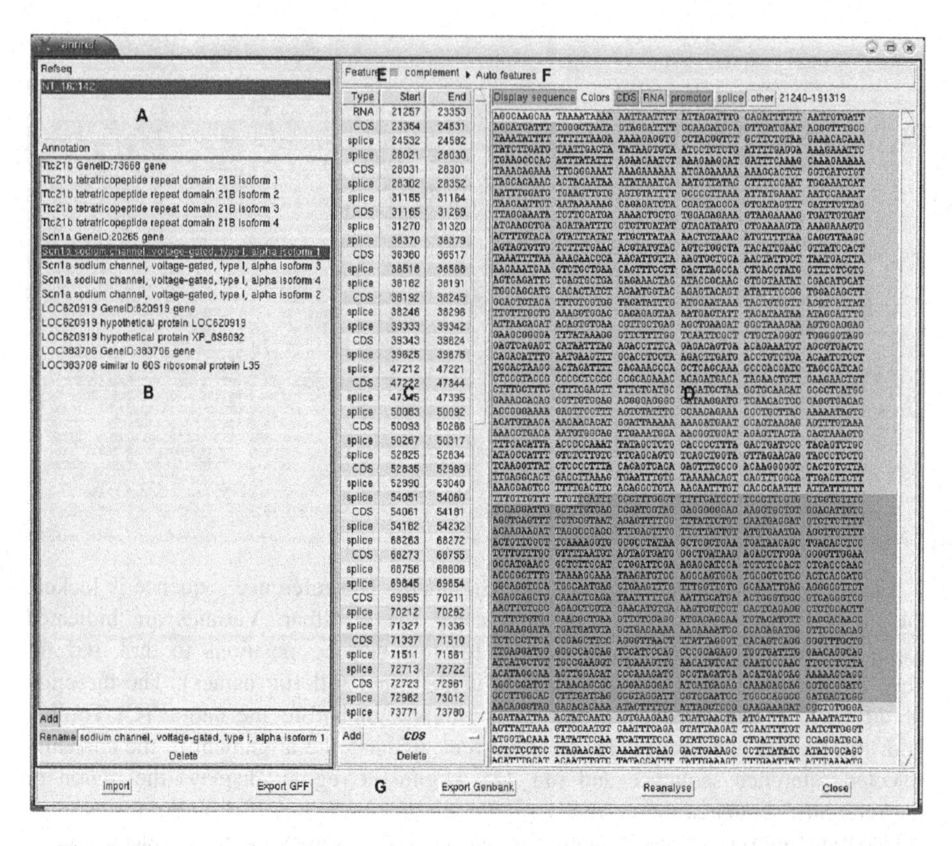

Fig. 4. The annotation window. (**A**) List of reference sequences in the current project, indicating the currently selected reference sequence. (**B**) Annotation entries. These can be genes or other regions of interest in the genomic sequence. Typing the name in the "Add" entry, and pressing Return can add new entries. The rename box and Delete button can be used respectively to rename or to delete an Annotation entry. (**C**) Table of features of the currently selected annotation entry. Each feature assigns a type to a region of the reference sequence. The features of one entry cannot overlap. (They can overlap features of different entries.) The type and position can be edited in the table. Features can be added or deleted using the buttons below. (**D**) Display of the reference sequence with the current annotation features indicated using colors. This display is only active when the "Display sequence" option is selected. Selecting a region in the display will show the positions in the info field above. Adding new features will use the selected region. (**E**) The radio button labeled complement is selected when the current annotation is on the complementary strand of the reference sequence. (**F**) The "Auto Features" can be opened to reveal some options to easily change features. The splice entries add a region of the given size labeled "splice" before and after splice donors and acceptors. "CDS start" changes the position of the coding region within an annotated

2. Annotate variants. Variant annotation is invoked by selecting the "Annotate SNPs" entry in the "Annotation" menu. It determines the type and impact of all variants on the genes present in the reference sequence. Information on the type and impact of the variants and their position relative to different elements of the annotation is added to the information field above the traces (**Fig. 1E**). If a variant has an impact on more than one annotation entry, each is shown on a new line. In the variant table (**Fig. 1D**), a symbol indicating the most severe impact on one of the annotations will be displayed (*see* **Note 10**).

3.5. Reporting

Three types of reports can be generated from the Analysis menu. The "SNP list" report generates a tab-separated file containing a list of all variants currently shown in the variant table. Each line shows the position, score, summary of genotypes, and so on, for each selected variant.

The "Genotype" report generates a table (in tab separated format) listing the predicted genotype for each sample for each selected variant (*see* **Note 2**). The third report type "Graphical reports" is used to generate printed reports featuring the trace displaying the variant, together with a comparison trace without the variant, which is interesting for diagnostic applications.

3.6. Data Editing

Some datasets perform poorly using default parameter settings. Reanalysis of the dataset with adjusted parameters is possible and provides analysis flexibility.

1. Changing sample names. The sample names assigned to each trace are important in analysis and reporting (*see* **Note 2**). The names are automatically generated based on the file name according to the "sample name patterns" (**Fig. 2B**). The assigned sample names can be changed in batch in the "Options" dialog, invoked from the "Window" menu. The "sample name patterns" field can be edited by changing one of the patterns or adding a better pattern. The button labeled "Recalculate sample names" will reassign new sample names to all traces in the project using the new

Fig. 4. *(Continued)* RNA to either a given start codon or start base. (**G**) Annotation can be exported in GFF or Genbank format. Annotation in GFF, Genbank, or EMBL formats can be imported. The file format is automatically recognized. The sequence in the Genbank or EMBL file has to be the same as, or part of the reference sequence. The Import option also accepts files in FASTA format. These will be regarded as mRNA sequences, and are processed using Spidey *(10)* to generate the annotation.

pattern settings. Sample names can also be edited manually by opening a run in the top left tree frame and selecting the relevant trace. This will invoke the trace tool bar where trace data can be edited. (The file name of the trace is kept in the label field.)

2. Changing clipping cutoff. Clipping parameter can be changed in the "Option" dialog to increase or decrease the stringency of clipping (*see* **Note 5**). Clipping can be turned of completely using the "Quality clipping" checkbox, in this case a large amount of false-positives are expected. The clipping can also be adjusted for each trace individually in the alignment window (**Fig. 3**), or in the trace display.
3. Changing the alignment. The alignment can be edited in the alignment window (**Fig. 3**). This is useful when trace sequences show large differences among each other and/or with the reference sequence. Clicking the "Edit" button allows editing of the alignment. Part of the alignment can then be selected using the mouse, and the left and right arrows will move the selected part in the given direction, pushing other bases as needed. The "Insert" button allows the insertion of a number of gap symbols (equal to the number of positions currently selected in the alignment). The "Save" button will make the changes permanent.
4. Changing annotation. Annotation of the reference sequences can be added or edited at any time (**Fig. 4**). All annotation of variants will be removed upon changing the annotation. This can be restored (according to the new annotation) using "Annotate SNPs" in the "Annotation" menu or the reanalyze button in the "Annotate refseq" dialog (**Fig. 4**).
5. Changing parameters such as the alignment or clipping does not have an immediate effect on the variants detected earlier and shown in the variant table. To become effective the list of variants has to be updated first by reanalyzing the data with the new changes. This can be done from the "Reanalyze" entries in the "Analysis" menu. The currently selected run, all runs, or a number of selected runs can be reanalyzed. If the runs contain variants that have already been checked and for which the status has been set, the user is prompted and can select to keep the assigned status. The scores, and so on, of the variants will be updated according to the new data.

4. Notes

1. novoSNP will use the base call information stored in the trace files, e.g., by the KB base caller in case of ABI files (Applied Biosystems, Foster City, CA). If no base call information is present in a trace file, novoSNP will run Phred ***(8–9)*** on this file. To use Phred, it has to be obtained separately (http://www.phrap.org) and installed in the directory extern\phred.
2. The sample name of a trace identifies the sample sequenced to obtain this trace. Forward and reverse reads from the same sample are linked through their identical sample name. Scoring of variants can be improved by checking for matching genotype predictions in forward and reverse reads of the same sample. Reports

such as the genotype report list the genotypes for each sample. The sample name of a trace is automatically derived from its file name. Different conventions on how to include the sample name in the file name can be handled by using the "sample name patterns" option. This option contains a list of patterns that will be matched against the filename. A number in the pattern will match a string of numbers in the filename; a letter matches any number of letters. The part of the filename matched by * will be the sample name. More complex patterns can be specified using regular expressions.

3. The novoSNP project file is actually a single file SQLite (http://www.sqlite.org/) database. This has several advantages. Changes are directly made in the database and thus directly changed on disc omitting the need for manual saving of data. The nvs files are portable, and can be moved between Linux and Windows versions without problems. More than one user can access the same project at the same time, although different users cannot change data at the same time.
4. If a file with reference sequences contains only one sequence, the reference sequence name will be based on the file name. If more than one reference sequence is used, the names will be the ones assigned to each sequence in the file. Right clicking on the reference sequence in the tree frame, and selecting "Rename refseq" in the pop up menu allows names to be changed afterwards.
5. Clipping determines which region of the traces is actually scanned for variants. The 5′- and 3′-ends of a trace are usually of poor quality and therefore not suited for reliable analysis and are excluded from analysis by clipping. The clipped parts are not really removed as the clipping position may be changed. Clipping is determined by the clipping cutoff parameter. Lowering this parameter will increase the stringency of clipping, so a smaller region of higher quality will remain. A higher cutoff will leave larger regions of lower quality. The clipping parameters can be changed when adding new runs, or in the Options dialog (Window – Options menu). The Options dialog also allows the reclipping of runs with the changed parameters.
6. Adding traces to a project invokes novoSNP to check for (1) trace sequences that extend beyond the reference sequence (endmatch). If this happens the user is prompted to take action. If the endmatch is caused by a reference sequence that is too short, the reference sequence should be corrected, as variants outside the reference sequence will not be scored. In cases where this type of problem can be expected, e.g., if the reference sequence consists of only the expected amplicon sequence, novoSNP can safely align them. (2) Short matches are very short alignments, usually caused by a random hit and these are skipped by default. (3) Incomplete matches occur when a large part of the trace can be aligned, but the beginning and/or end differs too much to be aligned. This is most likely caused by conservative clipping (i.e., bad quality sequence was not removed in initial clipping) and is aligned by default. The default actions can be changed using the "Add run settings" entry in the Window menu.

7. novoSNP uses a cumulative scoring scheme (for more details *see* Wecks et al., 2005). Several typical features seen in SNPs are independently scored for both forward and reverse reads. If the genotype predictions of forward and reverse reads match, an extra score is added. The final score of a SNP is the sum of all these subscores. As a result, variants in a region that was only sequenced in on direction can never accomplish the same high score value of those that were sequenced in both directions. novoSNP also detects both homozygous and heterozygous insertion/deletion variants, and assigns them a score.
8. The filter parameters determine which variants are shown in the bottom right variant table, and in which order. The parameters can be changed in the filter toolbar or in the filter dialog (Filter entry of the Window menu). Some parameters have fields to be filled out, if these are empty, the parameter will not be used in filtering. Other parameters can only be turned on or off or selected from a list. In the filter toolbar **(Fig. 1B)** we find (in the following order): show SNPs, show indels, show only variants of the currently selected run, sort field and order, minimum and maximum score, minimum and maximum position, include variants with status set to ok, not ok, uncertain and no status, minimum impact of SNP on annotation, and annotation item. In the filter dialog some extra options are possible: three fields for sorting can be selected, and a number of runs for which variants are shown can be selected.
9. The trace toolbar is used to change options concerning the trace display. If not visible, the "Trace toolbar" button can be used to select this toolbar. The first tool selects whether the start or the end of an indel variant is shown. Traces important to the variant detection are moved to the top by default. This can be turned off using the next option. The "match forward and reverse reads" menu determines whether matching forward and reverse reads are shown together or not. The grouping parameters determine if traces are grouped together by trace similarity near the variant position. In this case, only one representative trace for each group is shown, reducing the scrolling required to review all traces. The group cutoff parameter is the minimum distance between traces that will be grouped together. The lower this value, the more similar traces are in one group, selecting a value of 0 will remove all grouping. Grouping can also be turned off using the grouping menu. The menu allows ordering the ungrouped traces according to several parameters (name, direction, . . .). The following numerical entries apply to the display of each trace: the "Trace height" entry (H) sets the height (in pixels) of each trace display. The "Max height" field (Y) sets the maximum height for the data points displayed. If empty, the height of the highest peak displayed will be used as a maximum. This height is shown in the right top corner of each trace. Larger values for the "X scaling" field (X) will make displayed peaks broader. The annotation entries indicated on the reference sequence using color can be selected in the "Annot" menu. The "Q" button turns the display of quality values in the traces on or off, whereas the "C" button removes or adds traces that have been quality clipped

at the position of the variant. The last three buttons let you choose how a trace display reacts to the mouse. The default "M" lets the user drag the display left and right, keeping the traces anchored at the position of the variant. When "S" is selected, the start of quality (nonclipped) sequence can be selected in a trace; The button labeled "E" allows changing the end of quality sequence.

10. In the annotation of variants, novoSNP takes into account elements of gene structure (CDS and UTR): The impact of variants with regard to the different annotation entries is added, e.g., intronic (i), 3′ UTR (u), coding (C). The position relative to elements of the annotation is also given: the exon or intron number, the position within the exon/intron, the position within the RNA (indicated by cpos). Variants in a CDS will also be annotated with the impact on the resulting protein sequence, e.g., CS (coding silent), CM (coding missense), CF (coding frameshift). The position in the protein and the amino acid change are shown in the standard notation (e.g., p.Q3K).

Acknowledgments

This work was in part funded by the Special Research Fund of the University of Antwerp, the Fund for Scientific Research Flanders (FWO-V), the Interuniversity Attraction Poles program P5/19 of the Belgian Federal Science Policy Office.

References

1. Adams, M. D., Celniker, S. E., Holt, R. A., et al. (2000) The genome sequence of Drosophila melanogaster. *Science* **287,** 2185–2195.
2. Lander, E. S., Linton, L. M., Birren, B., et al. (2001) Initial sequencing and analysis of the human genome. *Nature* **409,** 860–921.
3. The C. elegans Sequencing Consortium. (1998) Genome sequence of the nematode C. elegans: a platform for investigating biology. *Science* **282,** 2012–2018.
4. Venter, J. C., Adams, M. D., Myers, E. W., et al. (2001) The sequence of the human genome. *Science* **291,** 1304–1351.
5. Waterston, R. H., Lindblad-Toh, K., Birney, E., et al. (2002) Initial sequencing and comparative analysis of the mouse genome. *Nature* **420,** 520–562.
6. Kwok, P. Y., Carlson, C., Yager, T. D., Ankener, W., and Nickerson, D. A. (1994) Comparative analysis of human DNA variations by fluorescence-based sequencing of PCR products. *Genomics* **23,** 138–144.
7. Weckx, S., Del-Favero, J., Rademakers, R., Claes, L., et al. (2005) novoSNP, a novel computational tool for sequence variation discovery. *Genome Res.* **15,** 436–442.
8. Ewing, B. and Green, P. (1998) Base-calling of automated sequencer traces using phred. II. Error probabilities.*Genome Res.* **8,** 186–194.

9. Ewing, B., Hillier, L., Wendl, M. C., and Green, P. (1998) Base-calling of automated sequencer traces using phred. I. Accuracy assessment. *Genome Res.* **8,** 175–185.
10. Wheelan, S. J., Church, D. M., and Ostell, J. M. (2001) Spidey: a tool for mRNA-to-genomic alignments. *Genome Res.* **11,** 1952–1957

22

Rapid Identification of Single Nucleotide Substitutions Using SeqDoC

Mark L. Crowe

Summary

Identification and characterization of nucleotide substitutions in DNA sequences for single nucleotide polymorphism or point mutation detection can be a time consuming and sometimes inaccurate process, particularly in relatively low-throughput situations where fully automated solutions may not be appropriate. SeqDoC provides a simple web-based application to simplify this identification process, by using direct subtractive comparison of the raw sequence traces to highlight differences characteristic of nucleotide substitutions.

Sequencing artefacts, such as variable peak separation and signal strength, are compensated for with moving window normalisation functions, whereas the signal to noise ratio of the comparison trace is greatly enhanced by applying an algorithm to emphasise features associated with nucleotide substitutions. Analysis of the output is simple and intuitive, permitting rapid identification of points of difference between the reference and test sequence traces.

Key Words: Chromatogram; electropherogram; SNP detection; mutagenesis; sequence analysis; sequence comparison.

1. Introduction

The identification of single nucleotide changes in DNA sequences is critical to many fields of biological research, from areas as diverse as the characterization of single nucleotide polymorphisms (SNPs) *(1,2)* to identifying specific alterations caused by random mutagenesis screens *(3,4)*. There are many approaches which can be used to compare DNA sequences for such nucleotide substitutions, ranging from the most basic, of comparing plain text sequence to

From: *Methods in Molecular Biology, vol. 396: Comparative Genomics, Volume 2*
Edited by: N. H. Bergman © Humana Press Inc., Totowa, NJ

a reference, to complex multifunctional analysis software. However, these can often suffer from a compromise between simplicity and efficiency, where the simpler methods may generate false-positive and false-negative results, whereas the more accurate methods can be overly complex and time-consuming for small-scale analysis.

SeqDoC is a web-based application, which provides a way of identifying nucleotide substitutions with high accuracy and good sensitivity through a very simple and intuitive interface *(5)*. It does this by generating a "difference profile" between pairs of sequence chromatogram traces, readily highlighting any changes characteristic of single nucleotide substitutions. The user then has only to examine the sequence traces at these specific points to determine the nature of the substitution; this intervention avoids the incorrect calls, which can result from fully automated software. Simplicity of use is enhanced the use of all predefined parameters, which have been optimized to produce the highest signal to noise ratio for features indicative of nucleotide substitutions. SeqDoC also compensates for sequencing artefacts including sequence compression, peak height variation, drop off of signal strength, and offset start positions of the sequences.

2. Materials

2.1. Input

Input for SeqDoC is two (or optionally more) Applied Biosystems (ABI) format sequence chromatogram files. Many other sequence formats (including SCF and MegaBACE files) can be converted to ABI format using free conversion software such as Chromas Lite (http://www.technelysium.com.au/chromas_lite.html). One file should be a known reference sequence, the other(s) are test sequences to be compared against the reference. SeqDoC directly compares the chromatogram traces themselves between the sequences being examined, so no manual correction of base calling is required prior to testing.

2.2. Access to Program

SeqDoC is primarily a web-based tool, and can be found at http://research.imb.uq.edu.au/seqdoc/. It can be accessed using any Internet-enabled computer, and works with all major operating systems (Windows, MacOS, and Linux) and web-browsers (including Firefox, Internet Explorer, Mozilla, and Netscape). There are no restrictions on the use of the web-based version.

2.3. Source Code

The source code for SeqDoC can be downloaded from the SeqDoC website (http://research.imb.uq.edu.au/seqdoc/) or Bioinformics.Org (http://bioinformatics.org/project/?group_id=481) for installation on a local server. In this case it will be necessary to install the GD graphics library (http://www.boutell.com/gd/) and the Perl GD, GD::Graph and ABI modules (http://search.cpan.org/dist/GD/, http://search.cpan.org/dist/GDGraph/, http://search.cpan.org/dist/ABI/). SeqDoC is made available under the terms of the GNU General Public License (http://www.gnu.org/licenses/gpl.html).

3. Methods

There are two versions of SeqDoC available at the website. The standard version should be used for direct comparison of two sequences, whereas multiple SeqDoC allows the user to simultaneously compare up to ten sequence traces against a single reference trace. The two versions use identical processing algorithms, and interpretation of output is the same for both. The interface differs only by the number of sequence chromatograms which can be uploaded. Multiple SeqDoC has two primary benefits for the user when three or more sequences are to be compared: first, all sequences are aligned to the same reference and are displayed on the same output page, simplifying interpretation of results; second, there can be a performance benefit of up to 30% decrease in processing time per sample (*see* **Note 1**).

3.1. Running SeqDoC

SeqDoC has no user-definable parameters, the only necessary steps being to upload a reference sequence file (*see* **Notes 2** and **3**) and one or more test sequences files, depending on which version of SeqDoC is being used. The file upload is carried out using the upload boxes on the webpage, and the user then simply clicks the "submit query" button to perform the comparison. Results will typically be returned after around 15 s for a single comparison, and proportionally longer for multiple comparisons (*see* **Note 4**).

The reference and test sequences should of course be of the same region of DNA, and should have been sequenced in the same orientation; relative peak heights will be inconsistent in samples sequenced in opposite directions because of variations in sequencing chemistry ***(6,7)*** and are likely to prevent usable results being returned. It is preferable, although not necessarily essential, that the sequences being compared have been sequenced using the same primer (*see* **Note 5**).

3.2. Output

The output provided by SeqDoC is three aligned images in a web browser window **(Fig. 1)** or, for multiple SeqDoC, three aligned images per test sequence. The top image of each three is the reference sequence trace and the bottom image is the test sequence. Both are displayed as png images and cannot be reconverted back to chromatograms. The "difference profile" is provided as an aligned png image between the two sequence traces.

All three traces are colored according to standard ABI chromatogram colors (A – green, G – black, C – blue, T – red). In the difference trace, values above zero indicates the channel which is stronger in the reference sample, while values below zero means that the signal for that channel is stronger in the test sequence. The actual Y-coordinate values in all three traces are arbitrary values resulting from the normalization routine and difference calculation algorithm.

Base position counts are included in both sequence trace images. These correspond directly to positions on the input sequence and, because the start of the test sequence may have been shifted to align the two traces, these position counts may not match directly between the aligned traces.

The output traces are subjected to normalization and alignment adjustment, and so will typically appear subtly different to the input traces; however, this will only affect peak spacing and height, and will not fundamentally alter the trace.

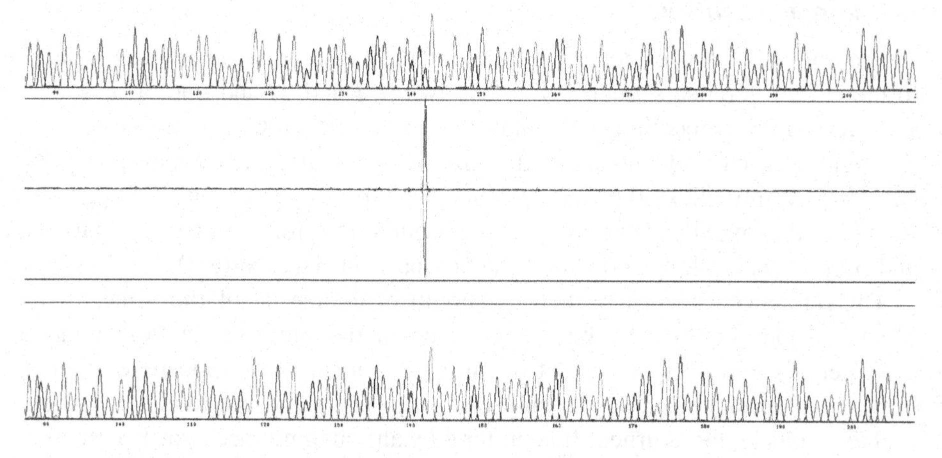

Fig. 1. Typical output from SeqDoC showing reference, difference and test traces. In this example, a C has been substituted by an A at position 142 (*see* **Note 13**).

3.3. Interpreting Output

3.3.1. Single Nucleotide Substitutions and Indels

SeqDoC does not provide any automatic classification of nucleotide substitutions beyond the algorithm for emphasising base substitutions over sequence noise; instead interpretation of the difference trace is left to the researcher. This interpretation is largely intuitive—the difference trace overall should be largely flat, with any major deviations indicative of a difference between the reference and test traces. Because the two sequence traces are aligned with each other and with the difference trace, the investigator need only look at the sequence peaks immediately above and below the point of difference to establish the nature of the difference.

The most frequently observed features are single nucleotide substitutions. These result in large bidirectional peaks in the difference profile as seen in **Fig. 1** and, in more detail, in **Fig. 2** (*see* **Note 6**). Examination of the reference and test traces at the point of this bidirectional peak will allow the researcher to identify the substitution as well as its coordinate positions within the original sequence traces. When a mixed peak occurs in the test sequence, such as that caused by a heterozygous SNP, a bidirectional peak will still be observed in the difference trace (*see* **Notes 7** and **8**).

In addition to highlighting base substitutions, SeqDoC can also be used to rapidly find single base insertions or deletions (indels). These are characterized by a sudden occurrence of major differences between the traces from the point of the indel (when nonmatching bases start to be compared) followed by a return to minimal difference after 8–10 bases once the alignment has been restored by the software (**Fig. 3**). Characterization of indels is similar to that for nucleotide substitutions; the investigator simply needs to examine the two sequence traces at the start of the region of misalignment to identify the inserted or deleted base.

3.3.2. Misalignments and Noise

The SeqDoC interface performs basic error checking on input sequences, and provides diagnostic messages where possible (for example, when a file does not appear to be in ABI format). However, no checking is carried out on the output and familiarity with some of the problems that can occur can save much time and effort when an unexpected result is obtained.

Probably the most frequent cause of failure is uploading two sequences that cannot be aligned. Usually this would occur because an incorrect sequence has been selected by accident, but it can also result from the start points of the

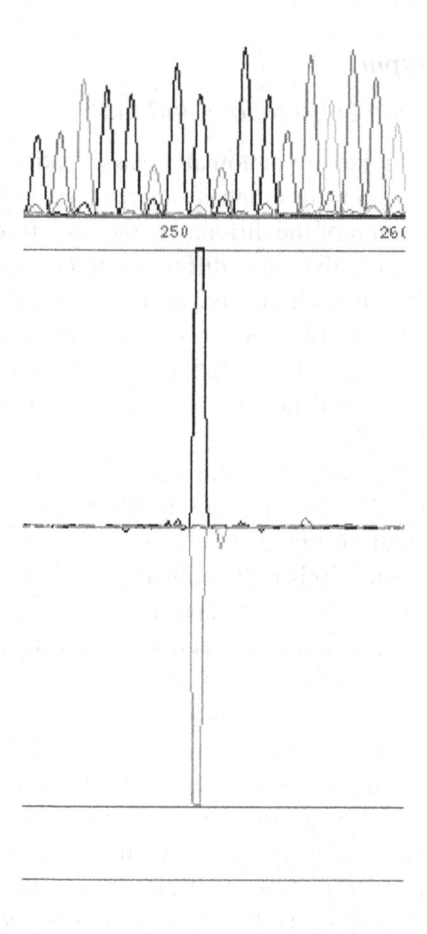

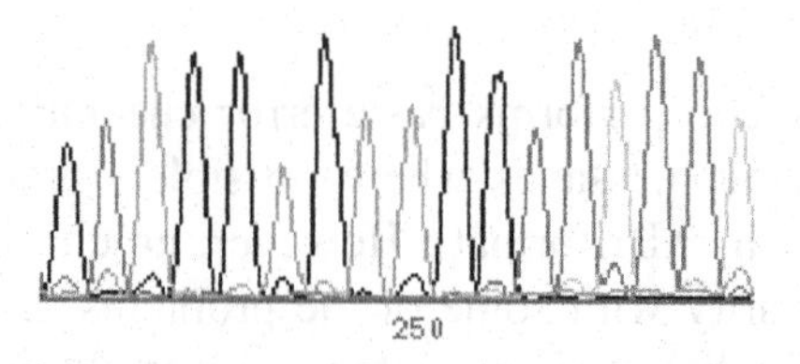

Fig. 2. The characteristic bidirectional peak caused by a single nucleotide substitution (in this case a G at position 251 in the reference sequence has changed to an A at position 249 in the test sequence) (*see* **Note 13**).

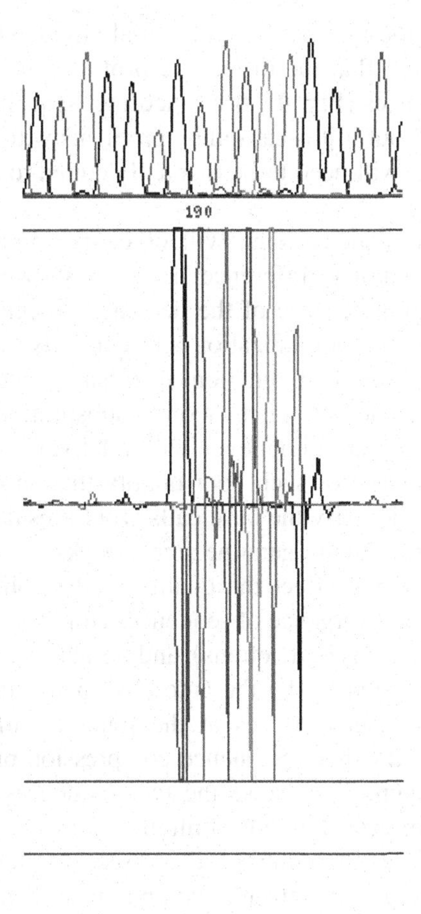

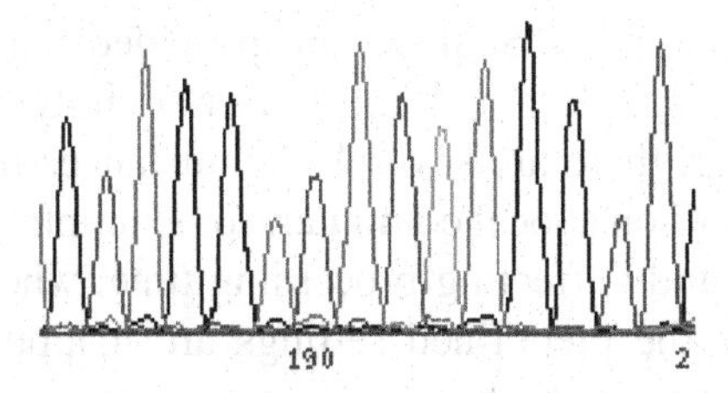

Fig. 3. The pattern resulting from a single base insertion/deletion. Here, a G at position 189 in the reference sequence is missing from the test sequence, causing the traces to shift temporarily out of alignment (*see* **Note 13**).

two sequences being too far apart for the initial alignment to succeed. In such cases, rather then being flat, the difference profile will be very noisy for the whole length of the traces (**Fig. 4**). If an incorrect sequence has been selected, this is easily resolved, but if it is caused by problems with the initial alignment, one or both sequences will need to be manually edited to better match the start position of the traces.

Poor quality input sequence data will of course produce reduced quality of output data with a noisy difference trace, as shown in **Fig. 5**, and will complicate interpretation because of the decrease in signal to noise ratio (*see* **Note 9**). Despite this, the normalization and corrective alignment algorithms built into SeqDoC, together with functions to enhance features of the difference profile which are characteristic of nucleotide substitutions, means that useful comparisons can still be carried out even with relatively poor quality sequence. Typically, only a sequence of such poor quality that an investigator could no longer unambiguously identify the base calls would result in a difference trace that was so noisy as to obscure genuine signal peaks.

Other issues which may affect the quality of data obtained from SeqDoC include dye-blob noise, divergence of sequences (for example by co-ligation, or as a result of large insertions or deletions), and sequence compression that is too extreme to be compensated for by the auto-alignment function (*see* **Note 10**). Examination of the sequence traces at the point of difference will readily identify the first two situations. Sequence compression problems are rare, but can be difficult to diagnose, because the two sequences are likely to appear identical yet still go irretrievably out of alignment (*see* **Note 11**).

Our website (http://research.imb.uq.edu.au/seqdoc/) features examples of a number of such problems with information on how to identify them and actions that can be taken to rectify them.

3.4. Installing SeqDoC locally

The web-based versions of SeqDoC are provided with predefined settings so as to provide the simplest and most user-friendly interface with broad applicability. However, there are several circumstances under which it can be useful to install SeqDoC on a local computer, for example where a slow network connection is significantly affecting processing time, where proprietary data is being used, or where the predefined settings are inappropriate for a specific analysis and need to be modified.

Settings which can be altered include increasing the window size for the initial alignment (to simplify comparison of samples sequenced with different primers), modifying the auto-alignment function to compensate for sequence

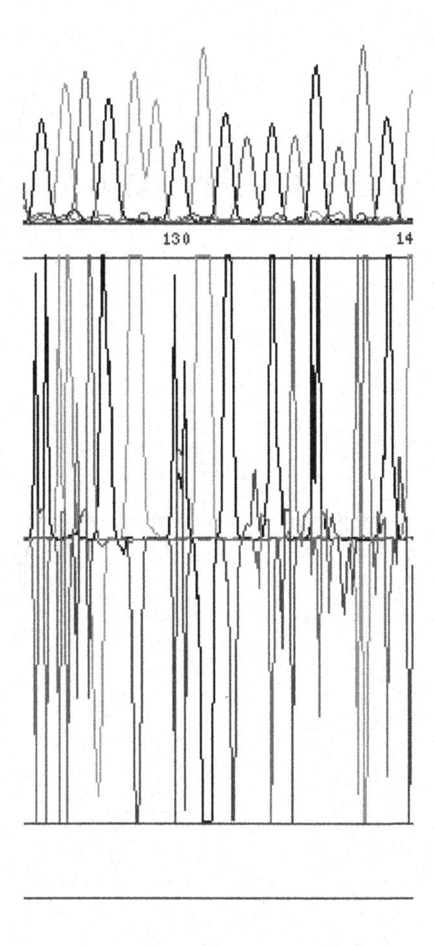

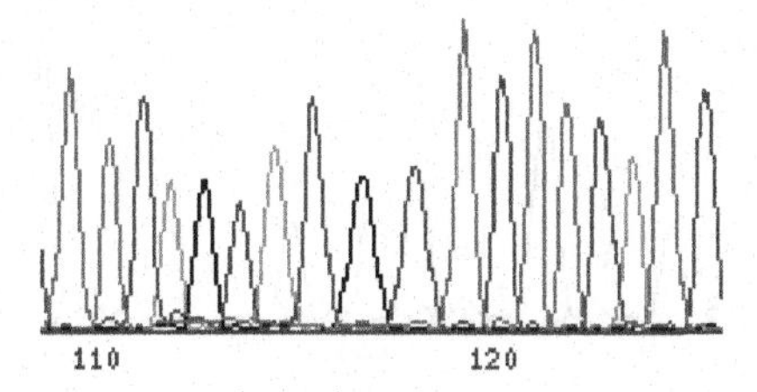

Fig. 4. Nonmatching sequences cannot be aligned, so give a very noisy difference trace (*see* **Note 13**).

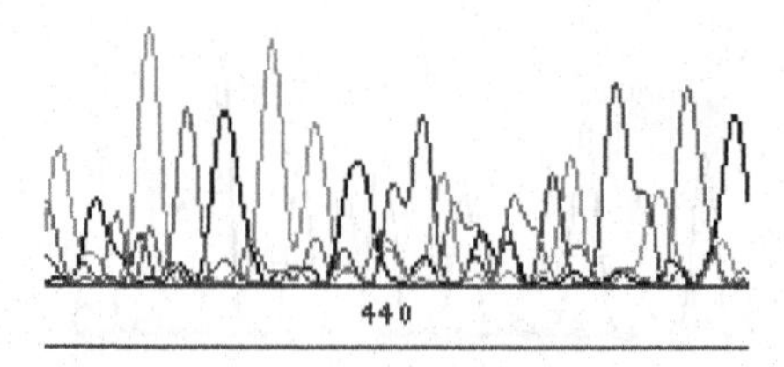

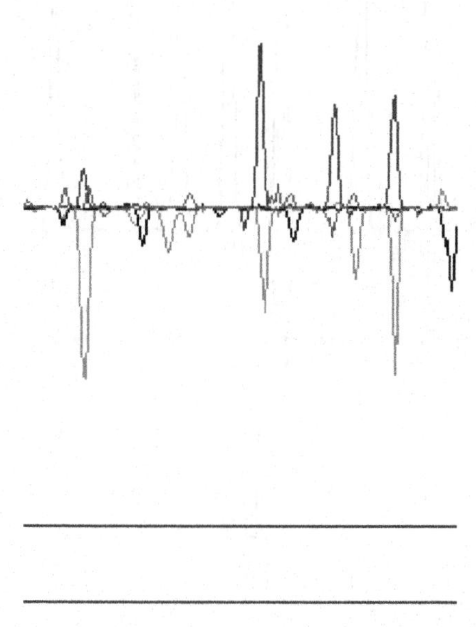

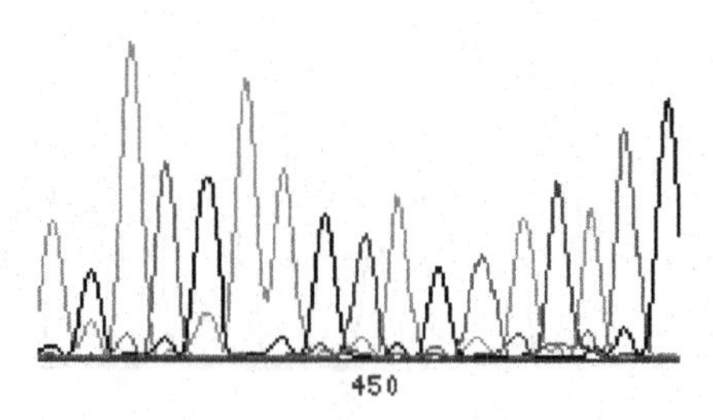

Fig. 5. Poor quality sequence data (top trace) results in a noisy difference trace (*see* **Note 13**).

compression, changing the difference trace calculations to increase the sensitivity of mixed peak detection, and even changing output trace colors to assist researchers with color blindness.

SeqDoC is a Perl program (http://www.perl.org/) and should work on any computer capable of running Perl (*see* **Note 12**). A number of accessory programs are also required to run SeqDoC—these are the GD graphics library and the Perl GD, GD::Graph and ABI modules (*see* references in **Subheading 2.**). In addition, web server software is required to use the web interface (the public SeqDoC version uses the Apache HTTP Server), although the code can be modified to generate a stand-alone application. Full instruction as to installation of these programs is outside the scope of this article, but interested researchers are welcome to contact the author for advice. This initial setup of SeqDoC requires a reasonable proficiency with building libraries and software from source, but subsequent modifications of parameters can be performed by anyone with basic Perl programming skills.

4. Notes

1. The increase in speed when using multiple SeqDoC is proportional to the number of test sequences. A two sequence comparison (i.e., with one reference and one test sequence) will take the same time using either version of SeqDoC, whereas the maximum 30% increase will be obtained when using 10 test samples.
2. For a comparison of two sequences, either can be assigned as the reference; the differences will be identified equally well. Processing of the two sequences is identical except that alignment alterations (insertion/deletion of sequence at the start of the trace for initial alignment and subsequent adjustments to peak spacing) are made only to the test sample.
3. For a multiple sequence comparison, the reference should be the sequence file from the wild-type/default sample if there is one, otherwise a good quality sequence trace with no mixed base calls (such as heterozygous SNPs). Although mixed bases in the reference do not have any specific impact on generation of the difference trace, they may complicate the subsequent analysis of the output.
4. The processing time will be significantly increased if a slow network connection is being used, because of the length of time taken to upload the files to the server.
5. Auto-start alignment can compensate for misalignments of up to around 20 nucleotides between the start positions of the two sequences, so nearly adjacent primers (such as in cloning vectors) should also be acceptable. Alternatively, initial manual alignment can be performed with chromatogram processing software, providing that the modified sequence traces are saved as ABI format files.
6. Bidirectional peaks in the difference trace are indicative of base substitutions because the value for one channel (that corresponding to the base in the reference sequence) will be very much higher in the reference than the test at that point in

the trace, whereas conversely the value for the channel corresponding to the base in the test sequence will be very much lower.

7. A bidirectional peak will be observed in the presence of a mixed base because, even though the base from the reference sequence is still present, it will be of reduced relative intensity in the test sequence. The magnitude of the difference trace peak for a 50/50 mixed base is theoretically approx 20% of that of a complete substitution. However this is subject to many other factors, and the magnitude of the difference peak is not a reliable method of determining the nature of the substitution.
8. If testing pooled DNA for SNPs, allele frequencies of around 20% or more can typically be detected. The default parameter settings of SeqDoC are optimized for substitutions, and consequently rarer alleles are not distinguished from noise (Dr. J. Logan, personal communication). The sensitivity of SeqDoC for such applications can be altered by installing a local version of the software and modifying the parameters to find the most suitable optimization settings.
9. Noise peaks do tend to appear different to the bidirectional peaks characteristic of nucleotide substitutions, in that they are often unidirectional or at least asymmetric, but they can also appear as bidirectional peaks very similar to genuine substitutions, and each of these will need individual examination.
10. Variation of gel conditions and DNA templates can, on occasion, cause the peak spacing interval to vary too greatly between the reference and test samples for the alignment algorithm to keep the two traces aligned. This may sometimes be corrected by modifying sequencing conditions, or can be compensated for by installing SeqDoC locally and modifying the alignment settings appropriately (Dr. A. Topf, personal communication).
11. Problems with sequence compression are probably best diagnosed by a preliminary comparison of the two sequences with a standard chromatogram viewer set to the same horizontal scale for both traces. A difference in base count of more than around 15% over the same size window may indicate sequence compression.
12. SeqDoC has been successfully installed and tested on Windows (using Cygwin: http://www.cygwin.com/) and on Linux operating systems, and has been reported also to work with ActiveState Perl under Windows.
13. All images in this article have been converted to greyscale, and many have been reduced in size. Full-size color versions are available on our website at http://research.imb.uq.edu.au/seqdoc/.

References

1. Collins, F. S., Guyer, M. S., and Chakravarti, A. (1997) Variations on a theme: cataloging human DNA sequence variation. *Science* **5343,** 1580–1581.
2. Twyman, R. M. (2004) SNP discovery and typing technologies for pharmacogenomics. *Curr. Top. Med. Chem.* **4,** 1423–1431.

3. Muñoz, I., Ruiz, A., Marquina, M., Barcelo, A., Albert, A., and Ariño, J. (2004) Functional characterization of the yeast Ppz1 phosphatase inhibitory subunit Hal3: a mutagenesis study. *J. Biol. Chem.* **279,** 42,619–42,627.
4. Guo, H. H., Choe, J., and Loeb, L. A. (2004) Protein tolerance to random amino acid change. *Proc. Natl. Acad. Sci. USA* **101,** 9205–9210.
5. Crowe, M. L. (2005) SeqDoC: rapid SNP and mutation detection by direct comparison of DNA sequence chromatograms. *BMC Bioinformatics* **6,** 133.
6. Parker, L. T., Deng, Q., Zakeri, H., Carlson, C., Nickerson, D. A., and Kwok, P. Y. (1995) Peak height variations in automated sequencing of PCR products using Taq dye-terminator chemistry. *Biotechniques* **19,** 116–121.
7. Zakeri, H., Amparo, G., Chen, S. M., Spurgeon, S., and Kwok, P. Y. (1998) Peak height pattern in dichloro-rhodamine and energy transfer dye terminator sequencing. *Biotechniques* **25,** 406–414.

23

SNPHunter

A Versatile Web-Based Tool for Acquiring and Managing Single Nucleotide Polymorphisms

Tianhua Niu

Summary

Sites in the DNA sequences where two homologous chromosomes differ at a single DNA base are called single nucleotide polymorphisms (SNPs). The human genome contains at least 10 million SNPs, making them the most abundant genetic "lampposts" for pinpointing causal variants underlying human diseases. SNP-related toolboxes and databases have become increasingly important for researchers to choose the most appropriate SNP set for attaining their research goals. This chapter introduces SNPHunter, a web-based software program that allows for SNP search (both *ad hoc* mode and batch mode), retrieval of SNP information, SNP management, automatic SNP selection based on customizable criteria including physical position, function class, flanking sequences at user-defined lengths, and heterozygosity from National Center for Biotechnology Information dbSNP. The SNP data extracted from dbSNP via SNPHunter can be exported and saved in plain text format for further down-stream analyses.

Key Words: Single nucleotide polymorphism; database; software; visualization; association studies.

1. Introduction

A single nucleotide polymorphism (SNP) is defined as a genomic locus where two or more alternative bases occur with appreciable frequencies (>1%). The human genome contains at least 10 million SNPs *(1)*. Among them, more than 5 million SNPs are expected to have minor-allele frequencies (MAFs) greater than 10% *(2)*.

From: *Methods in Molecular Biology, vol. 396: Comparative Genomics, Volume 2*
Edited by: N. H. Bergman © Humana Press Inc., Totowa, NJ

SNPs play a pivotal role in (i) candidate gene-based and genome-wide association studies, (ii) population genetics and molecular evolution studies, and (iii) fine mapping and positional cloning. SNPs account for 90% of interindividual variability, and for as many as 100,000 amino acid differences *(3)*. SNPs differ in frequencies, locations (intragenic vs intergenic), genetic context, and functional significance. Multiple linked SNPs can either be analyzed as individual markers, or can be analyzed as a haplotype, (defined as a combination of multiple marker alleles on a single chromosome) *(4)*.

Identifying SNPs underlying complex phenotypes remains a challenging enigma. The plethora of SNPs points out a major obstacle faced by scientists in planning costly population-based genotyping, which is to choose a SNP set that is a "tagging" SNP set consisting of representative SNPs that gleans the maximal information for a particular gene or genomic region, or a functional SNP set consisting of SNPs that affect either protein function or gene transcriptional activity or a combination of both.

The selection of SNPs to be genotyped relies increasingly on SNP database information rather than on expensive and time-consuming in-house SNP discovery studies. National Center for Biotechnology Information (NCBI) dbSNP (http://www.ncbi.nlm.nih.gov/SNP/) is the largest publicly accessible SNP database for genomic and cDNA sequence variations *(5)*, which contains 99.77% single nucleotide substitutions *(5)*. Further, NCBI dbSNP contains both disease-causing mutations as well as neutral polymorphisms *(5)*. Approximately 50% SNPs in NCBI dbSNP are validated *(5)*. NCBI dbSNP is designed to facilitate searches along five major axes of information: (i) sequence location, (ii) function class, (iii) cross-species homology, (iv) SNP quality or validation status, and (v) heterozygosity, i.e., proportion of heterozygous individuals in a population. The contents of dbSNP are also cross-linked to records in other databases such as NCBI MapViewer, Entrez Gene, and GenBank.

Because the budget for genotyping costs is not illimitable, scientists need tools to find the most appropriate set of SNPs from NCBI dbSNP, to evaluate their annotations, and to export them in formats suitable for subsequent analysis. SNPHunter *(6)* is a novel web-based tool that addresses such needs.

2. Software Requirement and Installation

2.1. Operating System Requirement

SNPHunter was written using Visual Basic .NET. Thus, it requires Windows Operating System (OS) (including Win98, Win2000, NT, and WinXP) with

.NET Framework (v1.0 or above) installed. Windows XP is recommended. NET Framework could be freely downloaded from Microsoft windows update.

2.2. Software Installation

SNPHunter package can be downloaded at the following website: http://www.hsph.harvard.edu/ppg/software.htm. Click on the clickable object "SNPHunter" of "SNPHunter for Windows (User Manual and Example files are included)", save the zipped folder "snphunter.zip" to your local directory. Unzip the "snphunter.zip" to get a file folder named "snphunter" with the following contents:

1. "example_files.zip". Unzip this zipped folder to extract its contents, which consists of a file folder named "example_files" containing six separate files: (i) test.summary.txt, (ii) test.gls, (iii) test.seq, (iv) test.sif, (v) test.sls, and (vi) test.snp.
2. "SNPHunter User Manual 1.7x.doc". This is the User's Guide to SNPHunter v1.7x.
3. SNPHunter.exe. This is the Windows OS executable. To launch the SNPHunter program. double click on the "SNPHunter.exe", and SNPHunter's main window will pop up.

3. Methods

SNPHunter is a web-based application designed to facilitate the acquisition and management of SNPs from NCBI dbSNP. SNPHunter is comprised of the following modules: (i) SNP Search Module, (ii) SNP Management Module, and (iii) Locuslink SNP Module. In the following, I provide a step-by-step protocol for using SNPHunter.

3.1. SNP Search Module

SNP Search Module provides the *ad hoc* mode SNP searching function. As depicted in **Fig. 1**, it provides two different options searching SNPs that meet the user's criteria: (i) "SNP Search" Panel; and (ii) "Direct SNP Search" Panel.

3.1.1. "SNP Search" Panel

In the "SNP Search" Panel, the gene symbol for the gene of user's interest can be input into the "Human Gene Symbol" box. Here, we plug in "CRP", which is the Locuslink Gene Symbol (i.e., Entrez Gene Symbol; note that NCBI Locuslink was superseded by the NCBI Entrez Gene, but this change does not affect SNPHunter's applicability—in the following Locuslink ID is defined as the GeneID of the NCBI Entrez Gene) for "C-reactive protein, pentraxin-related" for *Homo sapiens*. The user may choose customized function classes in "Gene Region" and/or customized 5′ and 3′ flanking sequence lengths in

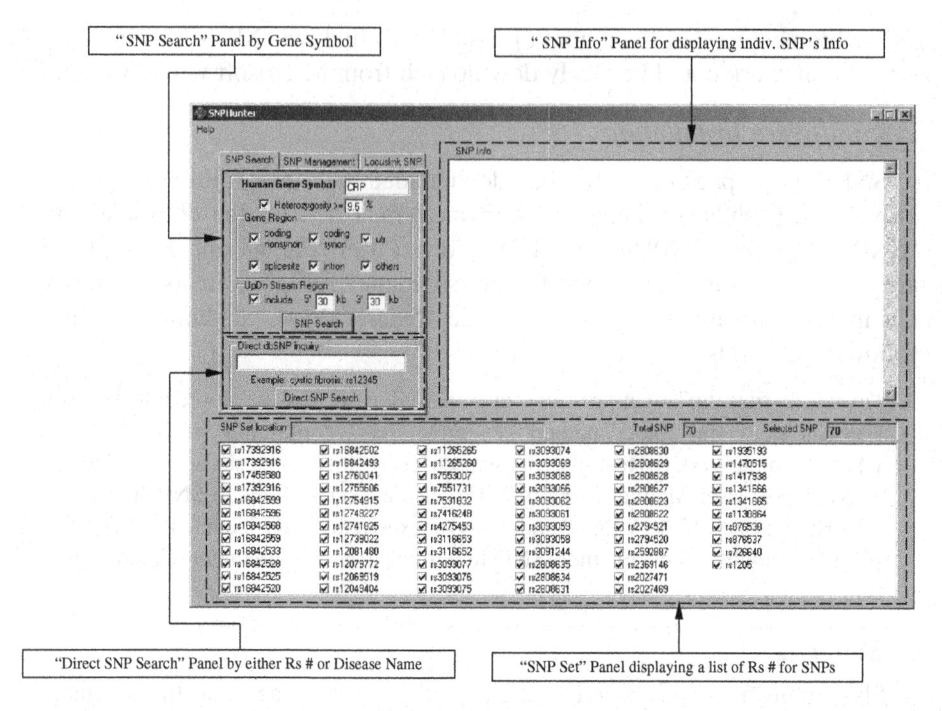

Fig. 1. The SNP Search Module. This Module consists of four panels: (i) "SNP Search" Panel by Gene Symbol; (ii) "Direct SNP Search" Panel by either NCBI dbSNP rs number or Disease Name; (iii) "SNP Set" Panel displaying a list of rs number(s) for SNP(s) identified; and (iv) "SNP Info" Panel displaying individual SNP's information for a SNP clicked on in the "SNP Set" Panel.

"UpDn Stream Region" (representing upstream/downstream region[s]) as the search criteria. In addition, by using the heterozygosity criteria, the user can restrict SNPs found to meet a customized heterozygosity threshold (e.g., 9.5%). After clicking on the "SNP Search" button, all SNPs that meet the user-specified search criteria will be displayed as a checklist of NCBI dbSNP rs number(s) in the "SNP Set" Panel **(Fig. 1)**. The user can check or uncheck a SNP by clicking on the checkbox in front of the corresponding SNP's rs number.

3.1.2. "Direct SNP Search" Panel

In the "Direct SNP Search" Panel, the user can directly search specific SNPs based on: (1) the name of the trait, such as "cystic fibrosis", or (2) a specific dbSNP rs number, such as “rs17173461” into the “Direct dbSNP inquiry” box. After clicking on “Direct SNP Search” button, all SNPs that meet the user’s search criteria will be listed in the “SNP Set” Panel.

3.2. SNP Management Module

In the preceding subsection, the user has found the SNP “rs17173461” in the SNP Search Module; now, the user can proceed to the “SNP Management” Module and retrieve its corresponding SNP summary information from NCBI dbSNP by clicking on the “Fetch SNP Info” button. After the retrieval process finishes (a status bar will notify the user about the status of the retrieval because this can be a time-consuming step for a long list of SNPs), the user can click on each SNP in the “SNP Set” Panel at the bottom, and the retrieved corresponding SNP summary information will be shown in the “SNP Info” Panel. By clicking on the “Save SNP Set” button, the user can save the retrieved SNP summary information into a file, which can be named as “rs17173461.sls” (“sls” is the file’s extension name that stands for “SNP list”). After saving this (.sls) file, the user can load this file into SNPHunter from the local directory by clicking on the “Load SNP Set” button. Afterward, the user can choose a particular length (e.g., 150 bp) for both 5′ upstream and 3′ downstream flanking sequences from the SNP site to retrieve the flanking sequences along with the SNP of interest, and the retrieved sequence file can be saved as “rs17173461.seq” (“seq” is the file’s extension name that stands for “Sequence”). Next, the user may click on the “Detect Nearby SNP” button to detect any potential SNPs in the flanking sequences contained in the “rs17173461.seq” file, and save the corresponding output file as “rs17173461.snp” (“snp” is the file’s extension name that stands for “SNP”).

The user can also work out the above steps using the example file called “test.sls” contained in the “example_files” folder. First, load the “test.sls” file by clicking on the “Load SNP Set” button. Second, the user click on the “Save Flanking Seq” button to retrieve the flanking sequence information based on a user-specified flanking sequence length (e.g., 250-bp), and the user may save the output file as “test2.seq”. Third, the user may click on the “Detect Nearby SNP” button and save the output file as “test2.snp”.

Overall, the operational order for “SNP Search” Module and “SNP Management” Module, as shown in **Fig. 2**, can be recapitulated as follows:

“SNP Search” Module, choose a “SNP Set” of intrest → “SNP Management” Module, click on the “Fetch SNP Info” button to fetch information for all the SNPs in the “SNP Set” from NCBI dbSNP → “SNP Management” Module, click

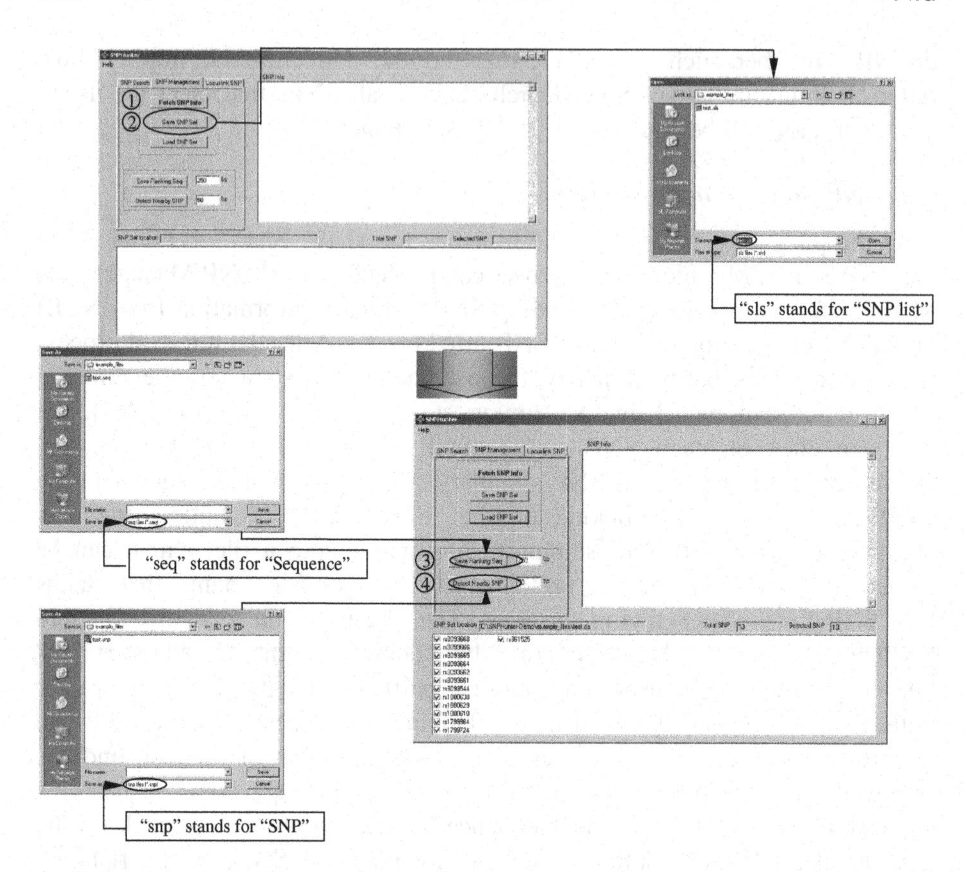

Fig. 2. The SNP Management Module. Once a SNP set is selected in the SNP Search Module, Four steps are involved as indicated: (1) the user can first click on the "Fetch SNP Info" button to fetch dbSNP information for all the SNPs in the SNP set; (2) the user can click on the "Save SNP Set" button to save the "SNP Info" file into a (.sls) file; (3) the user can click on the "Save Flanking Seq" button to save the "SNP Flanking Seq" file into a (.seq) file; and (4) the user can click on the "Detect Nearby SNP" button to save the "Nearby SNP" file into a (.snp) file.

on the "Save SNP Set" button to save the "SNP Info" file into a (.sls) file → "SNP Management" Module, click on the "Load SNP Set" button to load a (.sls) file → "SNP Management" Module, click on the "Save Flanking Seq" button to retrieve the flanking sequences of the SNP, and save the output file into a (.seq) file → "SNP Management" Module, click on the "Detect Nearby SNP" button to detect SNPs, and save the output file into a (.snp) file.

3.3. Locuslink SNP Module

Locuslink SNP Module provides the batch mode SNP searching function for multiple genes based on their Locuslink IDs (i.e., GeneIDs of NCBI Entrez Gene). It involves two steps: step 1, creating a "gene summary" file by using a preexisting "gene list" file; step 2, filter SNPs using the user-specified criteria.

3.3.1. Step 1: Creating a "Gene Summary" File

In step 1, the user loads a preexisting "gene list" file to retrieve the respective SNP information for each gene contained in the "gene list" file, and then, the retrieved information is saved into a new "gene summary" file with an extension name "summary.txt".

The user may use the file named "test.gls" as an example, which contains a single column and two lines (note that the file is just a plain text file).

4846

1234

Click the "Create Summary" button and load the file we just created, i.e., "test.gls", Now, a new text file named "test.summary.txt" is created.

3.3.2. Step 2: Filter SNPs

In step 2, the user needs to filter the SNPs using the user-specified criteria. Use the file "test.summary.txt" in the preceding subsection as the example. The user may click on the "Filter SNP" button, and a new "Filter SNP" window will pop up (**Fig. 3**). In this "Filter SNP" window, first, the user may load the summary file "test.summary.txt" by clicking on the "Load" button and the Locuslink IDs for the corresponding genes in this summary file are loaded. Next, the user can click on the Locuslink ID "4846" shown in the upper right corner, and a visual display of the SNPs located on this gene (Entrez Gene Symbol: NOS3) will be shown at the bottom of the "Filter SNP" window. Click on the "Out" button until all the SNPs (represented as colored vertical bars) can be seen. The height of each SNP bar indicates the heterozygosity value for that SNP, and the four dotted horizontal lines indicate heterozygosity values of 0, 0.25, 0.5, and 0.75, respectively. With regard to the SNP bar's color legends depicting SNP function class, "green" means "coding: synonymy unknown" or "synonymous;" "red" means "nonsynonymous"; "orange" means "all the others"; and "gray" means "unselected SNPs" (i.e., "Pick" field is "0"). In the graphical display, there is an inverse red triangle indicating the position of the SNP.

The user can freely pick or depick (i.e., filter) a SNP belonging to the table by manually setting the corresponding "Pick" value to "1" or "0", respectively.

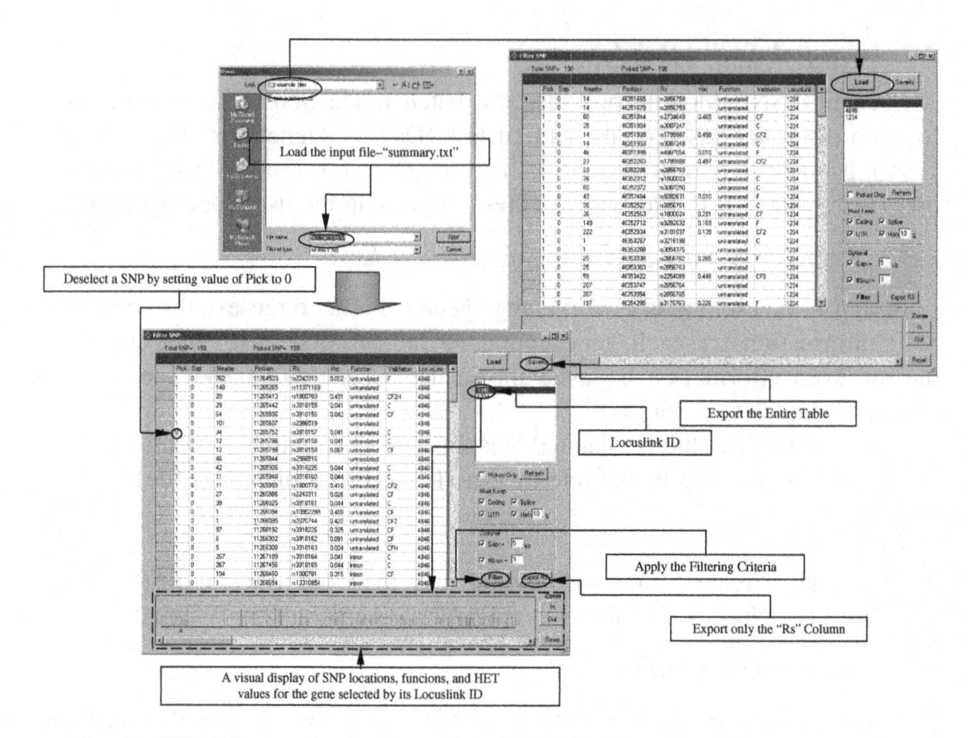

Fig. 3. SNP Filtering in the Locuslink SNP Module. The "Filter SNP" window provides a table corresponding to the gene summary file, a visual display of the SNP information for a gene with its Locuslink ID being clicked on, and various customizable SNP selection options.

When a SNP is depicked, the color of the corresponding SNP in the visual display will turn into gray. The user can then check the "Picked Only" checkbox and click on the "Refresh" button to refresh the table. Those depicked SNP(s) will be dropped from the table upon refreshing. There are two panels on the right-hand side that facilitate the user to filter SNPs according to predefined criteria: (i) "Must keep" panel. The user may keep those absolutely needed SNPs, such as "coding" or "splicing" SNPs. Once checked, those SNPs will *always* be kept regardless of other filtering procedures. (ii) "Optional" panel. This panel provides the user with the option to minimize inter-SNP gaps (with Gaps less or equal to a pre-specified length) and the option to select a minimal number of SNPs (with the total number of selected SNPs greater or equal to a pre-specified number) to meet the user's needs.

Once the SNP list has been properly filtered, there are two alternative choices of exporting the output table: (i) export only the dbSNP rs numbers of the selected SNPs by clicking on the "Export Rs" button (Here we can save the output file as "test_rs.sls".) (ii) export the complete table including all the columns out by clicking on the "SaveAs" button. Here, the user can save the output file as "test_completeTable.txt".

4. Exercises

Create a file folder called "exercises" within the "snphunter" folder. Save all the exercise files into this local folder.

4.1. SNP Search module and SNP Management Module

1. Select SNPs for the HNF4A gene. Set heterozygosity >= 0.095, select those SNPs belonging to two function classes "coding nonsyn" and "coding syn", include SNPs located also in 5 kb 5′ upstream region and 5 kb 3′ downstream region, fetch SNP information for the SNPs selected, and save the fetched information into the "HNF4A.sls" file; retrieve and save the 250-bp 5′ upstream and 3′ downstream flanking sequences for all selected SNPs into the "HNF4A.seq" file, and detect and save nearby SNPs for all selected SNPs into the "HNF4A.snp" file.
2. Generate the following list of SNPs indicated by their dbSNP rs numbers, and save it into a file named "SNP_list.txt". Then, rename the file name into "SNP_list.sls."

rs17115487
rs17101784
rs17567561

Load "SNP_list.sls" into the "SNP Management" Module of SNPHunter, fetch SNP information, and save the fetched information to override the "SNP_list.sls" file; retrieve and save the 250-bp 5′ upstream and 3′ downstream flanking sequences for all selected SNPs into the "SNP_list.seq" file, and detect and save nearby SNPs for all selected SNPs into the "SNP_list.snp" file.

4.2. Locuslink SNP Module

Generate the following list of genes indicated by their Locuslink ID (i.e., Entrez GeneID) numbers, and save it into a file named "GENE_list.txt". Then, rename the file name into "GENE_list.gls".

952
958
960

Create "cd.summary.txt". Use the "cd.summary.txt" to filter SNPs according to the following criteria:

1. Must keep coding and splicing SNPs as well as SNPs with heterozygosity values > 0.10. Keep SNP Gap (i.e. inter-SNP interval) <= 6 kb, and the total # SNPs must be at least eight.
2. Output the results into "cd_rs.sls" for the rs column only and "cd_completeTable.txt" for the complete table, respectively.

5. Notes

1. SNPs are either causal genetic variants or positional markers of a wide spectrum of human diseases. The SNP information retrieved by SNPHunter from NCBI dbSNP (function class, location, flanking sequences, nearby SNPs, and frequency) allows scientists to decide on the most optimal set of SNPs tailored for their study purpose. Besides SNP selection, SNPHunter can retrieve flanking sequences for a particular SNP, which are generally required to design primers for PCR-based SNP assays. Moreover, the graphical display in the "Filter SNP" window of SNPHunter provides a visually integrated representation of SNPs for a gene picked by the user.
2. To reduce an overwhelming number of SNPs to a reduced, manageable number, especially in the case of candidate gene studies, scientists often prioritize SNPs according to their functional significance ***(7,8)***. The Sorting Intolerant From Tolerant program ***(9)*** may help to discriminate between functionally neutral nonsynonymous SNPs (nsSNPs) and nsSNPs with deleterious effects. The "topoSNP" database provides a topographic mapping of nsSNPs onto known 3D structures of proteins ***(10)***, and the SNPs3D resource provides a way of identifying those nsSNPs that are likely to have a deleterious impact on molecular function in vivo ***(11)***.
3. It should be noted that there is no "panacea" solution for addressing the SNP selection problem, and SNP selection is always an iterative process that involves looping over several times to identify the most optimal set of SNPs.

Acknowledgments

The author is grateful to many SNPHunter users for sharing their constructive comments. The author would like to thank Drs. Lin Wang, Simin Liu, and Aditi Hazra for insightful discussions. This work was supported in part by National Institutes of Health grants R01 HG002518, R01 DK062290, R01 DK066401, and R01 HL073882.

References

1. Crawford, D. C., Akey, D. T., and Nickerson, D. A. (2005) The patterns of natural variation in human genes. *Annu. Rev. Genomics Hum. Genet.* **6**, 287–312.
2. Carlson, C. S., Eberle, M. A., Rieder, M. J., Yi, Q., Kruglyak, L., and Nickerson, D. A. (2004) Selecting a maximally informative set of single-nucleotide polymorphisms for association analyses using linkage disequilibrium. *Am. J. Hum. Genet.* **74**, 106–120.

3. Brookes, A. J. (1999) The essence of SNPs. *Gene* **234**, 177–186.
4. Niu, T. (2004) Algorithms for inferring haplotypes. *Genet. Epidemiol.* **27**, 334–347.
5. Sherry, S. T., Ward, M. H., Kholodov, M., et al. (2001) dbSNP: the NCBI database of genetic variation. *Nucleic Acids Res.* **29**, 308–311.
6. Wang, L., Liu, S., Niu, T., and Xu, X. (2005) SNPHunter: a bioinformatic software for single nucleotide polymorphism data acquisition and management. *BMC Bioinformatics* **6**, 60.
7. Emahazion, T., Feuk, L., Jobs, M., et al. (2001) SNP association studies in Alzheimer's disease highlight problems for complex disease analysis. *Trends Genet.* **17**, 407–413.
8. Schork, N. J., Fallin, D., and Lanchbury, J. S. (2000) Single nucleotide polymorphisms and the future of genetic epidemiology. *Clin. Genet.* **58**, 250–264.
9. Ng, P. C. and Henikoff, S. (2003) SIFT: predicting amino acid changes that affect protein function. *Nucleic Acids Res.* **31**, 3812–3814.
10. Stitziel, N. O., Binkowski, T. A., Tseng, Y. Y., Kasif, S., and Liang, J. (2004) topoSNP: a topographic database of non-synonymous single nucleotide polymorphisms with and without known disease association. *Nucleic Acids Res.* **32**, D520–D522.
11. Yue, P., Melamud, E., and Moult, J. (2006) SNPs3D: candidate gene and SNP selection for association studies. *BMC Bioinformatics* **7**, 166.

24

Identification of Disease Genes

Example-Driven Web-Based Tutorial

Medha Bhagwat

Summary

The National Center for Biotechnology Information (NCBI) has developed several web-based mini-courses (www.ncbi.nlm.nih.gov/Class/minicourses) illustrating the applications of NCBI resources. This chapter describes the problem-based minicourse called "Identification of Disease Genes." The mini-course guides us through one of the several ways to identify disease related genes, starting from the expressed sequence data such as that may have been obtained from patients. The chapter first provides an introduction to the human genome assembly and the resources such as the Basic Local Alignment Search Tool, Map Viewer, Single Nucleotide Polymorphism database, and Online Mendelian Inheritance in Man. The chapter then demonstrates the practical application of these resources to the identification of genes related to two diseases, hemochromatosis and sickle cell anemia. The chapter also provides links to the mini-course web pages and includes the screen images of the results of the applied steps.

Key Words: Disease genes; hemochromatosis; sickle cell anemia; BLAST; dbSNP; OMIM; Map Viewer; mini-courses; NCBI; EST.

1. Introduction to the National Center for Biotechnology Information Mini-Courses

National Center for Biotechnology Information (NCBI) provides several focused bioinformatics mini-courses (http://www.ncbi.nlm.nih.gov/Class/minicourses/). The mini-courses are either problem-based such as "Identification of Disease Genes" or NCBI resource-based such as "BLAST Quick Start." The courses are 2.5 h in length with the first 90 min devoted to an

From: *Methods in Molecular Biology, vol. 396: Comparative Genomics, Volume 2*
Edited by: N. H. Bergman © Humana Press Inc., Totowa, NJ

overview and an online demonstration of a problem or problem set by an instructor. This is followed by a 1-h hands-on session where students practice a similar problem or problem set to the one demonstrated at their own computers. The courses are taught on the National Institutes of Health (NIH) campus in Bethesda and at academic institutes in the United States.

2. Objective

This chapter describes the mini-course that focuses on the identification of a disease gene using NCBI's human genome assembly. The reference human genome assembly along with integrated maps, literature, and expression information comprises a powerful discovery system for exploring candidate human disease genes.

3. Genetics and Bioinformatics Background

3.1. Information Transfer Within a Cell

The pathway of genetic information transfer in a cell begins with the transcription of genes within a genome to produce mRNAs and ends with the translation of mRNAs to produce proteins. The sequence databases contain genomic, mRNA, and protein sequences representing all three stages in the pathway.

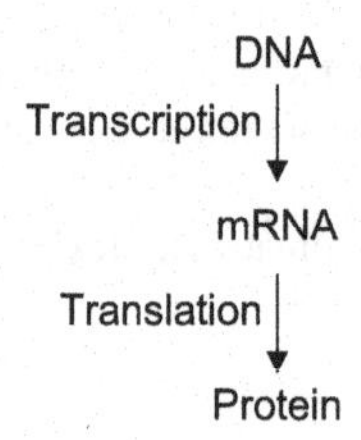

3.2. Data Available for Bioinformatics Analysis

One way to identify genes in a genome is to generate a cDNA library from the pool of RNA messages. To generate a cDNA library, the RNA messages from a tissue or from cells representing a developmental stage are copied into more stable cDNA molecules, which are then placed into an appropriate vector to generate a collection of cDNA clones (vector and the individual cDNA insert). The single pass, short 300–500 nucleotide sequences obtained from sequencing either end of the cDNA insert are called expressed sequence tags (ESTs).

For more background information about genetic terms, the user may refer to the following webpages:

Talking Glossary of Genetic Terms (http://www.genome.gov/glossary.cfm).

The NCBI Handbook Glossary (http://www.ncbi.nlm.nih.gov/books/bv.fcgi?rid=handbook.glossary.1237).

A Science Primer (http://www.ncbi.nlm.nih.gov/About/primer/index.html).

NCBI Bookshelf (http://www.ncbi.nlm.nih.gov/entrez/query.fcgi?db=Books).

One way to solve the problem of identifying genes responsible for a particular phenotype is to generate a cDNA library from patient tissues/samples and obtain a number of ESTs. Then, use the ESTs to determine the genes expressing them and to determine whether they contain any nucleotide variations or single nucleotide polymorphisms (SNPs) when compared to normal individuals. Sites of DNA sequences where individuals differ at a single nucleotide are called SNPs. We will obtain more information about the SNP database in the latter part of this chapter.

4. General Protocol and Required Resources

4.1. Outline of Steps

1. Compare the sequences of ESTs from a patient to the sequences of the human genome (using Basic Local Alignment Search Tool [BLAST]).
2. Identify the genes aligning to the ESTs and download their sequences (using Map Viewer).
3. Identify whether the EST sequences contain any known SNPs (using dbSNP).
4. Determine whether a gene variant is known to cause a phenotype (using Online Mendelian Inheritance in Man [OMIM]).

Thus, starting from the transcribed sequences derived from patients, we will obtain information about expressed genes and determine whether these genes contain known variations that lead to the disease phenotype.

4.2. Descriptions of Resources Used

NCBI assembles component sequences from the human genome sequencing project into longer sequences called contigs whose accession numbers begin with prefix "NT_". NCBI also performs a number of annotations on the assembly to identify genes, transcripts, clones, repeats, markers, and SNPs. NCBI releases the updated human genome assembly or the new "Build" periodically. For more information about the human genome assembly and annotation, *see* **ref. *1*** and the help document (http://www.ncbi.nlm.nih.gov/mapview/static/humansearch.html).

This problem based mini-course guides us through use of NCBI resources such as BLAST, Map Viewer, dbSNP, and OMIM as tools to identify disease genes *(2)*.

4.2.1. BLAST

BLAST provides a method for rapid searching of nucleotide and protein databases for similarities with a query nucleotide or protein sequence *(3,4)*. The human genome BLAST page at (http://www.ncbi.nlm.nih.gov/genome/seq/BlastGen/BlastGen.cgi?taxid=9606) provides centralized access to the NCBI human genome assembly and annotated transcript and protein sequences. The BLAST output links directly to the Human Genome Map Viewer, where database hits can be analyzed in their genomic context to see the relationship with other annotated features.

4.2.2. Map Viewer

The Map Viewer (http://www.ncbi.nlm.nih.gov/mapview/) allows us to view and search an organism's complete genome *(5)*. It shows integrated views of a collection of genetic, physical, and sequence maps for annotated genes, expressed sequences, SNPs, and other features, and, thus, is a valuable tool for the identification and localization of genes that contribute to human disease (as demonstrated in this mini-course).

4.2.3. dbSNP

NCBI's SNP database (http://www.ncbi.nlm.nih.gov/SNP/) contains both single nucleotide substitutions, and short deletion and insertions *(6)*. The data in dbSNP are integrated with other NCBI genomic data. SNPs are aligned to the human genome and the locations of SNPs with respect to the annotated genes and mRNAs are identified.

4.2.4. OMIM

OMIM (http://www.ncbi.nlm.nih.gov/entrez/query.fcgi?db=OMIM) is the database of human genes and genetic disorders developed and edited by Dr. Victor A. McKusick and his colleagues at Johns Hopkins and elsewhere, and adapted for the Internet by NCBI (*see* **Note 1** about Online Mendelian Inheritance in Animals) *(7)*.

5. Detailed Protocol: Problem 1

In this problem, we will use as an example the hemochromatosis disease, which is characterized by an iron overload. Consider that a researcher is working on the hemochromatosis disease and needs to obtain information about the gene(s) causing the phenotype. The following steps will describe the analysis of EST sequences that might have been obtained from a hemochromatosis patient.

It is recommended to follow the link to the "Identification of Disease Genes" through the mini-course webpage (http://www.ncbi.nlm.nih.gov/Class/minicourses/).

This page contains a link to a file containing the up-to-date screen images of each of the steps described below. Referring to the file is strongly recommended to follow the mini-course steps. However, a number of screen images are provided in this chapter as well for a reader to follow along. These screen images are from human Build 35.1.

5.1. Step 1: Compare ESTs to The Human Genome

One way to identify the genes expressing the ESTs is to compare their sequences using BLAST with the human genome assembly and the genes annotated on it. The specialized BLAST page for searching against the annotated human genome assembly is at (http://www.ncbi.nlm.nih.gov/genome/seq/BlastGen/BlastGen.cgi?taxid=9606) (*see* **Note 2**). The user may directly access the human genome BLAST page through the "Identification of Disease Genes" mini-course web page by clicking on the "BLAST (human genome)" link. We can concatenate a number of EST sequences to run the search as a batch. However, we will use only one EST sequence as a query for this analysis (*see* **Note 3**). Paste the EST sequence provided on the mini-course page in the query box of the BLAST page and select the "genome (reference only)" database from the pull down menu and use the default program MegaBlast *(8)* (*see* **Note 4**). Start the search by clicking on the "Begin Search" button and obtain the results by clicking on the "Format" button. The BLAST results page shows only one match to the contig sequence NT_007592.14 on chromosome 6 in the human genome Build 35.1. In certain cases, there may be multiple matches to the human genome assembly (*see* **Notes 5** and **6**).

The alignment of the query EST sequence (indicated by "query") and the matched sequence from chromosome 6 (indicated by "sbjct") shows that the EST sequence is only 99 % identical to the genomic sequence (**Fig. 1**). Note the location of the nucleotide that is different between the two sequences (a G to A variation at the nucleotide 16951392 of the contig NT_007592.14).

The difference may be due to a sequencing error in the low quality EST sequence or it may represent a real SNP in the human genome. For future reference, paste your results, such as the alignment and the nucleotide difference (sequence difference at the nucleotide 16951392 on NT_007592.14; G in the genomic and A in the query EST sequence), in the window provided in the mini-course webpage.

```
>ref|NT_007592.14|Hs6_7749    Homo sapiens chromosome 6 genomic contig
          Length=48945890

 Features in this part of subject sequence:
   hemochromatosis protein isoform 11 precursor
   hemochromatosis protein isoform 10 precursor

 Score =  525 bits (273),  Expect = 4e-147
 Identities = 275/276 (99%), Gaps = 0/276 (0%)
 Strand=Plus/Plus

Query  1         TGCCTCCTTTGGTGAAGGTGACACATCATGTGACCTCTTCAGTGACCACTCTACGGTGTC  60
                 ||||||||||||||||||||||||||||||||||||||||||||||||||||||||||||
Sbjct  16951164  TGCCTCCTTTGGTGAAGGTGACACATCATGTGACCTCTTCAGTGACCACTCTACGGTGTC  16951223

Query  61        GGGCCTTGAACTACTACCCCCAGAACATCACCATGAAGTGGCTGAAGGATAAGCAGCCAA  120
                 ||||||||||||||||||||||||||||||||||||||||||||||||||||||||||||
Sbjct  16951224  GGGCCTTGAACTACTACCCCCAGAACATCACCATGAAGTGGCTGAAGGATAAGCAGCCAA  16951283

Query  121       TGGATGCCAAGGAGTTCGAACCTAAAGACGTATTGCCCAATGGGGATGGGACCTACCAGG  180
                 ||||||||||||||||||||||||||||||||||||||||||||||||||||||||||||
Sbjct  16951284  TGGATGCCAAGGAGTTCGAACCTAAAGACGTATTGCCCAATGGGGATGGGACCTACCAGG  16951343

Query  181       GCTGGATAACCTTGGCTGTACCCCCTGGGGAAGAGCAGAGATATACGTACCAGGTGGAGC  240
                 |||||||||||||||||||||||||||||||||||||||||||||||| |||||||||||
Sbjct  16951344  GCTGGATAACCTTGGCTGTACCCCCTGGGGAAGAGCAGAGATATACGTGCCAGGTGGAGC  16951403

Query  241       ACCCAGGCCTGGATCAGCCCCTCATTGTGATCTGGG  276
                 ||||||||||||||||||||||||||||||||||||
Sbjct  16951404  ACCCAGGCCTGGATCAGCCCCTCATTGTGATCTGGG  16951439
```

Fig. 1. MegaBlast of the query EST against the human genome: alignment overview. The difference between the EST and genomic sequence (a G to A variation at the nucleotide 16951392 of the contig NT_007592.14) is highlighted by a rectangle.

5.2. Step 2: Identify the Genes Expressing the ESTs and Download Their Sequences

We will now take advantage of the NCBI annotation of the human genome assembly to identify the gene corresponding to the EST by using the Map Viewer. To visualize the BLAST hit on the genome using Map Viewer, click the "Genome View" button at the top of the BLAST results page, then on the Map element "NT_007592." Currently, four maps should be displayed (Model, RNA, Genes_seq, and Contig) **(Fig. 2)**.

The Genes_seq map shows the "known" genes annotated by alignment of EST and/or mRNA sequences to the assembly. The Contig map shows the assembled genome contig sequence in the region, the Model map shows the *Ab initio* model genes predicted by the NCBI's program Gnomon and the RNA map shows the alignments of the known alternatively spliced transcripts. For more information about the human genome assembly and annotation,

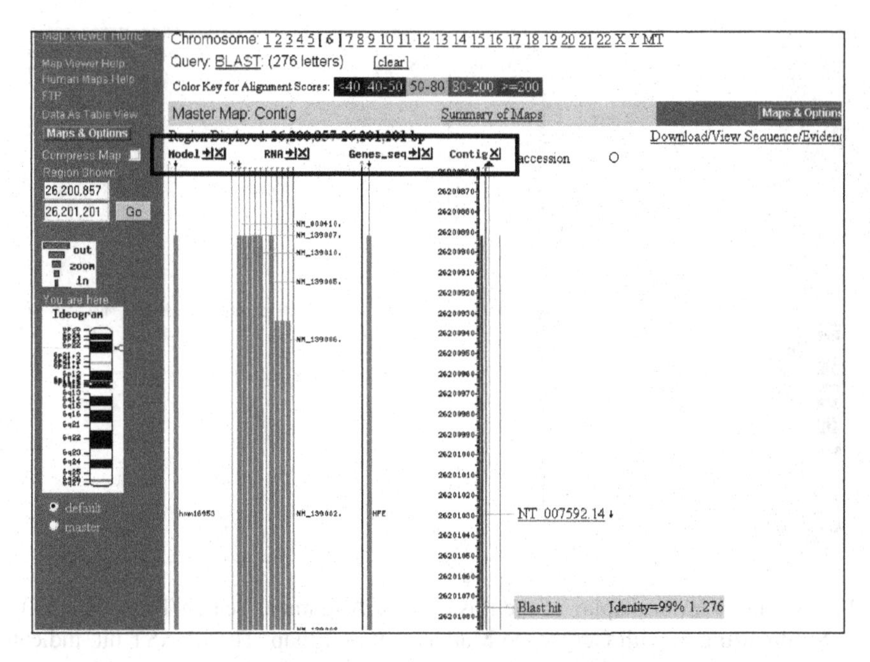

Fig. 2. Map Viewer display of the Basic Local Alignment Search Tool hit from **Subheading 5.2**. The four maps displayed in this view, Model, RNA, Gene-seq, Contig, are highlighted by a rectangle.

see **ref. *4*** and the help document (http://www.ncbi.nlm.nih.gov/mapview/static/humansearch.html).

Make the Genes_seq map the master map by clicking on the arrow at its top. The BLAST hit, indicated by the red bar, is in the region of one of the exons of the *HFE* gene annotated on the human genome **(Fig. 3)**.

The thick bars in the Genes_seq map indicate the exons and the thin lines joining them indicate introns of the gene. Zoom out several times until the user sees the entire *HFE* gene structure by clicking on the gray line and selecting option "Zoom out 2 times" from the menu that appears. The query EST represents a known gene, HFE. The orientation of the arrow next to the gene link indicates the orientation of the gene on the forward or the reverse strand. A gene annotated on the forward strand is indicated by an arrow pointing downward whereas a gene annotated on the reverse strand is indicated by an arrow pointing upward (*see* **Note 7**). The HFE gene is annotated on the forward strand of chromosome 6.

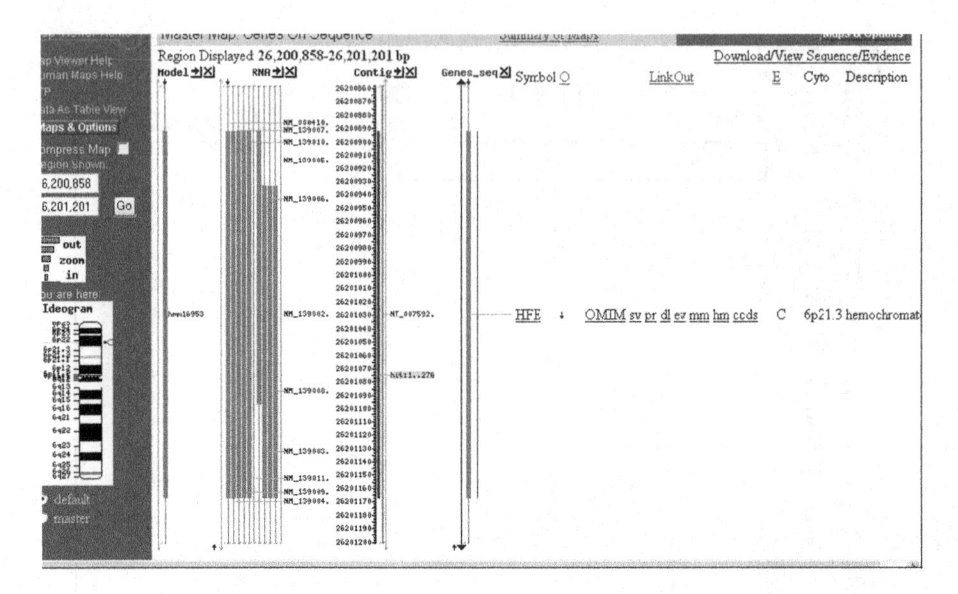

Fig. 3. Map Viewer display of the Basic Local Alignment Search Tool (BLAST) hit from **Subheading 5.** with Genes_seq Map as a master map. The BLAST hit, indicated by a bar on the right side of the Genes_seq map, is in the region of one of the exons of the *HFE* gene.

The right-most map is called the master map and links that give more information about map elements are provided next to it. For example, the current master map, Genes_seq map, has links to resources that provide more information about the *HFE* gene such as OMIM, sv (Sequence Viewer), pr (Reference Proteins), dl (Download Sequence), ev (Evidence Viewer), mm (Model Maker), and hm (Homologene). For more information, please refer to the Human Maps Help document http://www.ncbi.nlm.nih.gov/mapview/static/humansearch.html. We will use some of these links in this mini-course. Display the entire HFE gene sequence by clicking on the download "dl" link and then on "Display" on the next page (*see* **Notes 8** and **9**). Copy the sequence and paste it in the area provided on the mini-course page. Note the accession number of the longest transcript, NM_000410. We will use this information in the next step.

5.3. Step 3: Determine Whether the ESTs Contain Known SNPs

Go back to the Map Viewer report by clicking on the back button of the browser twice. Click the Maps and Options link.

Remove all the maps except the Genes_seq map by selecting the map under the "Maps Displayed" menu and clicking on "Remove." Now add the Variation map from the "Available maps" menu (by selecting the map and clicking Add). Make the Variation map the master map by selecting it and clicking the "Make Master/Move to Bottom" option. Then click "Apply." (The Mini-Course Map Viewer Quick Start describes the usage of the Map Viewer in detail.)

Now two maps are displayed, Variation (it is the rightmost and the master map) and Genes_seq (*see* **Fig. 4**). The master map provides detailed information for the map features, in this case SNPs. Zoom in on the blast hit area (bar on the right side of each map) by clicking on the map line next to it and choosing the appropriate zoom level. There are two SNPs in the area; rs1800562 and rs4986950 (*see* **Notes 10** and **11**).

Click any of the links and obtain information about the location and the nucleotide variation from the "Fasta sequence" and "Integrated maps" panels. The SNP, rs1800562, represents an A/G SNP **(Fig. 5)** at the nucleotide position 16951392 on the contig NT_007592.14 of the reference assembly (**Fig. 6**, *see* **Note 12**).

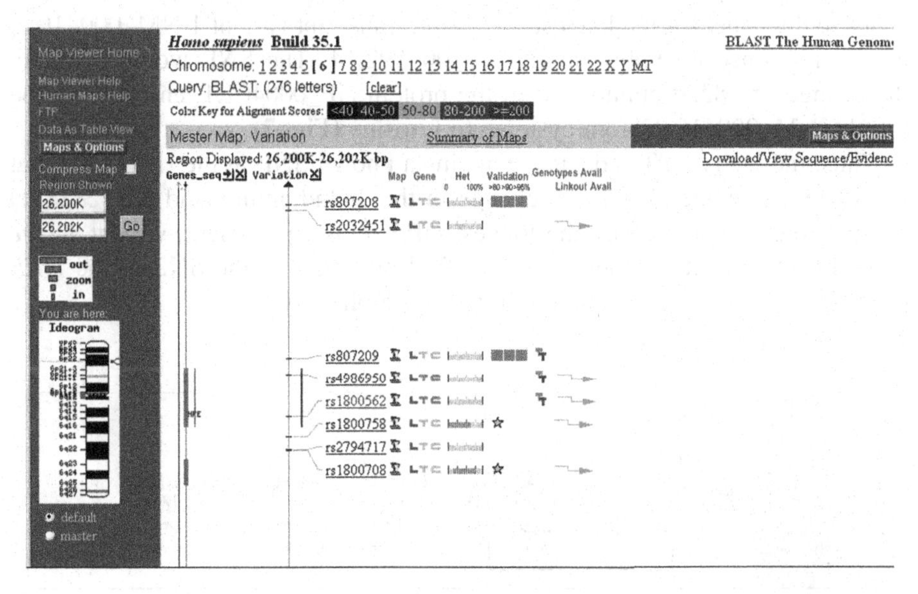

Fig. 4. Map Viewer display, containing the variation and Genes_seq maps, zoomed in the region of the Basic Local Alignment Search Tool (BLAST) hit in **Subheading 5.** There are two SNPs, rs1800562 and rs4986950, in the BLAST hit area indicated by a bar on the right side of each map.

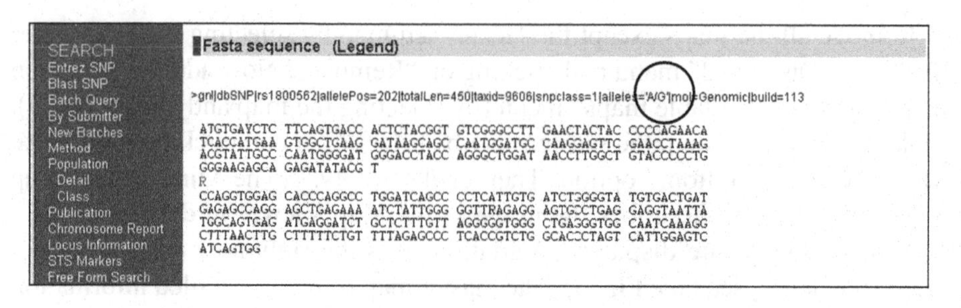

Fig. 5. Fasta sequence section of the SNP entry rs1800562. The A/G allele in the SNP, indicated in the definition line on the record, is highlighted by an oval.

This is the same nucleotide variation on the contig NT_007592.14 found in the BLAST result in **Subheading 5.1.** (16951392 G to A). To identify whether this change represents a change in an encoded amino acid, we will refer to the GeneView panel. This view shows the location of the SNP in the alternatively spliced products annotated on all the assemblies. It also provides information at the protein level; the amino acid number and the change in the sequence, if any. Refer to the panel for the longest transcript, transcript variant 1 NM_000410.2, on the reference assembly contig NT_007592.14. The SNP would result in the change of 282nd amino acid in the protein NP_000401.1, encoded by the mRNA NM_000410.2, from cysteine to tyrosine **(Fig. 7)**.

Thus, the query EST sequence contains a known SNP in the HFE gene that results in a cysteine to tyrosine change in the 282nd amino acid (Cys282Tyr) of the protein expressed by the longest HFE transcript variant, variant 1 (*see* **Note 13**). The next obvious step is to find out whether the SNP in the *HFE* gene is known to be associated with a disease phenotype.

Integrated Maps:

NCBI MapViewer: rs1800562 maps exactly once on NCBI human chromosome 6

Chromosome	Contig accession	Contig position	Chromosome position	Hit orientation	Group term	Group label	Contig label
6	NT_086686.1	25684223	25967140	plus strand	alt_assembly_2	Celera	Celera
6	NT_007592.14	16951392	26201120	plus strand	ref_haplotype	reference	reference

Fig. 6. Integrated maps section of the SNP entry rs1800562. The location of the SNP, nucleotide position 16951392 on the contig NT_007592.14 of the reference assembly, is highlighted by a rectangle.

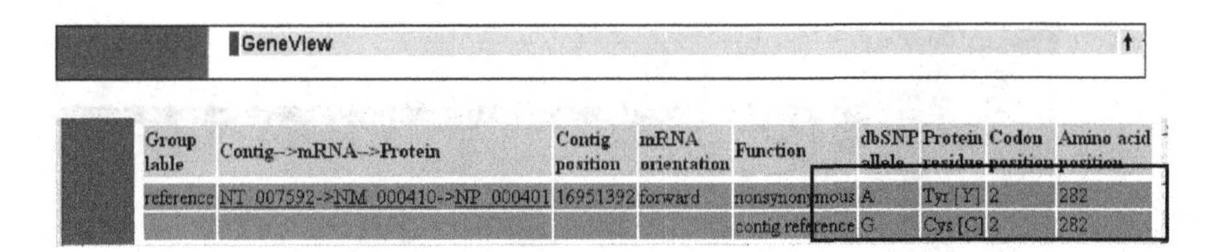

GeneView

Group lable	Contig-->mRNA-->Protein	Contig position	mRNA orientation	Function	dbSNP allele	Protein residue	Codon position	Amino acid position
reference	NT_007592->NM_000410->NP_000401	16951392	forward	nonsynonymous	A	Tyr [Y]	2	282
				contig reference	G	Cys [C]	2	282

Fig. 7. GeneView section of the SNP entry rs1800562 for the mRNA NM_000410 alignment on the reference assembly contig NT_007592. The resulting amino acid change, 282nd amino acid in the protein NP_000401.1, from cysteine to tyrosine, is highlighted by a rectangle.

5.4. Step 4: Determine Whether the HFE Gene Variant is Known to Cause a Disease Phenotype

To determine whether the Cys282Tyr amino acid change is linked to a phenotype, we will access the OMIM database. Go back to the Map Viewer report by clicking the back button of the web browser. Make the Genes_seq map a master map by clicking the arrow at the top of the Genes_seq map. Click on the OMIM link next to the *HFE* gene. This takes us to the OMIM report for the *HFE* gene. It describes the relationship between the mutations in the *HFE* gene and the hemochromatosis phenotype. Click the Allelic Variants "View list" in the side blue bar to get information about the mutant proteins from patients. One variant, Cys282Tyr, is reported to cause the hemochromatosis phenotype (*see* **Fig. 8**). The query EST contains a known variation that would lead to the expression of the Cys282Tyr variant protein associated with the hemochromatosis phenotype (*see* **Notes 14** and **15**).

5.5. Results for Problem 1

This Mini-Course describes the steps needed to identify the gene producing an EST obtained from a hemochromatosis patient, download the gene sequence, identify known SNPs in the gene, and find SNP-associated phenotypes.

Results of **Subheading 5.1.**: the query EST sequence was found to align to contig NT_007592.14 on chromosome 6 with one nucleotide difference (G to A with respect to the nucleotide 16951392 on the contig).

Results of **Subheading 5.2.**: The query EST was found to align to the *HFE* gene.

Results of **Subheading 5.3.**: The query EST sequence contains a known SNP (G/A with respect to the nucleotide 16951392 on contig NT_007592.14) that results in the Cys282Tyr change in the hemochromatosis protein expressed by the longest HFE mRNA variant.

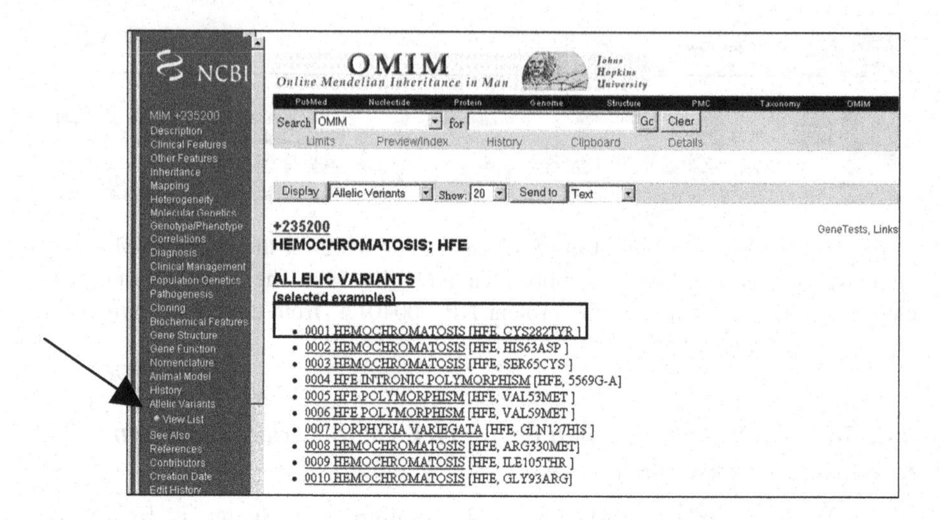

Fig. 8. Allelic variants list section from the Online Mendelian Inheritance in Man report for the *HFE* gene. The Cys282Tyr variant, highlighted by a rectangle, is reported to be associated with hemochromatosis.

Results of **Subheading 5.4.**: The Cys282Tyr change in the *HFE* protein is associated with hemochromatosis.

6. Detailed Protocol: Problem 2

For more practice, we will now perform a similar analysis using another EST sequence from a sickle anemia patient. Sickle cell anemia is a disease in which the red blood cells are curved in shape and have difficulty passing through small blood vessels. It is recommended to follow along from the webpage (http://www.ncbi.nlm.nih.gov/Class/minicourses/diseasegene2.html).

This page contains a link to a file containing the screen images of each of the steps described next. Referring to the file is strongly recommended to follow the mini-course steps. However, a number of screen images are provided in this chapter as well for a reader to follow along. These screen images are from human Build 35.1.

6.1. Step 1

Paste the EST sequence that is provided in the mini-course page into the query box of the human genome BLAST page (http://www.ncbi.nlm.nih.gov/genome/seq/BlastGen/BlastGen.cgi?taxid=9606). The user may directly access the human genome BLAST page through the "Identification of Disease Genes"

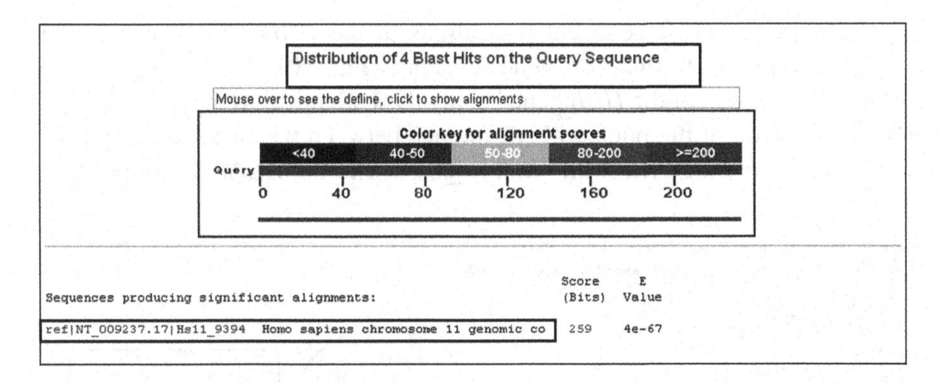

Fig. 9. MegaBlast of the query expressed sequence tags against the human genome: graphical overview. There are four hits to the contig sequence NT_009237.17 on chromosome 11 as highlighted by rectangles.

mini-course web page by clicking on the "BLAST (human genome)" link. Select the "genome (reference only)" database from the pull down menu, start the search by clicking on the "Begin Search" button and obtain the results by clicking on the "Format" button (*see* **Notes 2–4**). The BLAST results page shows four hits to the contig sequence NT_009237.17 on chromosome 11, in the human genome build 35.1, with varying percent identity (**Fig. 9**).

These multiple hits could arise from similarity to multiple gene family members and/or the query EST sequence originating from multiple exons (*see* **Notes 5** and **6**).

6.2. Step 2: Identify the Genes Expressing the ESTs and Download Their Sequences

To determine the gene expressing the EST in this case, it is much easier to view the BLAST hits in the Map Viewer. Click the "Genome View" button at the top of the BLAST results page, then on the Map element "NT_009237."

Currently, four maps are displayed; Model, RNA, Genes_seq, and Contig (*see* **Fig. 10**). Refer to **Subheading 5.1.** for more description of these maps. The four BLAST hits are indicated by the shaded areas on the right side of each map. Two of these align to the two exons of the *HBB* gene and two align to the two exons of the *HBD* gene (highlighted by the ovals). Note the percent identity of the BLAST hits (highlighted by the rectangles). The EST sequence is more similar to the HBB exons than to the HBD exons (100 and 99 % compared to 98 and 92 %, respectively). Thus, the query EST is probably expressed by the *HBB* gene but also aligns to the *HBD* gene because of its sequence similarity to the *HBB* gene sequence **(Fig. 10)**.

One of the exons of the *HBB* gene is only 99 % identical to the query EST. To note the location of the nucleotide difference between the two sequences, click on the corresponding "Blast hit" link to go back to the BLAST results page.

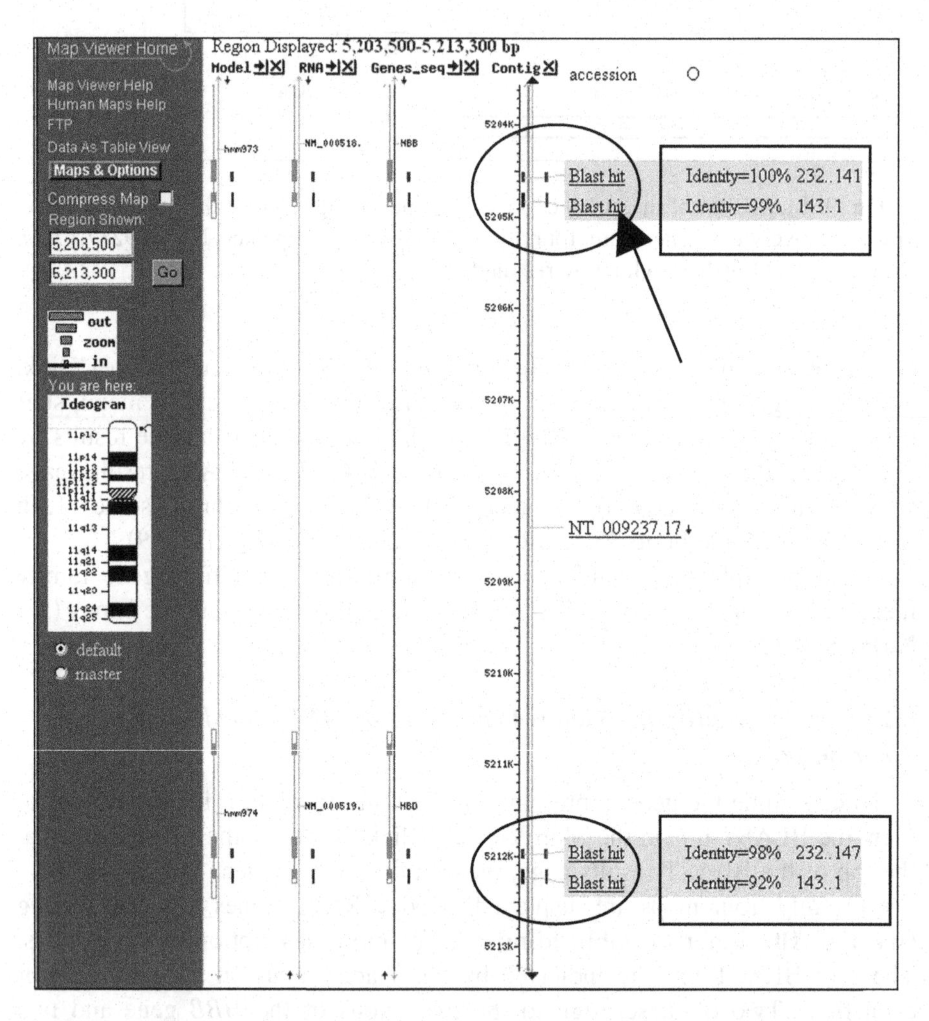

Fig. 10. Map Viewer display obtained from the Genome View button link on the BLAST results page of **Subheading 6.1.** The four BLAST hits are indicated by the shaded areas on the right side of each map. Two of these align to the two exons of the *HBB* gene and two align to the two exons of the *HBD* gene (highlighted by the ovals). Note the percent identity of the BLAST hits (highlighted by the rectangles).

Note that the alignment is on the minus (reverse) strand of the contig NT_009237.17 (**Fig. 11**). One nucleotide sequence is different between the query EST and the genomic sequence "Sbjct." Identify the nucleotide number at the site of difference with respect to the closest nucleotide number 4035482 of "Sbjct" contig NT_009237.17 and count downward by 10 (because the site of difference is 10 nucleotides away from 4035482 and the alignment is on the minus strand). Thus the query EST has a T to A variation with respect to the nucleotide 4035473 of the contig NT_009237.17. The difference could be due to a sequencing error in the low quality EST sequence or it may represent a real SNP in the human genome. For future reference, paste your results, such as the alignment and the nucleotide difference (sequence difference at the nucleotide 4035473 on NT_009237.17; A in the genomic and T in the query EST sequence), in the window provided in the mini-course webpage.

To download the HBB genomic sequence, go back to the Map Viewer report by clicking the "back" button on the browser once. Set the Genes_seq map as the master map by clicking on the arrow at its top. Because the gene of interest for further analysis is the *HBB* gene, we can remove the *HBD* gene from the

```
                                  Alignments

>ref|NT_009237.17|Hs11_9394 Homo sapiens chromosome 11 genomic contig
Length=49571094

 Features in this part of subject sequence:
   beta globin


 Score =  259 bits (140),  Expect = 4e-67
 Identities = 142/143 (99%), Gaps = 0/143 (0%)
 Strand=Plus/Minus

Query  1        ACATTTGCTTCTGACACAACTGTGTTCACTAGCAACCTCAAACAGACACCATGGTGCATC  60
                ||||||||||||||||||||||||||||||||||||||||||||||||||||||||||||
Sbjct  4035542  ACATTTGCTTCTGACACAACTGTGTTCACTAGCAACCTCAAACAGACACCATGGTGCATC  4035483

Query  61       TGACTCCTGTGGAGAAGTCTGCCGTTACTGCCCTGTGGGGCAAGGTGAACGTGGATGAAG  120
                |||||||||| |||||||||||||||||||||||||||||||||||||||||||||||||
Sbjct  4035482  TGACTCCTGAGGAGAAGTCTGCCGTTACTGCCCTGTGGGGCAAGGTGAACGTGGATGAAG  4035423

Query  121      TTGGTGGTGAGGCCCTGGGCAGG  143
                |||||||||||||||||||||||
Sbjct  4035422  TTGGTGGTGAGGCCCTGGGCAGG  4035400
```

Fig. 11. Alignment of the 99% identical Basic Local Alignment Search Tool (BLAST) hit in **Subheading 6.1.** The BLAST hit is on the minus (reverse) strand (highlighted by an oval) of the contig NT_009237.17. There is one nucleotide difference (highlighted by a rectangle) between the query EST and genomic sequences at nucleotide 4035473 of the contig NT_009237.17.

view by clicking on the gray line at the appropriate BLAST hit location and selecting the "Recenter" option (*see* **Fig. 12**).

The upward pointing arrow next to the *HBB* gene link shows the placement of the gene on the reverse strand of chromosome 11 (*see* **Note 7**).

Click on the "dl" link next to the HBB gene. Because the gene is on the reverse strand, select minus on the Stand pull down menu and click on the "Change Region/Strand" button. Display the gene sequence by clicking on the "Display" option (*see* **Notes 8** and **9**). Copy the sequence and paste it in the area provided in the Mini-Course page. You can adjust the nucleotide locations

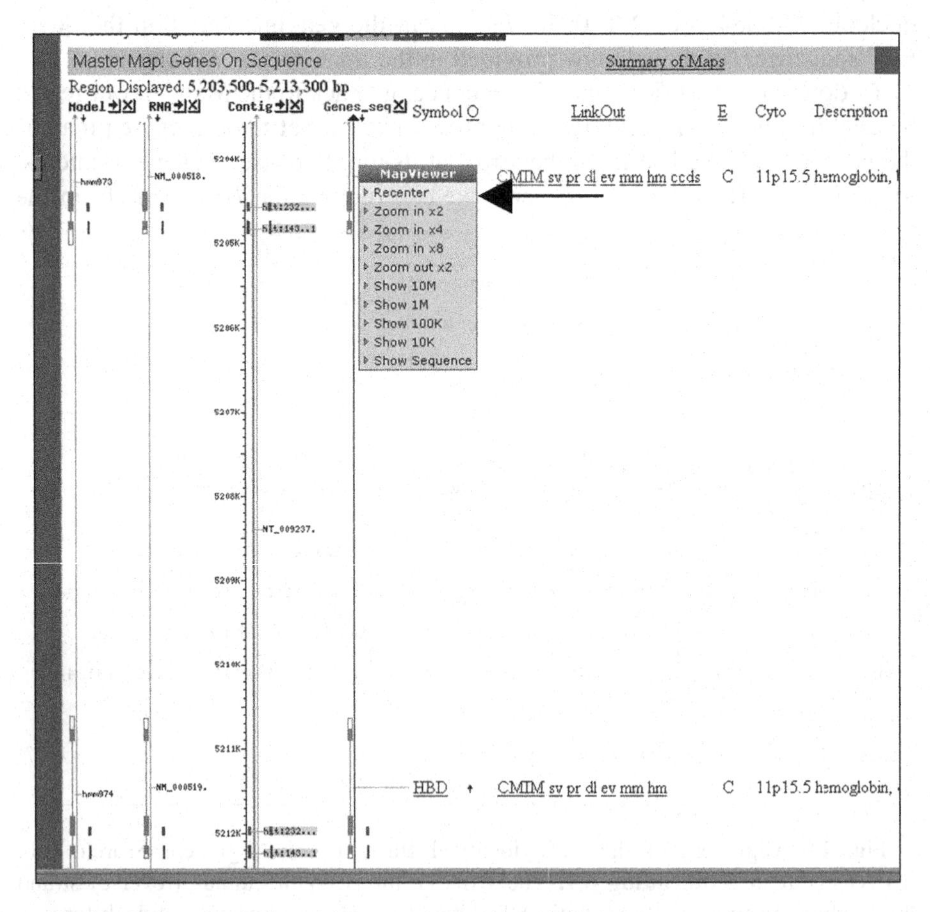

Fig. 12. Map Viewer display showing the recenter option by clicking on the gray line indicating the contig Map.

to download the upstream or downstream sequence by using the "adjust by" and "Change Region/Strand" option.

6.3. Step 3: Determine Whether the ESTs Contain Known SNPs

Go back to the Map Viewer report by clicking on the back button of the browser twice. Click the Maps and Options link.

Remove all the maps except the Genes_seq map and add the Variation map as the master map. Zoom in on the blast hit area (represented by the bars) by clicking on the thin gray line next to it and choosing the appropriate zoom level. There are three SNPs in the area; rs713040, rs334, rs11549407 (*see* **Notes 10** and **11**; **Fig. 13**).

Click on any of the links and obtain information about the location and the nucleotide variation from the "Fasta sequence" and "Integrated maps" panels. The SNP, rs334, represents an A/T SNP **(Fig. 14)** at the nucleotide position 4035473 on the contig NT_009237.17 (**Fig. 15** and **Note 12**).

This is the same nucleotide variation on the contig NT_009237.17 found in the BLAST result in **Subheading 6.1.** (4035473 T to A). Next, to identify whether this change represents a change in an encoded amino acid, we will refer to the GeneView panel. This view shows the location of the SNP in the alternatively spliced products annotated on all the assemblies. It also provides information at the protein level; the amino acid number and the change in the

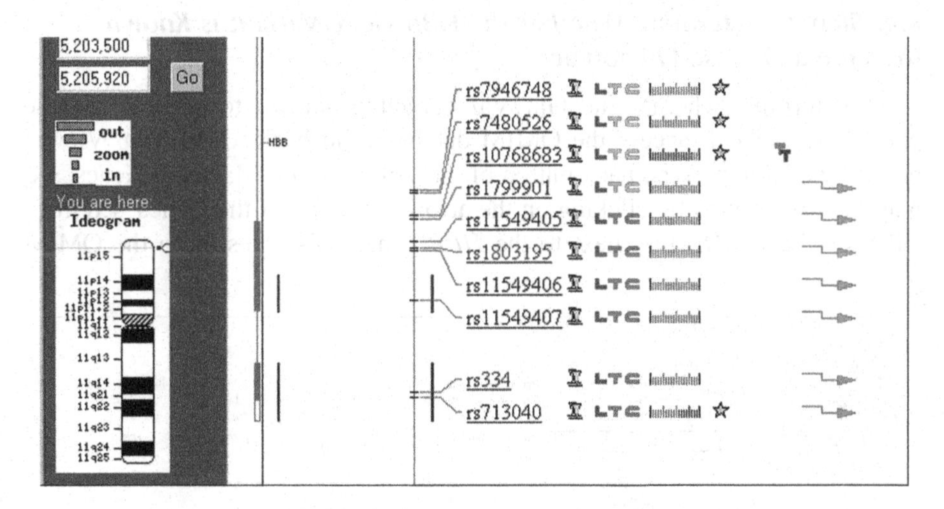

Fig. 13. Map Viewer, displaying the variation and Genes_seq maps, zoomed in the region of BLAST hit in **Subheading 6.** There are three SNPs in the BLAST hit area indicated by the bars on the right side of each map; rs713040, rs334, rs11549407.

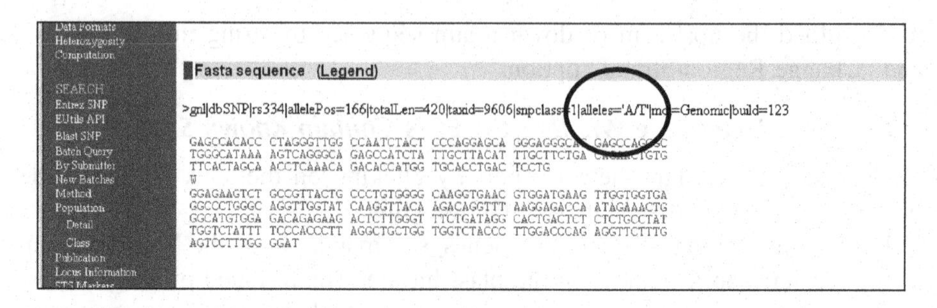

Fig. 14. Fasta sequence section of the SNP entry rs334. The SNP contains an A/T SNP, indicated in the definition line of the record, highlighted by an oval.

sequence, if any. Refer to the panel for the transcript, NM_000518, on the reference assembly contig, NT_009237.17. The SNP would result in the change at the seventh amino acid in the protein NP_000509.1, encoded by the mRNA NM_000518.4, from glutamate to valine (*see* **Fig. 16**).

Thus, the query EST sequence contains a known SNP in the HBB gene that results in a glutamate to valine change in the seventh amino acid (Glu7Val) of the β-globin protein (*see* **Note 13**). The next obvious step is to find out whether the SNP in the *HBB* gene is known to be associated with a disease phenotype.

6.4. Step 4: Determine Whether the HBB Gene Variant is Known to Cause a Disease Phenotype

To determine whether the Glu7Val variant is known to cause a disease phenotype, we will access the OMIM database. Go back to the Map Viewer report by clicking on the back button of the web browser. Make the Genes_seq map the master map by clicking on the arrow at the top of the Genes_seq map. Click on the OMIM link next to the *HBB* gene. This takes us to the OMIM

Integrated Maps:

NCBI MapViewer: rs334 maps exactly once on NCBI human chromosome 11

Chromosome	Contig accession	Contig position	Chromosome position	Hit orientation	Contig Allele	Group term	Group label	Contig label	Neighbor SNP	SNP_flank position
11	NT_086780.1	860802	5173605	minus	A	alt_assembly_2	Celera	Celera	view	165
11	NT_009237.17	4035473	5204808	minus	A	ref_haplotype	reference	reference	view	165

Fig. 15. Integrated maps section of the SNP entry rs334. The location of the SNP, nucleotide position 4035473 on the contig NT_009237.17 of the reference assembly, is highlighted by a rectangle.

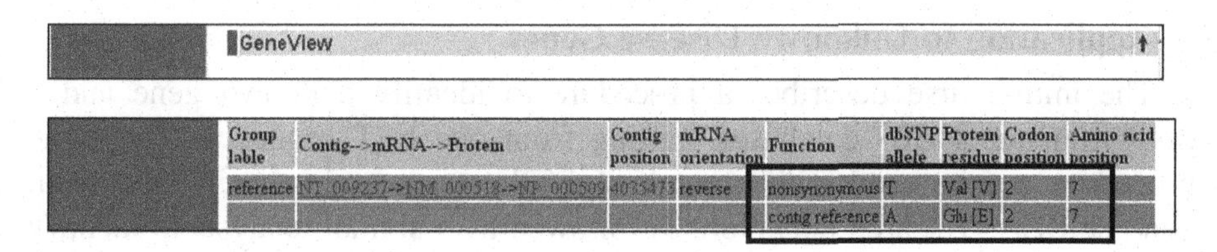

Fig. 16. GeneView section of the SNP entry rs334 for the mRNA NM_000518 alignment on the reference assembly contig NT_009237. The SNP results in the change at the seventh amino acid in the protein NP_000509.1 from glutamate to valine.

report for the *HBB* gene that details how variants (HBB gene variants) the *HBB* gene are associated with various phenotypes. As mentioned in the OMIM report under the "Psuedogenes" section, the allelic variants are listed for the mature HBB (β-globin) protein which lacks the initiator methionine. The SNP database reports them for the precursor protein. Hence, the allelic variants in the OMIM report are off by one amino acid compared to the variants in the SNP report (*see* **Note 14**). Thus, the Glu7Val variant in the SNP report corresponds to the Glu6Val variant in the OMIM report. Access the allelic variants list by clicking on the "View list" in the blue side bar. The Glu6Val variant, called hemoglobin S, is reported to cause the sickle cell anemia phenotype. The query EST contains a known variation that leads to the expression of the Glu7Val variant protein associated with the sickle cell anemia phenotype (*see* **Note 15**).

6.5. Results for Problem 2

This mini-course describes steps to identify the gene expressing the ESTs obtained from a sickle cell anemia patient, download the gene sequence, identify known SNPs in the gene and find SNP-associated phenotypes.

Results of **Subheading 6.1.**: the query EST sequence was found to align to the contig NT_009237.17 on chromosome 11 with one nucleotide difference (T to A with respect to the nucleotide 4035473 on the contig).

Results of **Subheading 6.2.**: the query EST was found to be expressed by the *HBB* gene.

Results of **Subheading 6.3.**: the query EST sequence contains a known SNP (T/A with respect to the nucleotide 4035473 on contig NT_009237.17).

Results of **Subheading 6.4.**: the Glu7Val change in the *HBB* protein is associated with sickle cell anemia.

7. Application to Unknown Disease Genes

The mini-course describes a procedure to identify a known gene and a SNP from the NCBI databases starting from one EST sequence. The same procedure can be used with a batch of EST sequences in the initial human genome BLAST search (*see* **Note 3**) followed by a similar analysis to identify genes corresponding to them. Some ESTs may be produced by known genes (as described in the mini-course) and some may be produced by novel genes not yet annotated on the Genes_seq map. The Model map may be useful to identify the novel genes. Also, some ESTs may contain new SNPs, which can be deposited in dbSNP. By comparing the DNA sequence from patients and normal individuals, it can be discerned whether the novel SNP and/or the novel gene are associated with the disease.

8. Further Analysis

The mini-course "Correlating Disease Gene and Phenotype" elucidates the biochemical and structural basis for the function of the mutant proteins and their relationship to the particular phenotype.

9. Notes

1. Online Mendelian Inheritance in Animals is a database of genes, inherited disorders and traits in animal species (other than human and mouse) authored by Professor Frank Nicholas of the University of Sydney, Australia, with help from many collaborators over the years.
2. In addition to the human genome, a number of other genomes are available as BLAST databases. A complete list is available under the Genomes panel on the BLAST page (http://www.ncbi.nlm.nih.gov/BLAST/).
3. MegaBlast also accepts a batch of query sequences. Each query sequence must have a unique identifier written on a separate line before the sequence and the identifier line should begin with a greater than (">") sign. For example,

   >identifier1
   atgcggctta...
   >identifier2
   ttggcatactg...
   >identifier3
   ggatcgatcag...
4. Since the human Build 36, NCBI also provides access to the previous assembly release (Build) as a BLAST database. More information about each build is provided in the release notes at http://www.ncbi.nlm.nih.gov/genome/guide/human/release_notes.html. You may choose to run the BLAST search against the previous build by using the appropriate option in the database field.

5. In **Subheading 5.1.**, there is only one hit to the database sequence. Multiple hits are possible for several reasons such as finding similarity to other gene family members. For example, refer to **Subheading 6.1.** The query EST sequence shows similarity to two members of the globin family, HBB (encoding β-globin), and HBD (encoding delta globin). The gene encoding the EST may be the one with high similarity, HBB in this case (**Fig. 10**).
6. BLAST may align a single EST in two or more segments if the EST sequence spans two or more exons. For example, refer to **Subheading 6.1.** The query EST sequence aligns to two exons of both the *HBB* and *HBD* genes (**Fig. 10**).
7. The orientation of the gene on the chromosome can also be discerned from the placement of the blue bars representing the gene structure with respect to the gray line. If the gene is placed on the forward strand then the blue bars representing the gene structure are drawn to the right of the gray line (as for the *HFE* gene in **Subheading 5. (Fig. 3)**). If the gene is placed on the reverse strand, then the blue bars are drawn to the left of the gray line (as for the *HBB* gene in **Subheading 6. (Fig. 12)**).
8. If the gene of interest is on the reverse strand, then change the Strand pull down menu to minus and click on the "Change Region/Strand" option before displaying the sequence (refer to **Subheading 6.2.**).
9. The user can also adjust the nucleotide locations to download the upstream or downstream sequence by using the "adjust by" and "Change Region/Strand" options.
10. When a single nucleotide polymorphism is submitted to dbSNP, an identifier with prefix "ss" is assigned to the entry. It is possible that multiple laboratories may submit information on the same SNP as new techniques are developed to assay variation, or new populations are typed for frequency information. Each of these SNP entries is assigned a unique identifier with prefix "ss". When two or more submitted SNP records refer to the same location in the genome, a Reference SNP record is created, with an "rs" prefix on the identifier, by NCBI during periodic "builds" of the SNP database. This reference record provides a summary list of submitted "ss" records in dbSNP.

    For example, the Reference SNP record from **Subheading 5.**, rs1800562, contains three submitted SNP records; ss2420669, ss5586582, and ss24365242 in the dbSNP build 125. The Reference SNP record from **Subheading 6.**, rs334, contains six submitted SNP records; ss335, ss1536049, ss4397657, ss4440139, ss16249026, and ss24811263 in the dbSNP build 125.
11. When the "variation" map is selected as the master map in the Map Viewer (**Figs. 4** and **13**), the user is presented with a graphical summary of several properties of the SNP, such as the quality of the SNP computed from mapping, location in the gene region, marker heterozygosity and validation information. For more information, refer to http://www.ncbi.nlm.nih.gov/SNP/get_html.cgi?whichHtml=verbose.

    For example, the green triangle next to rs1800562 indicates that this marker is mapped to a unique position in the genome (**Fig. 4**). The highlighted L, T, and C symbols indicate that the SNP is in the Locus, Transcript, and the coding region of the gene. For the SNP rs807209, only the L symbol is highlighted indicating that the SNP is not in the transcript or the coding region but is in the locus region of the gene. A locus in this report is defined as any part of the marker position on sequence map within a 2000-base interval 5′ of the most 5′ feature of the gene (CDS, mRNA, gene), or the marker position within a 500-base interval 3′ of the most 3′ feature of the gene.

12. NCBI has additional human genome assemblies such as the assembly submitted by Celera and chromosome 7 assembly from the Center for Applied Genomics, TCAG. The assembly information is provided under the contig label heading in the Integrated Maps panel. On the Celera assembly, the SNP in the *HFE* gene in **Subheading 5.**, rs18000562, is at the nucleotide 25684223 on the contig NT_086686.1 (*see* **Fig. 6**). On the Celera assembly, the SNP in the *HBB* gene in **Subheading 6.**, rs334, is at the nucleotide 860802 on the contig NT_086780.1 (*see* **Fig. 15**).
13. The GeneView panel shows the locations of the SNPs on the genomic assemblies with respect to the genes, their alternatively spliced mRNAs and encoded proteins. The view is color coded for quick identification of the location and to show whether the change is synonymous (not altering the amino acid translation) or nonsynonymous (altering the amino acid translation). For example, nonsynonymous SNPs are represented in red, synonymous in green and those in introns are in yellow. A link to the "Color Legend" is provided next to the Gene Model under the GeneView panel.
14. Some OMIM entries report the allelic variants for the mature protein, whereas dbSNP reports variants for the precursor protein. Thus, for the same SNP, amino acid numbering for the allelic variant may be different in these databases. For example, refer to **Subheading 6.6.** OMIM reports allelic variants for the beta globin mature protein (after removal of the initiator methionine). Thus, the Glu7Val change reported in dbSNP is the same as Glu6Val allelic variant reported in the OMIM database.
15. The OMIM report and thus its allelic variants list are manually derived from publications. dbSNP contains the SNPs reported by the submitters. Currently, a link is provided from dbSNP to OMIM if the amino acid number of the allelic variant in the OMIM report matches the number of the changed amino acid due to a SNP in dbSNP. Since the sources of the two databases, OMIM and dbSNP, are different, each may contain information not found in the other.

References

1. Kitts, P. (2002–2005) Genome assembly and annotation process, in *The NCBI Handbook,* (McEntyre, J. and Ostell, J., eds.), National Library of Medicine (US), NCBI, Bethesda, MD.

2. Wheeler, D. L., Barrett, T, Benson, D. A., et al. (2006) Database resources of the National Center for Biotechnology Information. *Nucleic Acids Res.* ***34,*** D173–D180.
3. Altschul, S. F., Madden, T. L., Schaffer, A. A., et al. (1997) Gapped BLAST and PSI-BLAST: a new generation of protein database search program. *Nucleic Acids Res.* **25,** 3389–3402.
4. Madden, T. (2002–2005) The BLAST sequence analysis tool, in *The NCBI Handbook,* (McEntyre, J. and Ostell, J., eds.), National Library of Medicine (US), NCBI, Bethesda, MD.
5. Dombrowski, S. M. and Maglott, M. (2002–2005) Using the Map Viewer to Explore Genomes, in *The NCBI Handbook* (McEntyre, J. and Ostell, J., eds.), National Library of Medicine (US), NCBI, Bethesda, MD.
6. Kitts, A. and Sherry, S. (2002–2005) The single nucleotide polymorphism database (dbSNP) of nucleotide sequence variation, in *The NCBI Handbook,* (McEntyre, J. and Ostell, J., eds.), National Library of Medicine (US), NCBI, Bethesda, MD.
7. Maglott, D., Amberger J. S., and Hamosh, A. (2002–2005) Online Mendelian Inheritance in Man (OMIM): a directory of human genes and genetic disorders, in *The NCBI Handbook,* (McEntyre, J. and Ostell, J., eds.), National Library of Medicine (US), NCBI, Bethesda, MD.
8. Zhang, Z., Schwartz, S., Wagner, L., and Miller, W. (2000) A greedy algorithm for aligning DNA sequences. *J. Comput. Biol.* **7,** 203–214.

25

Variable Number Tandem Repeat Typing of Bacteria

Siamak P. Yazdankhah and Bjørn-Arne Lindstedt

Summary

Analysis of bacterial genomes revealed a high percentage of DNA consisting of repeats, in which DNA motifs existed in multiple copies. Study of these DNA motifs has resulted in the development of variable number tandem repeat (VNTR) or multilocus variant-repeat analysis (MLVA) assays, which have shown to be valuable bacterial typing methods, especially in relation to disease outbreaks. The VNTR-based assay is based on direct PCR amplification of a specific locus, which is well defined. The range and polymorphism index of each locus can be calculated. This chapter describes the VNTR analysis of *Neisseria meningitides*-based on separation in low resolution media agarose, and VNTR analysis of *Salmonella enterica* subsp. *enterica* serovars Typhimurium-based on high resolution capillary electrophoresis.

Key Words: Molecular typing; bacteria; VNTR; epidemiology; outbreak.

1. Introduction

The availability of the whole genome of a number of bacteria has opened the way for the evaluation of repetitive DNA motifs and their application to epidemiological investigation. Large proportions of genomic DNA consist of repeats, where DNA motifs exist in multiple copies. The repeats may vary in size, location, complexity, repeat mode, and are designated as belonging to different repeat classes such as direct repeats, dyad repeats, and inverted repeats *(1)*. Short DNA tandem repeats have been recognized as an important mechanism for controlling the expression of some bacterial outer membrane proteins *(2,3)*. Furthermore, short DNA sequence repeats may be involved

From: *Methods in Molecular Biology, vol. 396: Comparative Genomics, Volume 2*
Edited by: N. H. Bergman © Humana Press Inc., Totowa, NJ

in coding or in the promoter area of genes associated in the biosynthesis of surface antigens *(4)*.

Epidemiological analysis of infections, caused by pathogenic bacteria, is dependent on the precise identification at the strain level. During the last 10 yr, hypervariable minisattelite regions have been utilized to detect DNA fingerprints in the human gnome *(5)* and other eukaryotes like *Candida albicans (6)*, *Candida crusei (7)*, *Leishmania infantrum (8)*, *Sphaeropsis sapinea (9)*. During the last 6 yr VNTR has been used for molecular typing of several bacterial species including *Bacillus anthracis (10)*, *Yersinia pestis (11)*, *Mycobactrium tuberculosis (12)*, *Haemophilus influenza (13)*, *Francisella tuleransis (14)*, *Xyllela fastidosa (15)*, *Legionella pneumophilia (16)*, *Staphylococcus aureus (17)*, *Salmonella enterica* subsp. *enterica* serovars *typhimurium (18)*, *Escherichia coli O157 (19)*, and *Neisseria meningitidis (20,21)*.

The genome of a particular bacterium can be screened for repetitive DNA with the Tandem Repeats Finder *(22)*, which is available for free at www.tandem.bu.edu/tools.html.

The program generates an output file giving the repeat location, the repeat segment size, the nucleotide composition, and the copy number of the array. PCR primers can be designed from the sequences flanking the repeats of the loci. Primers may be designed by free software available at www.biotools.umassmed.edu/biopass/primer3_www.cgi. In a VNTR typing assay, the length of PCR products is measured. From this length, the number of units are deduced for each locus, and this list of values for a given strain is the genotype. It should be emphasized that some loci may have low/no polymorphism in a genetically diverse set of isolates *(20)*. Therefore, each locus may be evaluated for polymorphism before it can be used in genotyping. PCR may be run either singly, using one primer set or multiplex using more than one primer sets. The PCR-products may be pooled together into a single tube for each sample and thoroughly mixed. The PCR-mix may be run either on an agarose gel or by capillary electrophoresis.

An Internet-based resource (www.minisattelites.u-psud.fr) has recently been developed, which takes advantage of available genome sequences to help develop and run tandem-repeat based strain identification *(23)*. This database allows finding of VNTR regions in sequenced organisms, comparing the VNTR loci between isolates or species and identification of similar and dissimilar VNTR regions. The website makes possible searches for primer matches in whole genome sequences or restricted sequences associated with minisatellites. This website, as an online resource, greatly facilitates the work for development of VNTR-based typing systems.

We present, in this chapter, the gel-based VNTR typing system for *Neisseria meningitides* and multiple-locus variant analysis of *Salmonella typhimurium* using capillary electrophoresis.

2. Materials

2.1. Gel-Based VNTR for N. meningitidis

1. Chocolate agar plate for growth of the bacteria (Oxoid), store at 4 °C.
2. TE-buffer: 1 *M* Tris-EDTA, pH 8.0, store at 4 °C.
3. dH_2O: store at 4 °C.
4. PCR-primers: the oligionucleotide primers (**Table 1**) for the PCR reactions can be ordered from any commercial primer supplier, store at −20 °C.
5. PCR kit (Applied Biosystems): the kit contains premixed dNTPs, DNA polymerase, PCR-buffer, ddH_2O, store at −20 °C.
6. Gene Amp 9700 PCR system (Applied Biosystems).
7. 5X TBE buffer: TBE powder (AppliChem) is dissolved in dH_2O, autoclaved at 121 °C for 15 min, store at 4 °C.
8. Loading dye: dissolve 100 mg Evans blue (Sigma) in 10 mL dH_2O and 75 mL glycerol (Sigma). Store at 4 °C.
9. Agarose: SeaKem® ME agarose (Cambrex Bio Science), store at room temperature.
10. Gel Electrophoresis apparatus and Power Supplier Pack 3000 (Bio-Rad) and gel tray.
11. DNA molecular weight marker: 4 μL 1 kb DNA-ladder (InVitrogen), 20 μL loading dye and 76 μL dH_2O, store at 4 °C.
12. Ethidium-bromide (Sigma): prepare 10 mg/mL by mixing undiluted ethidium-bromide in dH_2O, store at room temperature.
13. Photography apparatus: Polaroid MP4+(Polaroid). Polaroid Film should be stored at 4 °C.
14. Ultraviolet (UV)-illumination apparatus: TM36 Model (KEBO-LAB, Norway)
15. Software (Applied Maths BVBA, Sint-Martens-Latem, Belgium) for data analysis and generation of dendrogram.

Table 1
PCR Primers for VNTR Analysis of *Neisseria meningitidis*

Primer	Primer sequence (5′–3′)	Annealing temperature
VNTR01F	GACGGGTCAAAAGACGGAAG	59 °C
VNTR01R	GGCATAATCCTTTCAAAACTTCG	
VNTR02F	CTCCCCGATAGGCCCCGAAATACC	57 °C
VNTR02R	AAAGCGGCGGGAATGACGAAGAGT	
VNTR08F	GACCCTGACCGTCGGAAACC	57 °C
VNTR08R	ATACCGCCCTGCTGTTGTGC	

Table 2
PCR Primers for VNTR Analysis of *Salmonella typhimurium*

Primer name	(Dye name) Sequence (5′–3′)	Annealing temperature[a]
STTR3-	(HEX)-CCCCCTAAGCCCGATAATGG	61 °C
STTR3-R F	TGACGCCGTTGCTGAAGGTAATAA	
STTR5-F	(HEX)-ATGGCGAGGCGAGCAGCAGT	58 °C
STTR5-R	GGTCAGGCCGAATAGCAGGAT	
STTR6-F	(6FAM)-TCGGGCATGCGTTGAAA	54 °C
STTR6-R	CTGGTGGGGAGAATGACTGG	
STTR9-F	(6FAM)-AGAGGCGCTGCGATTGACGATA	54 °C
STTR9-R	CATTTTCCACAGCGGCAGTTTTTC	
STTR10-F	(TET)-CGGGCGCGGCTGGAGTATTTG	62 °C
STTR10-R	GAAGGGGCCGGGCAGAGACAGC	

[a] For multiplex-PCR by using Qiagen kit, one can use annealing temperature 63 °C, independent.

2.2. Capillary Electrophoresis-Based VNTR for S. typhimurium

1. Luria-Bertani (LB) agar plate (Difco).
2. ddH_2O.
3. Oligonucleotide primers: the oligionucleotide primers (**Table 2**) for the PCR reactions can be ordered from any commercial primer supplier. The forward (F) primer is 5′-labeled with a fluorescent dye (HEX, 6FAM, or TET).
4. Multiplex PCR-kit (Qiagen) the kit contains pre-mixed dNTPs, and DNA polymerase, PCR amplification buffer, and ddH_2O.
5. ABI-310 Genetic Analyzer, POP4-polymer for 310 genetic analyzer, 10X Genetic Analyzer buffer with EDTA, 310 capillary 47 cm (all Applied-Biosystems), Geneflo-625-TAMRA internal size marker (CHIMERx), Formamide (Applied-Biosystems).

3. Methods

3.1. Gel-Based VNTR for N. meningitidis

3.1.1. Growth of Bacteria and DNA Isolation

1. Plate meningococcal isolates on chocolate agar and incubate overnight at 37 °C in a 5% CO_2 atmosphere.
2. Remove a loopful (10 μ) of bacterial growth from the plate and suspend in 100 μL of 1 *M* TE buffer, pH 8.0, in an Eppendorf tube.

3. Boil the bacterial suspension for 10 min.
4. Centrifuge the boiled suspension at 6000 rpm for 10 min.
5. Transfer the supernatant to a new tube and use for further analyses.
6. The supernatant containing DNA may be stored at −70 °C.

3.1.2. PCR, PCR-Product Mixture (Pool), Gel Electrophoresis

1. Make a list of samples, including negative and positive controls.
2. Prepare a reaction mixture of a volume 49 µL (*see* **Note 1**). The reaction mixture contains: 10X PCR buffer 5 µL (*see* **Note 2**), deoxyribonucleoside triphosphate (dATP, dTTP, dCTP, and dGTP), 200 m*M*: 4 µL (*see* **Note 2**), Primer forward 100 pM: 1 µL (*see* **Note 2**), Primer reveres 100 pM: 1 µL (*see* **Note 2**), Ampli*Taq* polymerase 2.5 U/µL: 0.5 µL (*see* **Note 3**), dH_2O: 37.5 µL (*see* **Note 4**). Depending on the number of samples including negative and positive samples, prepare a master-mixture (*see* **Note 5**) and dispense 49 µL to each sterile microfuge tube, which is appropriate to your thermocycler. Label the microfuge tubes.
3. Add 1 µL of supernatant containing DNA (*see* **Note 6**).
4. Turn the thermal cycler, place the microfuges and run: initial denaturing at 94 °C for 5 min, and 32 cycles of: denaturing: 94 °C for 1 min, annealing: °C depending on primer set (*see* **Table 1**) for 1 min, extension: 72 °C for 1 min 30 s, and finally an extension step of 72 °C for 5 min.
5. Mix 5 µL of each of the PCR products, resulting from the amplification of the three VNTR for each isolate, together into a single tube. Using the micropipet with a clean tip and add 5 µL of loading dye to the each PCR-product pool (*see* **Note 7**).
6. Seal both ends of the gel tray with tape or stoppers.
7. Make sure that one comb (*see* **Note 8**) is in place at the negative electrode (black end of the gel).
8. Mix 4 g SeaKem ME agarose in 200 mL 1X TBE buffer and boil in a microwave until all agarose is melted. Leave at room temperature for 20–30 min (*see* **Note 9**).
9. Pour the melted agarose into the gel space. Let the agarose harden, which should takes 1 h at room temperature (*see* **Note 10**).
10. When the gel has hardened, remove the tape or stoppers.
11. Load both the pooled PCR-reactions and standard DNA markers samples into the gel (*see* **Note 11**).
12. Pour 1X TBE buffer carefully so it fills the electrophoresis apparatus and just cover the gel.
13. Run the gel at 40 V for 18 h (overnight), at room temperature (*see* **Note 12**).
14. Turn off the power supply, disconnect the electrodes, and remove the top of the electrophoresis apparatus.
15. Place gel in staining tray and stain the gel in ethidium bromide for 10 min, destain in water for 20 min. Both steps occur at room temperature (*see* **Note 13**).
16. Take a picture of the gel under UV illumination (*see* **Note 14**).

3.1.3. Data Analysis

The gel photograph can be analyzed using the BioNumerics Software (Applied Maths BVBA, Sint-Martens-Latem, Belgium) and a dendrogram can be constructed using the Dice coefficient of similarity and cluster analysis with the unweighted pair-group method with arithmetic averages. Both the position tolerance and the optimization can be set at 1%, as recommended in the BioNumerics Software manual.

3.2. Capillary Electrophoresis (S. typhimurium)

3.2.1. Bacterial Growth and Lysis

1. Streak bacteria on LB agar plates and grow at 37 °C overnight in an incubator.
2. Transfer fresh *S. typhimurium* cultures, grown overnight on a LB plate, with an inoculation loop, to a microcentrifuge tube containing 300 μL of ddH_2O.
3. Vortex the cells briefly until they are evenly suspended.
4. Boil the bacterial suspension for 15 min and then centrifuge at 960*g* for 3 min.
5. Transfer the supernatant to a new tube for further analysis.
6. The supernatant containing DNA may be stored at −70 °C.

3.2.2. Multiplexed PCR Reactions

1. Multiplex the primers in two solutions using the Qiagen multiplex-PCR kit in a total of 50 μL as described by manufacturer (*see* **Notes 15** and **16**):

   a. Multiplex 1 (M1): 10 pmol of the STTR3 and STTR6 primer pairs.
   b. Multiplex 2 (M2): 10 pmol of the STTR5, STTR9 and STTR10 primer pairs.

2. Amplify both multiplex solutions with 2 μL of the supernatant containing DNA (*see* **Notes 17** and **18**) and under the following cycling conditions on a GeneAmp 9700 thermal cycler:

   1. Initial denaturing at 95 °C for 15 min, and then 25 cycles of:
   2. Denaturing: 94 °C for 30 s.
   3. Annealing: 63 °C for 90 s.
   4. Extension: 72 °C for 90 s, and ending an extension step of 72 °C for 10 min.

3.2.3. Sample Pooling

1. Before capillary electrophoresis pool the two multiplexed PCR reactions as follows: mix 10 μL of M1 + 2.5 μL of M2, and then add 87.5 μL of ddH_2O to a total of 100 μL. This constitutes the run-mix (RM).

3.2.4. Capillary Electrophoresis

1. Prepare the samples as follows before capillary electrophoresis: 1 μL of the RM is mixed with 1 μL of the internal size-standard, Geneflo-625-TAMRA (CHIMERx), and 12 μL of formamide (*see* **Note 19**).
2. Denature the samples at 94 °C for 2 min and then cool to room temperature.
3. For capillary electrophoresis the ABI-310 Genetic Analyzer is used. The electrophoresis is run at 60 °C for 35 min using POP4-polymer with an injection voltage of 15 kV for 5 s, and a running voltage of 15 kV (*see* **Note 20**).

For further information please *see* **Notes 21, 22,** and **23**.

3.2.5. Data Analysis

The resulting electropherogram will typically look like **Fig. 1** when all loci are present. Each peak is a PCR product and is equivalent to a "band" on an agarose gel. Loci STTR6 and STTR9 labeled with the dye 6FAM are depicted in blue, loci STTR3 and STTR5 labeled with the dye HEX are depicted in black, and locus STTR10 labeled with the dye TET is depicted in green. Sizes in basepairs are automatically recorded by the ABI-310 instrument using GeneScan (Applied-Biosystems) software. The internal size-standard, labeled with the dye TAMRA, is depicted in red. This pattern will vary dependent on the number of tandem repeats at each VNTR locus in the strains under study, and give each isolate its own distinct DNA-fingerprint. Variation in size according to multiples of the repeat is interpreted as separate alleles.

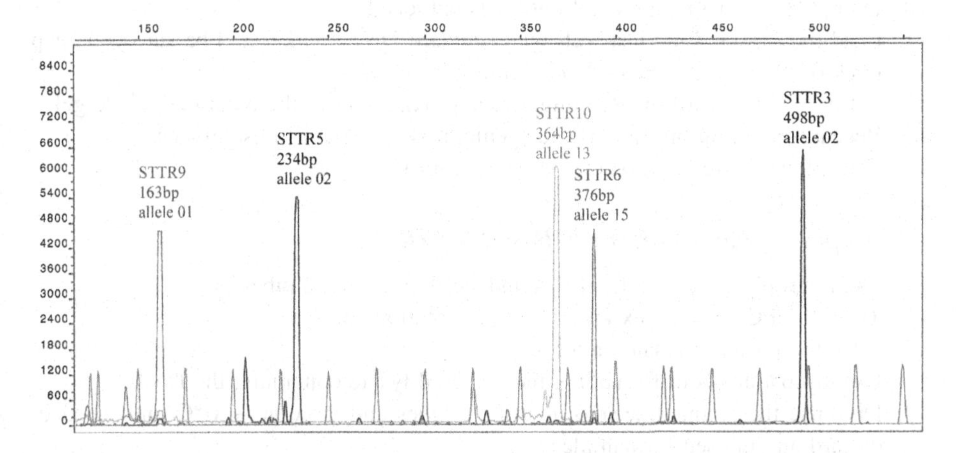

Fig. 1. Capillary-electrophoresis based VNTR of S. Typhimurium. Blue, black, green peaks represent different VNTR loci. Red peaks represents the internal standard.

BioNumerics v3.5 (Applied-Maths) can be used to enter PCR amplicon sizes, or sizes converted to allele numbers, recorded at each locus, and to draw dendrograms in which the pattern similarities are visualized, and matching profiles can rapidly be identified. The MLVA profiles can rapidly be shared among laboratories using email.

4. Notes

4.1. Gel-Based VNTR

1. Use the micropipet and sterile tips, sterile microfuges.
2. The 10X PCR buffer, dNTP and primers are held at room temperature for 30 min and then placed on ice.
3. Keep the enzyme on the ice.
4. Keep the water at room temperature.
5. Keep the master-mix on ice.
6. Use separate room for adding DNA. Thaw the supernatant containing DNA at room temperature prior adding to reaction mixture, keep the samples on ice prior running PCR. High concentrations of DNA may result in no PCR-products. Dilute DNA in dH_2O
7. Because loading dye contain Evans blue (toxic and may cause cancer) and glycerol (harmful), it is important to use gloves. PCR product may be reduced to 2.5 μL if the PCR products have a high intensity.
8. Use a comb allowing application of at least 25 μL volume.
9. Do not leave the melted agarose at room temperature longer than 30 min and be sure the agarose has not begun to harden.
10. Do not touch or move the gel until it is hardened.
11. Load the first and the last well with standard DNA marker and be sure you keep track of which samples are loaded into which wells.
12. Let the gel run until the dye migrates approximately to the other end of the gel.
13. Because ethidium bromide is mutagenic, it is important to use gloves.
14. Use safety glasses to protect your eyes from UV light.

4.2. Capillary Electrophoresis-Based VNTR

15. Use micropipets, sterile filter tips and sterile microfuge tubes.
16. Thaw the PCR-multiplex kit and primer solutions on ice.
17. Keep the bacterial lysate on ice.
18. Use a separate room for adding the bacterial lysate containing the DNA.
19. Dispense the formamide into microfuge tubes and store at –20 °C until needed, discard any unused formamide.
20. Use chemically resistant disposable gloves, lab coat, and safety glasses when handling the POP4 polymer.

4.3. General Notes

21. Extra bands. Some VNTR profiles can be difficult to read because of extra bands/peaks occurring in the gel/electropherogram. These bands can have several causes. They may be real bands indicating a duplicated segment of DNA, or in the case of locus STTR10 in *S. thyphimurium* the presence of more than one *pSLT* plasmid. These are however, rare problems. The extra bands may also be artefactual bands, which may result from:

    a. The differential migration of the two DNA strands in denaturing acrylamide gels.
    b. The "+A" activity of the *Taq* polymerase generating x + 1 bands.
    c. Slippage of the *Taq* polymerase during polymerization generating "stutter" or "shadow" bands.
    d. Nondenatured secondary structures adopted by the PCR products.

    Using capillary electrophoresis, problem A is not seen as only one strand is labeled with fluorescence and can thus be visualized. In the case of *N. meningitides* and the *S. typhimurium* VNTR method, using the previously mentioned protocol, it has also been shown that problems B and D do not result in misinterpretation. Problem C does however occur in some isolates, and will manifest on the gel/electropherogram as either a row of minor intensity peaks each representing a repeat unit of the VNTR locus forming a "saw" pattern before the correct band/peak or a larger band/peak that appears some VNTR repeat units short of the correct band/peak. In the case of capillary electrophoresis, it is relatively easy to find the correct peak in the "saw" pattern of artefactual bands, as the correct band will be the last and most intense (i.e., highest peak) in a row of small peaks leading up to it. In the case of a larger more predominant band, the correct band can be found by running a dilution of the PCR products, as the incorrect band will in this assay be greatly reduced in intensity or even disappear.
22. Loss of some peaks. Some isolates will be negative for some of the VNTR loci used in the *S. typhimurium* method. This is normal, and especially locus STTR10 and locus STTR6 are prone to show no amplification in a subset of isolates. This is still valuable information and is regarded as a "0" allele in comparisons. Loss of bands is a rare problem in *N. meningitides*.
23. Loss of all or almost all peaks. If all bands (gel-based method)/peaks (including the red peaks from the internal size-marker) (capillary electrophoresis-based method) are absent, this points to a problem with the electrophoresis instrument. In the case gel-based method the thermocycler should also be inspected for a possible malfunction.

    In the case of capillary electrophoresis-based method, the capillary, the running buffer and POP4-gel should be replaced. If the problems persist, the readings for the laser (capillary electrophoresis-based method) and the power supply (in both methods) should be checked to ensure they are working within operational parameters.

If all bands/peaks are absent, the cause of this problem may be faulty PCR amplification, and all conditions and mixes used for the PCR amplification should be checked, and new primer mixes should be prepared. The Thermocycler should also be inspected for a possible malfunction.

If just one or two peaks appear, this indicates that the isolate in testing is not a *N. meningitides* or *S. typhimurium* isolate and should be inspected by other methods.

References

1. Lindstedt, B. A. (2005) Multiple-locus variable number tandem repeats analysis for genetic fingerprinting of pathogenic bacteria. *Electrophoresis* **26,** 2567–2582.
2. Jordan, P., Snyder, L. A., and Saunders, N. J. (2003) Diversity in coding tandem repeats in related *Neisseria* spp. *BMC Microbiol.* **3,** 23.
3. Martin, P., van de Ven, T., Mouchel, N., Jeffries, A. C., Hood, D. W., and Moxon, E. R. (2003) Experimentally revised repertoire of putative contingency loci in *Neisseria meningitidis* strain MC58: evidence for a novel mechanism of phase variation. *Mol. Microbiol.* **50,** 245–257.
4. Hammerschmidt, S., Muller, A., Sillmann, H., et al. (1996) Capsule phase variation in *Neisseria meningitidis* serogroup B by slipped-strand mispairing in the polysialyltransferase gene (*siaD*): Correlation with bacterial invasion and the outbreak of meningococcal disease. *Mol. Microbiol.* **20,** 1211–1220.
5. Pati, U. and Pati, N. (2000) Lipoprotein(a), atherosclerosis, and apolipoprotein(a) gene polymorphism. *Mol. Genet. Metab.* **71,** 87–92.
6. Lunel, F. V., Licciardello, L., Stefani, S., et al. (1998) Lack of consistent short sequence repeat polymorphisms in genetically homologous colonizing and invasive *Candida albicans* strains. *J. Bacteriol.* **180,** 3771–3778.
7. Shemer, R., Weissman, Z., Hashman, N., and Kornitzer, D. (2001) A highly polymorphic degenerate microsatellite for molecular strain typing of *Candida krusei. Microbiology* **147,** 2021–2028.
8. Bulle, B., Millon, L., Bart, J. M., et al. (2002) Practical approach for typing strains of *Leishmania infantum* by microsatellite analysis. *J. Clin. Microbiol.* **40,** 3391–3397.
9. Burgess, T., Wingfield, M. J., and Wingfield, B. W. (2001) Simple sequence repeat markers distinguish among morphotypes of *Sphaeropsis sapinea. Appl. Environ. Microbiol.* **67,** 354–362.
10. Fouet, A., Smith, K. L., Keys, C., et al. (2002) Diversity among French *Bacillus anthracis* isolates. *J. Clin. Microbiol.* **40,** 4732–4734.
11. Klevytska, A. M., Price, L. B., Schupp, J. M., Worsham, P. L., Wong, J., and Keim, P. (2001) Identification and characterization of variable-number tandem repeats in the *Yersinia pestis* genome. *J. Clin. Microbiol.* **39,** 3179–3185.

12. Skuce, R. A., McCorry, T. P., McCarroll, J. F., et al. (2002) Discrimination of *Mycobacterium tuberculosis* complex bacteria using novel VNTR-PCR targets. *Microbiology* **148,** 519–528.
13. van Belkum, A., Scherer, S., van Leeuwen, W., Willemse, D., van Alphen, L., and Verbrugh, H. (1997) Variable number of tandem repeats in clinical strains of *Haemophilus influenzae*. *Infect. Immun.* **65,** 5017–5027.
14. Johansson, A., Goransson, I., Larsson, P., and Sjostedt, A. (2001) Extensive allelic variation among *Francisella tularensis* strains in a short-sequence tandem repeat region. *J. Clin. Microbiol.* **39,** 3140–3146.
15. Coletta-Filho, H. D., Takita, M. A., de Souza, A. A., Aguilar-Vildoso, C. I., and Machado, M. A. (2001) Differentiation of strains of *Xylella fastidiosa* by a variable number of tandem repeat analysis. *Appl. Environ. Microbiol.* **67,** 4091–4095.
16. Pourcel, C., Andre-Mazeaud, F., Neubauer, H., Ramisse, F., and Vergnaud, G. (2004) Tandem repeats analysis for the high resolution phylogenetic analysis of *Yersinia pestis*. *BMC Microbiol.* **4,** 22.
17. Sabat, A., Krzyszton-Russjan, J., Strzalka, W., et al. (2003) New method for typing *Staphylococcus aureus* strains: multiple-locus variable-number tandem repeat analysis of polymorphism and genetic relationships of clinical isolates. *J. Clin. Microbiol.* **41,** 1801–1804.
18. Lindstedt, B. A., Heir, E., Gjernes, E., and Kapperud, G. (2003) DNA fingerprinting of *Salmonella enterica* subsp. *enterica* serovars Typhimurium and Enteritidis with emphasis on phage type DT104 based on variable number tandem repeat loci. *J. Clin. Microbiol.* **41,** 1469–1479.
19. Lindstedt, B. A., Heir, E., Gjernes, E., Vardund, T., and Kapperud, G. (2003) DNA fingerprinting of Shiga-toxin producing *Escherichia coli* O157 based on multiple-locus variable-number tandem-repeats analysis (MLVA). *Ann. Clin. Microbiol. Antimicrob.* **2,** 12.
20. Yazdankhah, S. P., Lindstedt, B. A., and Caugant, D. A. (2005) Use of variable-number tandem repeats to examine genetic diversity of *Neisseria meningitidis*. *J. Clin. Microbiol.* **43,** 1699–1705.
21. Yazdankhah, S. P., Kesanopoulos, K., Tzanakaki, G., Kremastinou, J., and Caugant, D. A. (2005) Variable-number tandem repeat analysis of meningococcal isolates belonging to the sequence type 162 complex. *J. Clin. Microbiol.* **43,** 4865–4867.
22. Benson, G. (1999) Tandem repeats finder: a program to analyze DNA sequences. *Nucleic Acids Res.* **27,** 573–580.
23. Denoeud, F. and Vergnaud, G. (2004) Identification of polymorphic tandem repeats by direct comparison of genome sequence from different bacterial strains: a web-based resource. *BMC Bioinformatics* **5,** 4.

26

Fluorescent Amplified Fragment Length Polymorphism Genotyping of Bacterial Species

Meeta Desai

Summary

High-resolution and reproducible whole genome methodologies are needed as tools for rapid and cost-effective analysis of genetic diversity within bacterial genomes. These should be useful for a broad range of applications such as identification and subtyping of microorganisms from clinical samples, for identification of outbreak genotypes, for studies of micro- and macrovariation, and for population genetics.

Fluorescent amplified fragment length polymorphism analysis is one such technique that has been used successfully for studying several bacterial genera. It combines the principle of restriction fragment length polymorphism with the capacity to sample bacterial genomes by selective amplification of a subset of DNA fragments generated by restriction endonucleases, thereby sampling multiple loci distributed throughout the genome. Typically, the genomic DNA is digested with two restriction endonucleases, followed by ligation of double-stranded oligonucleotide adaptors to ends of restriction fragments. Subsets of these fragments are amplified by PCR using fluorescently labeled primers complementary to the adaptor sequences. Amplified fragments are resolved by electrophoresis on an automated DNA sequencer and precisely sized using internal size standards in each sample.

Key Words: High-resolution; whole genome; subtyping; genetic diversity; fluorescent amplified fragment length polymorphism.

1. Introduction

With increasing amounts of DNA sequence information available, the most direct and informative approach to studying DNA-sequence polymorphisms to assess bacterial species is sequence-based comparisons of genomes. However, it

From: *Methods in Molecular Biology, vol. 396: Comparative Genomics, Volume 2*
Edited by: N. H. Bergman © Humana Press Inc., Totowa, NJ

is currently not feasible to sequence whole bacterial genomes on a large scale, and hence sequencing is directed at very small regions of one or a few genes. Therefore, whole genome analysis techniques that detect genetic polymorphisms between bacterial strains are becoming very popular with researchers. Each genotyping method attempts to identify unique characteristics of a strain which differentiates it from other strains. This could be unique DNA sequence/s, or a group of DNA fragments of unique size or sizes. Amplified fragment length polymorphism (AFLP), first described by Vos et al. *(1)* for analysis of plant genomes using radioactive labeled primers is one such technique that has been adapted successfully for DNA fingerprinting of bacterial species. Fluorescent amplified fragment length polymorphism (FAFLP) uses primers that are fluorescently labeled and simple modifications make it possible to tailor the technique for molecular epidemiological studies of bacterial genomes of any size. The sources of AFLP polymorphisms could be point mutations within the restriction site, resulting either in loss or gain of a restriction site, or insertion or deletion between the priming sites, thus resulting in polymorphic fragments of varying sizes.

FAFLP usually involves the digestion of genomic DNA with a restriction endonuclease with a six-base recognition sequence (rare-cutter) followed by digestion with an endonuclease with a four-base recognition sequence (frequent-cutter) (**Fig. 1**). This is followed by ligation of the restricted fragments to double-stranded oligonucleotide adaptors, which consist of a core sequence and an enzyme-specific recognition-site sequence, to create target sites for primer annealing. A subset of these restriction fragments is amplified using primers that have sequences complementary to the ligated adaptor-restriction site. One of the primers is labeled with fluorescent dye and can be detected on an automated DNA sequencer.

FAFLP has numerous advantages over other DNA fingerprinting techniques. With or without prior knowledge of the genome sequence, it assesses the whole genome for both conserved and rapidly evolving sequences in a relatively unbiased way. The number of data points (fragments) obtained for comparative purposes between isolates is significantly greater (as much as 10 times more, compared to pulsed-field gel electrophoresis) thus making it more discriminatory than PFGE *(2)*. The FAFLP results are highly reproducible because of stringent PCR cycling parameters, and the software, using an internal size standard in each lane, precisely and accurately sizes the amplified fragments *(3,4)*. Using selective primers that have one or two additional nucleotides on the 3′-end, the number of amplified fragments can be varied to suit the needs of the investigation. It is also possible to

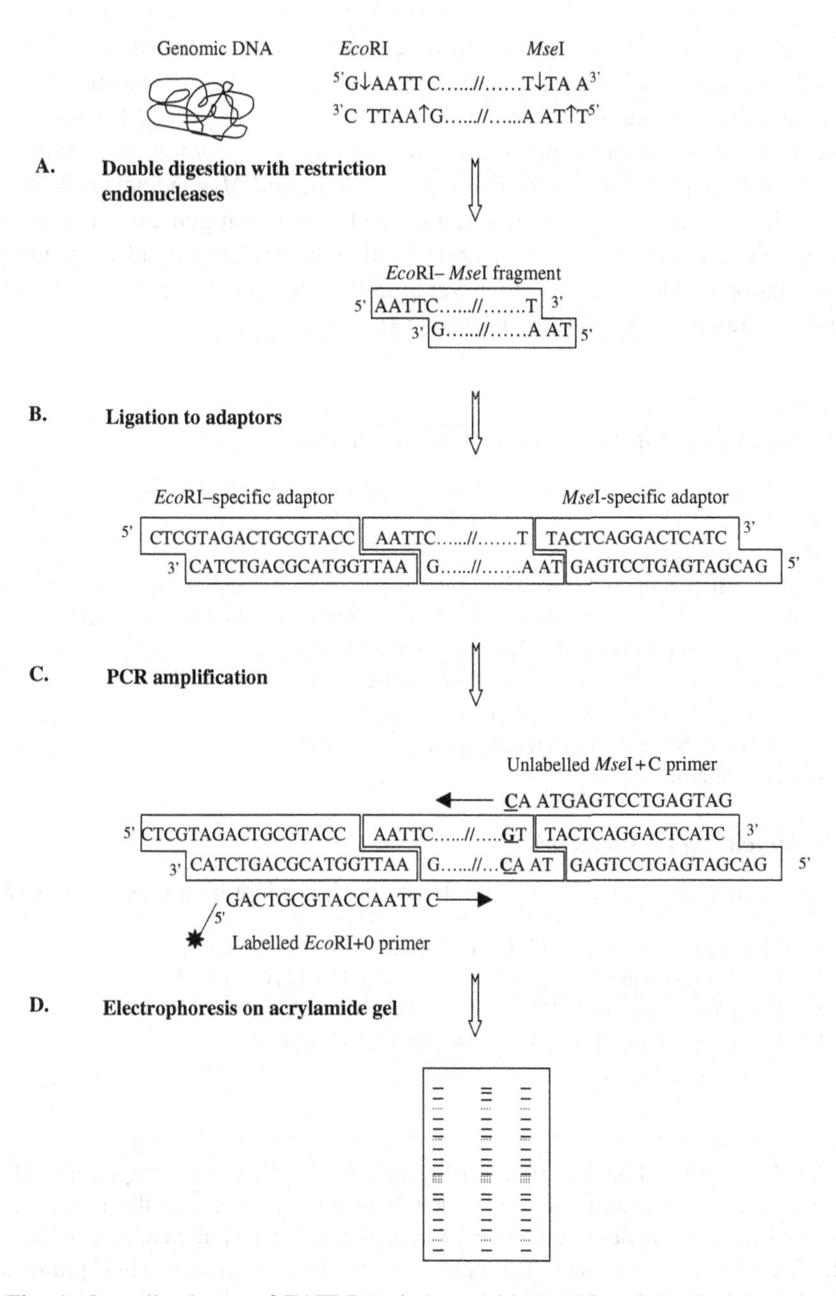

Fig. 1. Overall scheme of FAFLP technique with *Eco*RI and *Mse*I endonucleases. Total genomic DNA with *Eco*RI and *Mse*I recognition sequences indicated. (**A**) The DNA is double-digested with *Eco*RI and *Mse*I, generating *Eco*RI-*Mse*I restriction

automate the whole procedure using capillary sequencers along with the fluorescently labeled primers, thus making it appealing to researchers with large numbers of samples. FAFLP has been successfully used for subtyping several bacterial genera including *Enterococcus, Streptococcus, Staphylococcus, Campylobacter, Escherichia, Salmonella*, and *Mycobacteria* ***(2–8)*** and for studies of micro- and macrovariation and population genetics of *Neisseria* species ***(9)***. It has also been demonstrated that FAFLP can identify unique combinations of precisely sized marker amplified fragments within and between clusters, thus defining strain genotypes ***(4)***.

2. Materials

2.1. Digestion With Restriction Endonucleases

1. Restriction endonucleases (*see* **Note 1**): (New England BioLabs, NEB). Store at −20°C.
   *Mse*I: restriction-site T↓TA↑A
   *Eco*RI: restriction-site G↓AATT↑C
2. 10X Specific restriction buffer 2 for *Mse*I (supplied by the manufacturer).
3. 100X BSA (supplied by the manufacturer with *Mse*I).
4. 30 mg/mL Aqueous DNase-free ribonuclease A. Store at −20°C.
5. 0.5 *M* Tris-HCl, pH 7.2. Store at room temperature.
6. 0.5 *M* NaCl. Store at room temperature.
7. Water (*see* **Note 2**).

2.2. Annealing of Adaptors

The sequences of the *Eco*RI and *Mse*I-specific adaptors are as follows ***(1)***:

1. *Eco*RI adaptor strand 1: 5′-CTC GTA GAC TGC GTA CC-3′.
2. *Eco*RI adaptor strand 2: 5′-AAT TGG TAC GCA GTC TAC-3′.
3. *Mse*I adaptor strand 1: 5′ GAC GAT GAG TCC TGA G-3′.
4. *Mse*I adaptor strand 2: 5′-TAC TCA GGA CTC ATC-3′.

Fig. 1. fragments (as example). **(B)** Ligation of *Eco*RI-specific adaptor and *Mse*I-specific adaptor to ends of the restriction fragments, resulting in the formation of restricted-ligated fragments which serve as template for amplification for PCR. **(C)** Selective amplification of these fragments with one-base selective *Mse*I+C primer and FAM-labeled nonselective *Eco*RI+0 primer pair (as example) by Touchdown PCR cycling conditions (*see* **Subheading 3.**). **(D)** Resolution of amplified fragments is achieved on an automated DNA sequencer.

2.3. Ligation

1. T4 DNA ligase (NEB). Store at −20°C.
2. 10X Ligase reaction buffer (supplied by the manufacturer).
3. 2 μ*M* *Eco*RI-specific adaptor.
4. 20 μ*M* *Mse*I-specific adaptor.
5. Water.

2.4. PCR Amplification

1. 1 μ*M* Fluorescently labeled *Eco*RI primer (*Eco*RI+0): 5′-6FAM-GAC TGC GTA CCA ATT C-3′ (*see* **Notes 3** and **4**).
2. 5 μ*M* Unlabeled one-base selective *Mse*I primer (*Mse*I+**C**): 5′-GAT GAG TCC TGA GTA A**C**-3′ (*see* **Note 4**).
3. 10X *Taq* DNA polymerase supplied with Taq polymerase buffer and 50 m*M* $MgCl_2$.
4. 10 m*M* Each dNTPs: 10 μL of 100 μ*M* of each of the four dNTPs, and 60 μL of water. Aliquots are stored at −20°C.
5. Water.

3. Methods

The success of the technique depends on the quality of DNA, choice of enzymes used for restriction of the genomic DNA, the sequences of the adaptors and primers, and the PCR cycling conditions used for amplification of the restricted fragments. Extraction of genomic DNA from bacterial cells should be performed using established protocols, and the DNA quantified using a spectrophotometer by standard methods *(10)* or by gel comparison with known molecular weight standards. Although it has been shown that the input template DNA concentration does not affect AFLP data, except in the case of very high or very low DNA concentrations *(1)*, the quality of the DNA is very important (*see* **Note 5**). The choice of endonucleases and primers depends on the G+C content of the bacterial genome. Amplification of the restriction fragments using selective nucleotides at the 3′-end ensures that the primers anneal precisely to randomly chosen sequences immediately internal to a restriction endonuclease site. This decreases the number of amplified fragments comprising an FAFLP profile thus ensuring better resolution of amplified fragments across the gel. Separation of amplified fragments can be achieved by capillary electrophoresis or on slab-based systems using the fluorescent primers.

Endonuclease and primer combinations used in FAFLP reactions for a lot of bacterial genera and species have been standardized in our laboratory. For this chapter, the most extensively used endonuclease combination, *Eco*RI and

*Mse*I, has been described with the following primers: a FAM-labeled nonselective *Eco*RI adaptor-specific forward primer (termed *Eco*RI+0) and a one-base selective nonlabeled *Mse*I adaptor-specific reverse primer (*Mse*I+C).

3.1. Sequential Digestion With Two Restriction Endonucleases (see Note 6)

1. Aliquot the appropriate volume of genomic DNA (approx 500 ng) into 0.5-mL reaction tubes, depending on the concentration of DNA. Add water to bring the final volume to 17 μL, if necessary.
2. In a separate tube, prepare an *Mse*I enzyme master mix that contains the following reagents for each reaction: 2 μL of 10X *Mse*I reaction buffer, 0.2 μL of 100X BSA, 0.3 μL of 30 mg/mL Ribonuclease A, and 0.5 μL of 10 U/μL *Mse*I enzyme (*see* **Note 7**).
3. Mix the reaction thoroughly (*see* **Note 8**).
4. Aliquot 3 μL of the master mix into each of the tubes with genomic DNAs.
5. Mix thoroughly and incubate at 37°C for 2 h.
6. In a separate tube, prepare an *Eco*RI enzyme master mix that contains the following reagents for each reaction: 0.5 μL of water, 1.7 μL of 0.5 *M* Tris-HCl (pH 7.2), 2.1 μL of 0.5 *M* NaCl, and 0.5 μL of 10 U/μL *Eco*RI enzyme.
7. Add 4.8 μL of the *Eco*RI master mix to each of the *Mse*I digest.
8. Mix the reaction thoroughly and incubate at 37°C for a further 2 h.
9. Inactivate the endonucleases by heating at 65°C for 10 min.
10. Immediately cool the reactions on ice.

3.2. Annealing of Adaptors

1. The two individual strands (strand 1 and 2) of each of the adaptors are reconstituted to make a stock solution of 100 μ*M* each.
2. To make up a 2-μ*M* *Eco*RI adaptor solution, add 2 μL of *Eco*RI adaptor strand 1, 2 μL of *Eco*RI adaptor strand 2, and 96 μL of water.
3. Mix the reagents thoroughly.
4. Place the tubes in a block-based programmable thermocycler or a heating block set at 95°C for 10 min.
5. Slow cool to room temperature. This is achieved by leaving the tubes in a block-thermocycler or heating block and switching off the instrument until the tubes reach room temperature (*see* **Note 9**).
6. Aliquot the annealed adaptors and store at −20°C.
7. To make up a 20-μ*M* *Mse*I adaptor solution, add 20 μL of *Mse*I adaptor strand 1, 20 μL of *Mse*I adaptor strand 2, and 60 μL of water.
8. To anneal the two adaptor strands, follow **steps 3–6** as for the *Eco*RI adaptor previously described.

3.3. Ligation of Annealed Adaptors

1. Following digestion with the two endonucleases, ligation is carried out using the previously annealed adaptors.
2. In a separate tube, prepare a ligation master mix that contains the following reagents for each reaction: 9.9 μL of water, 5.0 μL of 10X T4 ligase reaction buffer, 0.5 μL of 2 μ*M* *Eco*RI adaptor, 0.5 μL of 20 μ*M* *Mse*I adaptor, 0.1 μL of 400 U/μL T4 DNA Ligase. Mix thoroughly.
3. Add 25 μL of the previously described master mix to the double-digested DNA in **Subheading 3.1.**
4. Mix the reaction thoroughly.
5. Incubate the reaction mixture at 12°C overnight (*see* **Note 10**).
6. Inactivate the reaction by heating at 65°C for 10 min.
7. Cool the reaction on ice. Once used for PCR, this double-digested ligated DNA can be stored at −20°C.

3.4. PCR Amplification of Double-Digested/Ligated Reaction Mix

1. PCR reactions are performed in 25 μL final volume.
2. Prepare a PCR-enzyme master mix that contains the following reagents for each reaction: 15.0 μL of sterile water, 2.5 μL of 10X *Taq* reaction buffer, 1.25 μL of 50 m*M* $MgCl_2$, 0.5 μL of 10 m*M* dNTPs mix, 2.0 μL of 1 μ*M* labeled *Eco*RI primer, 1.0 μL of 5 μ*M* nonlabeled *Mse*I selective primer, and 0.25 μL of 5 U/μL *Taq* DNA polymerase.
3. Mix the reagents thoroughly.
4. Aliquot 22.5 μL of the master-mix into thin-walled PCR tubes.
5. Add 2.5 μL of the double-digested/ligated reaction mix into individual tubes. Mix thoroughly.
6. Touchdown PCR (*see* **Note 11**) is performed in a thermal cycler, and the cycling conditions used for amplification are as follows: initial denaturation at 94°C for 2 min (1 cycle) followed by 30 cycles of denaturation at 94°C for 20 s, primer annealing step for 30 s, and extension at 72°C for 2 min. The primer annealing temperature for the first cycle is 66°C, and for the next nine cycles, it is decreased by 1°C at each cycle. The annealing temperature for the remaining 20 cycles is 56°C. This is followed by a final extension at 60°C for 30 min.
7. Amplification products are separated by electrophoresis on a denaturing polyacrylamide gel (*see* **Subheading 3.5.**). FAFLP reactions can be stored at −20°C for at least 6–9 mo.

3.5. Fragment Separation

This can be achieved on various platforms available in the market. In our laboratory, we have primarily used the ABI Prism 377 automated gel-based DNA sequencer, but the Beckman Coulter capillary electrophoresis for fragment separation has also been used (*see* **Note 12**). FAFLP reaction mixtures for running on the ABI 377 sequencer contained 1 μL of the PCR sample, 1.25 μL of formamide, 0.25 μL of loading solution (dextran blue in 50 m*M* EDTA), and 0.5 μL of fluorescently labeled internal size standard (GeneScan-2500 or GeneScan-500 labeled with the red dye, ROX). The sample mixture was heated at 95°C for 2 min, cooled on ice, and immediately loaded onto the gel. The running electrophoresis buffer was 1X TBE, and the electrophoresis conditions were 2 kV at 51°C for 17 h, with the well-to-read distance of 48 cm *(8)*.

3.6. FAFLP Fragment Analysis

Amplified fragments (AFs) separated on the ABI automated 377 sequencer are sized with GeneScan 3.1.0 software (Applied Biosystems), and the GenoTyper 2.5 software (Applied Biosystems) is used to generate a table in a binary format (with presence "1" or absence "0" of fragments) as a text file (tab-delimited) in MS Excel. These data can be imported into suitable software packages for further analysis. For example, Bionumerics (Applied Maths) can be used for obtaining dendrograms to demonstrate relationships between the isolates.

4. Notes

1. The appropriate combination of endonucleases is the primary determinant of the number and size range of amplified fragments, and the optimum distribution of these fragments on the polyacrylamide gel. In today's genomic era, published genome sequences can be used to model accurate *in silico* FAFLP analysis using online tools (e.g., http://193.129.245.227/aflp.html) *(3,4)*. This gives a rough indication of the number and sizes of amplified fragments that can be obtained using various endonuclease combinations. The combination *Eco*RI and *Mse*I has been successfully used in our laboratory for *Staphylococcus* spp., *Streptococcus* spp., *Salmonella* spp., *Neisseria* spp., and *Escherichia coli*. The other enzyme combination *Hind*III and *Hha*I has also been applied for FAFLP analysis of *Salmonella* spp. and *Campylobacter* spp. *(4,8)*.
2. Throughout the text, water refers to sterile water of molecular biology grade.
3. One of the primers, generally the one specific to the rare-cutting enzyme, *Eco*RI, is labeled at the 5′-end with a fluorescent dye, depending on the sequencer used for fragment separation. For the ABI instrumentation, either capillary or slab gel, 5-carboxyfluorescein (FAM, "blue" color), N,N,N′,N′-tetramethyl-6-carboxyrhodamine (TAMRA, "yellow"), or 2′,4′-dimethoxy-5′,7′-dichloro-6-carboxyfluorescein (JOE, "green") can be used. In our laboratory, our primers were labeled with the blue dye, FAM. The internal size standard is labeled with 6-carboxy-x-rhodamine (ROX, "red") (Applied Biosystems). For the Beckman

capillary sequencer, the fluorescent dyes Cy5 (color "blue") or Cy5.5 (color "green") can be used for labeling the primer and appropriate internal size standards labeled with fluorescent dye D1 (color "red") are used for sizing.

4. The amplification of the restriction fragments can be made selective by the use of primers that have between one and three selective nucleotides on the 3′-end. Typically for smaller genomes such as prokaryotes, the total number of selective bases should be restricted to one or two. The number of selective nucleotides in the primers should be such that 50–100 fragments evenly distributed over a size range of 100 to 600 bp are generated per sample. The selective primer is directed at the endonuclease restriction site, e.g., the *Mse*I restriction site (5′-TTAA-3′), whereby it contains one or two additional nucleotides at its 3′-end. For example, a one-base selective primer, *Mse*I+C will have a core sequence (complementary to the adaptor sequence) followed by 5′-TTAAC-3′, where the first four bases are complementary to an *Mse*I restriction site while the last nucleotide C at the 3′-end is selective. Hence, only fragments with the sequence 3′-AATTG-5′ will be amplified. Assuming random base distribution, the addition of one selective nucleotide theoretically reduces the number of fragments amplified by a quarter, thus reducing the complexity of the resultant FAFLP pattern. A non-selective primer, e.g., *Eco*RI+0 will have a core sequence followed by the six bases (5′-GAATTC-3′) complementary to the *Eco*RI restriction site.
5. The purity of genomic DNA is very important for successful FAFLP reactions. The first step of FAFLP reactions is the digestion of the DNA with two restriction endonucleases. The restriction endonucleases are sensitive to the presence of inhibitors like negatively charged polysaccharides, and organic and inorganic molecules in the DNA preparations. These impurities can result in partial digestion, resulting in nonspecific fragments detectable after amplification. Over the years, we have used several methods for obtaining pure bacterial DNA, ranging from manual extraction protocols like cetyl trimethyl ammonium chloride method ***(11)***, various commercial kits available on the market to using robotic extraction instruments such as the MagNa Pure LC instrument and the associated MagNa Pure DNA isolation Kit I or Kit III (Roche Diagnostics Ltd.).
6. The original method and subsequent researchers recommend double-digestion of genomic DNA with the two endonucleases. However, extensive experiments carried out in our laboratory have found that the best results are obtained by sequential digestion with the two endonucleases rather than double-digestion. The two endonucleases, *Eco*RI and *Mse*I, require varying salt concentrations in their respective buffers for optimal activity. This is achieved by digesting the endonuclease requiring lower salt concentration first (*Mse*I) followed by adjusting the salt concentration of the reaction (using small volumes of Tris-HCl and NaCl) for the second endonuclease (*Eco*RI) requiring the higher salt concentration. This results in significant decrease of partial digestion of genomic DNA that results in

FAFLP profiles with very low (if any) background peaks. The endonucleases are also inactivated prior to digestion, to obtain clean FAFLP patterns.

7. Once a DNA extraction protocol has been standardized for a particular genus, the concentration of DNA for each bacterial preparation will be approximately the same. This will eliminate the need to add varied volumes of DNA (i.e., **Subheading 3.1., step 1**) for each of the bacterial strains. The necessary volume of water required to make up 500 ng of genomic DNA in a final volume of 17 μL can then be added to the master mix itself, thus reducing an additional pipetting step in the protocol.
8. While preparing master mix, add the reagents and mix thoroughly by briefly vortexing for a few seconds followed by a spin at 1000*g* for 15 s to collect the reaction mix to the bottom of the tube. All enzymatic reactions must be set up on ice.
9. It is very important to have equimolar concentrations of each of the two strands of the adaptor. Great care should be taken while pipetting precise volumes as excess of one of the strands (the "lower" strand) in the annealed adaptor can result in that strand acting as a primer in subsequent reactions. This can result in FAFLP profiles with fragments that are sized (n+1) bp, (n+2) bp, and so on, where "n" is the expected size of that fragment. The resulting electropherograms will display two or three additional peaks for each expected peak. These additional peaks were also observed when the concentration of the adaptors in the ligation reaction was increased 10-fold. We have also found that while annealing the adaptors, the slow cool step after the denaturation step is very important.
10. Following digestion, ligation is carried out in the same tube using the appropriate ligation buffer supplied with the ligase enzyme. Ligation is carried out at lower temperatures (12–16°C) overnight. However, it was found that 3–4 h was sufficient to achieve efficient ligation of adaptors to the digested sticky-ends generated by the two restriction endonucleases. The best results were obtained by using 400 U of T4 DNA Ligase per FAFLP reaction. The ligase is inactivated after ligation, and reaction mixes can be stored at −20°C. PCR from such ligations yielded reproducible FAFLP results even after 6 mo.
11. Touchdown PCR cycling conditions are used whereby higher than optimal annealing temperature is used for the first few cycles followed by 20 cycles at the optimum annealing temperature. This high-stringency PCR ensures amplification of only fragments that are complementary to the adaptor-restriction fragment sequence. For samples where the input DNA was very low, two rounds of PCR may be required. The first round includes a touchdown PCR (as described in **Subheading 3.4., step 6**) using both nonselective primers i.e. *Eco*RI+0 and *Mse*I+0. For the second round PCR, use the amplification product from the first round PCR as template (1–2 μL in a 25- μL PCR reaction) and use the primers *Eco*RI+0 and *Mse*I+C.

12. On the ABI 377 automated DNA sequencer, amplification products are separated on a denaturing polyacrylamide gel using Long Ranger™ Singel™ packs (Cambrex BioSciences), containing all ingredients necessary for 5% Long Ranger acrylamide solution. Electrophoresis gels could be run for longer periods with better resolution of the fragments in the lower size range by using the easy-to-use Singel packs.

Acknowledgments

The author thanks Catherine Arnold for critically reading the manuscript.

References

1. Vos, P., Hogers, R., Bleeker, M., et al. (1995) AFLP: a new technique for DNA fingerprinting. *Nucleic Acids Res.* **23,** 4407–4414.
2. Desai, M., Tanna, A., Wall, R., Efstratiou, A., George, R., and Stanley, J. (1998) Fluorescent amplified-fragment length polymorphism analysis of an outbreak of group A streptococcal invasive disease. *J. Clin. Microbiol.* **36,** 3133–3137.
3. Arnold, C., Metherell, L., Clewley, J. P., and Stanley, J. (1999) Predictive modelling of fluorescent AFLP: a new approach to the molecular epidemiology of *E. coli*. *Res. Microbiol.* **150,** 33–44.
4. Desai, M., Logan J. M., Frost, J. A., and Stanley, J. (2001) Genome sequence-based FAFLP of *Campylobacter jejuni* and its implications for sero-epidemiology. *J. Clin. Microbiol.* **39,** 3823–3829.
5. Antonishyn, N. A., McDonald, R. R., Chan, E. L., et al. (2000) Evaluation of fluorescence-based amplified fragment length polymorphism analysis for molecular typing in hospital epidemiology: comparison with pulsed-field gel electrophoresis for typing strains of vancomycin-resistant *Enterococcus faecium*. *J. Clin. Microbiol.* **38,** 4058–4065.
6. Grady, R., Blanc, D., Hauser, P., and Stanley, J. (2001) Genotyping of European isolates of methicillin-resistant *Staphylococcus aureus* by fluorescent amplified-fragment length polymorphism analysis (FAFLP) and pulsed-field gel electrophoresis (PFGE) typing. *J. Med. Microbiol.* **50,** 588–593.
7. Sims, E. J., Goyal, M., and Arnold, C. (2002) Experimental versus in silico fluorescent amplified fragment length polymorphism analysis of *Mycobacterium tuberculosis*: improved typing with an extended fragment range. *J. Clin. Microbiol.* **40,** 4072–4076.
8. Lawson, A. J., Stanley, J., Threlfall, E. J., and Desai, M. (2004). Fluorescent amplified fragment length polymorphism subtyping of multiresistant *Salmonella enterica* serovar Typhimurium DT104. *J. Clin. Microbiol.* **42,** 4843–4845.
9. Goulding, J. N., Stanley, J., Olver, W., Neal, K. R., Ala'Aldeen, D. A., and Arnold, C. (2003) Independent subsets of amplified fragments from the genome of

Neisseria meningitidis identify the same invasive clones of ET37 and ET5. *J. Med. Microbiol.* **52,** 151–154.
10. Maniatis, T., Fritsch, E. F., and Sambrook, J. (1982) *Molecular Cloning: A Laboratory Manual*. Cold Spring Harbor Laboratory, New York.
11. Wilson, K. (1987) *Preparation of Genomic DNA From Bacteria, Unit 2.4.1. Current Protocols in Molecular Biology*. John Wiley and Sons, Inc., New York.

27

FLP-Mapping

A Universal, Cost-Effective, and Automatable Method for Gene Mapping

Knud Nairz, Peder Zipperlen, and Manuel Schneider

Summary

Genetic mapping with DNA sequence polymorphisms allows for map-based positional cloning of mutations at any required resolution. Numerous methods have been worked out to assay single nucleotide polymorphisms (SNPs), the most common type of molecular polymorphisms. However, SNP genotyping requires customized and often costly secondary assays on primary PCR products. Small insertions and deletions (InDels) are a class of polymorphisms that are ubiquitously dispersed in eukaryotic genomes and only about fourfold less frequent than SNPs. InDels can be directly and universally detected as fragment length polymorphisms (FLPs) of primary PCR fragments, thus eliminating the need for an expensive secondary genotyping protocol. Genetic mapping with FLPs is suited for both small-scale and automated high-throughput approaches. Two techniques best suited for either strategy and both offering very high to maximal fragment-size resolution are discussed: Analysis on nondenaturing Elchrom gels and on capillary sequencers. Here, we exemplify FLP-mapping for the model organisms *Drosophila melanogaster* and *Caenorhabditis elegans*. FLP-mapping can, however, be easily adapted for any other organism as the molecular biology involved is universal. Furthermore, we introduce FLP mapper, a JAVA-based software for graphical visualization of raw mapping data from spreadsheets. FLP mapper is publicly available at http://bio.mcsolutions.ch.

Key Words: Genetic mapping; sequence polymorphisms; InDels; model organisms; *Drosophila melanogaster; Caenorhabditis elegans*; mapping software; *FLP mapper*.

From: *Methods in Molecular Biology, vol. 396: Comparative Genomics, Volume 2*
Edited by: N. H. Bergman © Humana Press Inc., Totowa, NJ

1. Introduction

Small insertions/deletions (InDels) constitute a ubiquitous and frequent class of biallelic sequence polymorphisms in eukaryotic genomes. In complex genomes such as rice and human, InDels constitute roughly 20% of the sequence polymorphisms and occur as frequently as once every 1000 bp to about once every 5000 bp *(1–3)*. Similar frequencies have been observed in lower metazoans. Some studies reach highly divergent conclusions about InDel frequency, e.g., varying between 7 and 37% in Arabidopsis. In part, high estimates may be attributed to sequencing errors, as about two-third of InDels are 1-bp polymorphisms that have a low confirmation rate *(2,4)*.

Genotyping with InDels offers the main advantage that the measured output is the size of a primary PCR-product rather than its sequence. Fragment length polymorphisms (FLPs) can be easily measured on standard lab equipment at standardized conditions. The determination of sequence differences (single nucleotide polymorphisms [SNPs]) on the other hand involves either secondary, often costly and complex enzymatic reactions on the primary product (e.g., primer extension and primer ligation-based techniques, SNPs on chips *[5]*) or the analysis of different properties of homo- and heteroduplexes of primary PCR-products at partially denaturing conditions (e.g., SSCP and DHPLC *[6]*). The latter principle generally needs optimization for every single PCR product and in part specialized machinery.

Microsatellites (STRs) constitute a FLP subclass that has been widely used for genetic mapping by the same methodology *(7)*. They differ from InDels in that they are comprised of repetitive sequence, which may cause PCR problems leading to "stuttering" products. Besides, STRs are far less common than InDels—best illustrated by the current numbers of entries in the dSNP database for human polymorphisms (http://www.ncbi.nlm.nih.gov/SNP/) containing about 33 million SNPs, 3 million InDels, but only about 5000 STRs.

Thus, FLP-mapping combines the advantages of genetic mapping with STRs and SNP-mapping, respectively: allele calling is based on robust methodology on the primary PCR product and InDel density on average may be sufficient to even map down to a gene.

FLP-mapping is both suited to a small-scale approach in labs equipped with basic instrumentation (i.e., gel apparatus and thermocycler) and to high-throughput procedures including automated PCR-setup in 96- or 384-well format, multiplexing strategies, fragment analysis on capillary sequencers, and software-supported allele calling. Here, we exemplify both strategies by providing protocols for high-throughput fragment analysis and for low and medium throughput genotyping on high-resolution nondenaturing gels.

Besides, we present *FLP mapper*, a software that is designed to graphically display the genotypes of recombinant chromosomes.

2. Materials (see Note 1)

2.1. Links to SNP and InDel Databases

Human:

http://snp.cshl.org/
http://hgvbase.cgb.ki.se/cgi-bin/main.pl?page=data_struct.htm

Mouse:

http://www.informatics.jax.org/javawi2/servlet/ WIFetch?page=snpQF

Caenorhabditis elegans:

http://genomebiology.com/2005/6/2/R19 (additional data files 12 to 17)
http://www.wormbase.org/db/searches/strains
http://genomeold.wustl.edu/projects/celegans/index.php?snp=1

Drosophila:

http://genomebiology.com/content/supplementary/gb-2005–6–2-r19-s18.pdf
http://www.fruitfly.org/SNP/
http://flysnp.imp.univie.ac.at/

Arabidopsis:

http://www.arabidopsis.org/servlets/Search?action=new_search&
type=marker

Zebrafish:

http://zfin.org/cgi-bin/webdriver?MIval=aa-crossview.apg&OID=ZDB-REFCROSS-010114–1
http://www.sanger.ac.uk/Projects/D_rerio/DAS_conf.shtml

2.2. Worm Lysis

1. Worm lysis buffer (10X): 500 m*M* KCl, 100 m*M* Tris-HCl pH 8.2, 25 m*M* $MgCl_2$, 4.5% NP-40 (=IGEPAL CA630, Sigma-Aldrich), 4.5% Tween-20 (Sigma-Aldrich), 0.1% Gelatin (Sigma-Aldrich). Store at room temperature.
2. Proteinase K (100X, from *Tritirachium album*, Sigma-Aldrich): 10 mg/mL in water. Store at −20 °C.

2.3. Fly Lysis

1. Squishing buffer (10X): 100 m*M* Tris-HCl pH 8.2, 10 m*M* EDTA pH 8.0, 2% Triton X-100 (Sigma-Aldrich), 250 m*M* NaCl. Store at room temperature.

2. Proteinase K (100X, from *Tritirachium album*, Sigma-Aldrich): 20 mg/mL in water. Store at −20 °C.
3. Vibration mill (Retsch MM30).

2.4. PCR

1. Taq polymerase (5 U/μL, Euroclone Life Sciences) and 10X buffer without $MgCl_2$: 50 m*M* KCl, 160 m*M* $(NH_4)_2SO_4$, 670 m*M* Tris-HCl pH 8.8, 0.1% Tween-20. Add $MgCl_2$ from a 50-m*M* stock solution.
2. Phusion polymerase (2 U/μL, Finnzymes) and 5X buffer: HF buffer containing 7.5 m*M* $MgCl_2$ is provided with the enzyme.
3. 2 m*M* dNTPs: made from 100 m*M* dATP, dTTP, dGTP, dCTP stock solutions (Roche).
4. Primers: working concentration 10 pmol/μL, made from 100 pmol/μL stock solution.
5. Thermoycler: should be equipped with a 96-well thermo-block (e.g., MJR, Applied Biosystems, Biometra).

2.5. Capillary Electrophoresis

1. Denaturation buffer: HiDi formamide (Applied Biosystems).
2. Size standard: LIZ500 (Applied Biosystems).
3. Running parameters for fragment analysis on an ABI3730 capillary sequencer (Applied Biosystems): oven temperature 66 °C, buffer temperature 35 °C, prerun voltage 15 kV, prerun time 180 s, injection voltage 2 kV, injection time 10 s, first readout 200 ms, second readout 200 ms, run voltage 15 kV, voltage number of steps 10, voltage step interval 50 s, voltage tolerance 0.6 kV, current stability 15 mA, ramp delay 500 s, data delay 1 s, run time 850 s.
4. Capillary length: 36 cm.
5. Polymer: POP7 (Applied Biosystems).

2.6. Gel Electrophoresis

1. TAE buffer (40X): per liter 145.37 g Tris(hydroxymethyl) aminomethane ("Tris"), 11.16 g $Na_2EDTA \times 2\ H_2O$, 34.4 mL glacial acetic acid. Concentration of 1X TAE is 30 m*M*.
2. SYBR green I/SYBR gold (Invitrogen) staining solution: 1:10,000 dilution in 10 m*M* TAE. Stock solution shall be stored in 5-μL aliquots at −20 °C (*see* **Note 2**).
3. Ethidium bromide (EtBr, Sigma-Aldrich) staining solution: 0.4 μg/mL EtBr in water. Store 10 mg/mL stock solution in light-protected bottle at 4 °C (*see* **Note 3**).
4. Loading buffer (6X): 30% (v/v) glycerol, 0.25% (w/v) xylene cyanol, 0.25% (w/v) bromphenol blue. Store at 4 °C.
5. Gels: Elchrom Spreadex precast gels that are nondenaturing, but offer very high to maximum resolution.
6. Submarine gel system: Elchrom provides the specialized SEA2000 system for its gels featuring a thermostated water bath. However, Elchrom gels can be run on any

gel platform with at least 600 mL buffer volume and a width of at least 15 cm (*see* **Note 4**).
7. Transilluminator: should be equipped with a 254-nm ultraviolet source for achieving maximum sensitivity of SYBR-stained gels.

2.7. Software

1. Allele calling on capillary sequencer-generated electropherograms: GeneMapper (Applied Biosystems).
2. Visualization of raw mapping data: FLP mapper. Our software is specifically designed to extract and graphically display the genotype of recombinant chromosomes from Excel spreadsheets. Two algorithms are implemented: (1) deduction of the genotype of the recombinant chromosome in the presence of an unrecombined counter-chromosome (e.g., suited to *Drosophila* and mammalian genetics). (2) Visualization of both recombinant chromosomes of self-fertilizing organisms (e.g., *C. elegans*).

The program can be downloaded freely from http://bio.mcsolutions.ch and runs on all operating systems supporting Java Runtime Environment.

3. Methods

3.1. Establishing a FLP Map

1. From a suitable database (**Subheading 2.1.**) choose candidate InDels/FLPs in the region of interest. For chromosomal linkage analysis at least one (central) FLP per chromosome is required, for mapping at highest resolution FLPs have to be spaced appropriately in a subchromosomal region. One-basepair FLPs do not pose a resolution problem on sequencers, but may approach the limits of resolution on Elchrom gels (**Subheading 3.6.**). Besides, unverified 1-bp FLPs frequently are because of sequencing artefacts and, thus, have a lower confirmation rate than larger FLPs *(2,4)*.
2. Design primers spanning the polymorphism to amplify fragments in the size range of 50 to 350 bp at an annealing temperature of 56–60 °C. This task can be performed by the program Primer3, which is available as an online service on the Internet at: http://frodo.wi.mit.edu/cgi-bin/primer3/primer3_www.cgi (*see* **Note 5**). Primers may be "pig-tailed" to control Taq polymerase-catalyzed addition of adenosines on PCR-products (**Subheading 3.5.**). For FLPs being resolved on a sequencer one primer has to be fluorescently labeled. One-basepair FLPs that have to be separated on Elchrom gels shall have an approximate length of 100 bp.
3. Genotype parental chromosomes and verify the FLP on the system of choice. If FLPs shall be resolved on sequencers and primer costs shall be kept low, it is advisable to first confirm the FLP with unlabeled primers on high-resolution Elchrom gels (**Subheading 3.6.**).

4. From the verified FLPs choose assays that are evenly spaced on the genetic map at the required resolution. For example, if a resolution of <= 1 cM is demanded, take one FLP every 2 cM.
5. Genotype other chromosomes used for the mapping experiment.

3.2. Worm Lysis

1. Collect individual worms in microfuge tubes or in wells of a 96-well microtiter plate each containing 10 µL 1X lysis buffer (1.0 µL 10X buffer, 0.1 µL proteinase K, 8.9 µL water).
2. Shock-freeze in liquid nitrogen.
3. Digest worm suspension by placing tubes/plates at 60 °C for 60 min. Heat-inactivate Proteinase K for 30 min at 90 °C (*see* **Note 6**). Cool down to 4 °C. Store at −20 °C. Avoid more than 10 freeze–thaw cycles.
4. Dilute suspension 10-fold prior to use for PCR. Use 2 µL diluted extract per PCR reaction (*see* **Note 7**).

3.3. Fly Lysis

3.3.1. Bulk Preparation

1. Place single flies into wells of a 96-well deep-well plate (Eppendorf) filled with each 200 µL squishing buffer (20 µL 10X buffer, 2 µL Proteinase K, 178 µL water).
2. Add one tungsten carbide bead (Qiagen) per well.
3. Seal deep-well plate with a rubber mat (Eppendorf) and clamp it into the vibration mill.
4. Shake for 20 s at 20 strokes per second.
5. Transfer suspension into a 96-well PCR-plate (MJR) avoiding debris (*see* **Note 8**).
6. Incubate plate for 30 min at 37 °C and subsequently for 20 min at 95 °C to heat-inactivate Proteinase K (*see* **Note 6**). Cool down to 4 °C. Store at −20 °C.
7. Dilute suspension 20-fold prior to use for PCR. Use 2 µL diluted extract per PCR reaction (*see* **Note 7**).

3.3.2. Single Preparation

1. Place a single fly into a microfuge tube and add 50 µL squishing buffer (5 µL 10X buffer, 0.5 µL Proteinase K, 44.5 µL water).
2. Squish the fly with the pipet tip or, alternatively, use a pestle (Treff-Lab) fitting the microfuge tube to grind the fly.
3. Incubate tube for 30 min at 37 °C and subsequently for 20 min at 95 °C to heat-inactivate Proteinase K (*see* **Note 6**). Cool down to 4 °C and add 150 µL water. Store at −20 °C.
4. Dilute suspension 20-fold prior to use for PCR. Use 2 µL diluted extract per PCR reaction (*see* **Note 7**).

3.4. PCR

3.4.1. Using Taq Polymerase

1. Mix 2 μL DNA (diluted single worm lysate or single fly extract) with 23 μL PCR reaction mix (2.5 μL 10X buffer, 1 μL $MgCl_2$, 0.5 μL dNTPs, 1 μL forward-primer, 1 μL reverse-primer, 16.95 μL water, and 0.05 μL Taq) in an appropriate microfuge tube or PCR-plate (MJR) on ice. Multiplexing two to three different primer pairs in one reaction is usually possible without further optimization.
2. Cycling parameters: 2 min 95 °C, 10 cycles: 20 s 95 °C, 20 s 61 °C (−0.5 °C for each cycle), 45 s 72 °C, 24 cycles: 20 s 95 °C, 20 s 56 °C, 45 s 72 °C, final extension: 10 min 72 °C, hold at 12 °C.

3.4.2. Using Phusion Polymerase

1. Mix 2 μL DNA (diluted single worm lysate or single fly extract) with 23 μL PCR reaction mix (5 μL 5X buffer, 2.5 μL dNTPs, 1 μL forward-primer, 1 μL reverse-primer, 13.25 μL water, and 0.25 μL Phusion) in an appropriate microfuge tube or PCR-plate (MJR) on ice. Multiplexing two to three different primer pairs in one reaction is usually possible without further optimization.
2. Cycling parameters: 30 s 98 °C, 30 cycles: 10 s 98 °C, 20 s 60 °C, 15 s 72 °C, final extension: 10 min 72 °C, hold at 12 °C.

3.5. Capillary Electrophoresis

Taq polymerase tends to catalyze the addition of an adenosine to the 3′-end of PCR-products. This activity may yield two products exhibiting a size difference of 1 bp. A-addition could constitute a genotyping problem especially for 1 bp FLPs and for small FLPs causing polymerase "stuttering." Two different strategies are feasible to control A-addition: first, the non-fluorescent primer can be "pig-tailed" on the 5′-end, which leads to nearly 100% A-addition on the fluorescently labeled strand *(8)*. Second, A-addition can be prevented by using Polymerases like Pfu (Promega) or Phusion (**Subheadings 2.4.** and **3.4.2.** [*see* **Note 9**]). An example of a 1-bp FLP amplified with either Taq or Phusion polymerase is given in **Fig. 1A,B**.

1. Add 2 μL PCR-reaction (performed with one labeled primer) to 198 μL water. Dilution is necessary to reduce signal intensity and to reduce protein content that potentially could harm capillaries.
2. Mix 2 μL of the diluted PCR-reaction with 10 μL HiDi formamide containing LIZ500 size standard (25 μL LIZ500 in 10 mL HiDi) in 96-well sequencing-plates (Applied Biosystems).
3. Denature by heating mixtures to 95 °C for 2 min.
4. Separate fragments on an ABI3730 capillary sequencer using POP7 polymer according to running parameters given in **Subheading 2.5**.

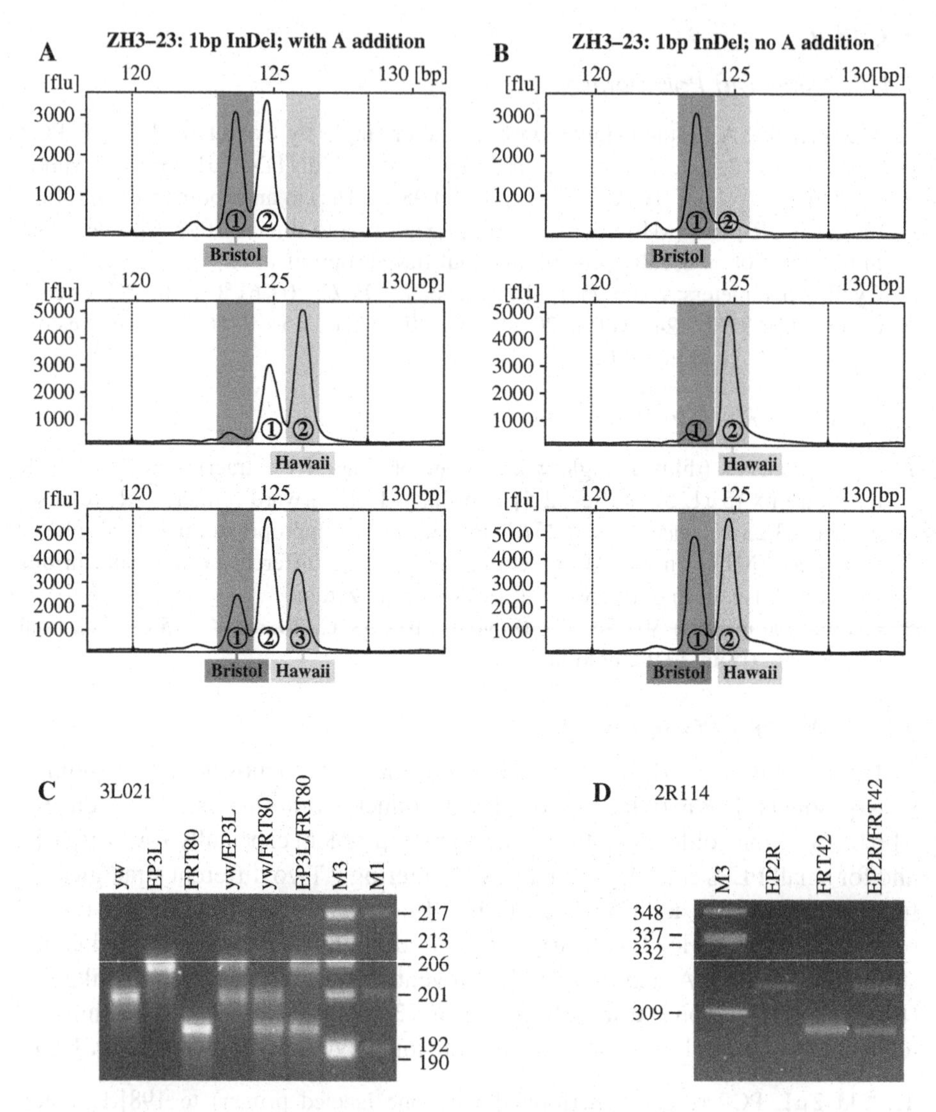

Fig. 1. Resolution of FLPs on sequencers and on nondenaturing gels. *ZH3-23*, a 1-bp FLP between the *Caenorhabditis elegans* strains Bristol and Hawaii, amplified with Taq polymerase (**A**) and Phusion polymerase (**B**), respectively, resolved on an ABI3730 sequencer and displayed by GeneMapper software (Applied Biosystems). Each homozygous and the heterozygous genotypes are shown. To perform unambiguous allele-calling when Taq catalyzed A-addition occurs, the peak without A-addition for the smaller and the peak with A-addition for the larger fragment have to be assigned (**A**). Without A-addition each electropherogram is unambiguous (**B**). FLPs *3L021* and

5. Analyze data and perform allele calling using GeneMapper Software. In “Tools – Options – Analysis” choose “Duplicate homozygous alleles” function. Export raw genotyping data in Microsoft Excel-compatible form by the “Export Table” function as “Tab-delimited text (.txt)”. Export the following data: Sample name, marker, allele 1, allele 2.

3.6. Gel Electrophoresis

For low- to medium-throughput genotyping applications, fragment analysis on nondenaturing gels is ideal because of convenience and technical simplicity. However, because of relatively low resolution conventional agarose gels are only suited for the minority of FLPs exhibiting large size differences. Elchrom Spreadex gels are as easy to handle as agarose gels, but offer higher resolution, under ideal conditions down to the limit of 1 bp (*see* **Fig. 1C**). Allele calling has to be performed manually. Thus, this fragment separation technique is most adequate for small genotyping projects and for FLP assay validations with nonfluorescent primers.

1. Choose Elchrom Spreadex gels (http://elchrom.com/public/index.php?article=144) with appropriate number of slots and offering optimal resolution for the fragments of choice. Take gels with 8-cm length for 1-bp FLPs.
2. Unpack gel and place it into submarine gel unit filled with TAE (*see* **Note 4**). If the gel has been stored at 4 °C allow for a 30 min equilibration. The optimal buffer temperature is 55 °C, but runs at lower temperatures are possible.
3. Per slot load a mixture of 2-μL PCR reaction, 1 μL loading buffer, and 2 μL water.
4. Running conditions depend on the gel used and the resolution requirements. A typical run would be for 1.5–2.5 h at 120 V at 55 °C.
5. Transfer gel into SYBR green/SYBR gold or EtBr staining solution, respectively, and float them for 30 or 45 min in an appropriate gel tray. Up to four gels may be stained simultaneously in 50 mL (*see* **Notes 2** and **3**).
6. Destain for 30 min in water. SYBR stained gels may also be destained in destaining solution (Elchrom).
7. Place gel onto a transilluminator equipped with a 254-nm ultraviolet source and record gel image.

Fig. 1. *2R114* amplified from respective homozygous and heterozygous *Drosophila* strains *yw, EP3L (EP3104), FRT80B, EP2R (EP0755)*, and *FRT42D* and resolved on nondenaturing Elchrom Spreadex gels. The 5-, 6-, and 11-bp *3L021* (**C**) and the 12-bp *2R114* (**D**) FLPs are clearly separated. Resolution of 1-bp FLPs is possible for smaller, ca. 100 bp fragments. Size standards M3 and M1 are from Elchrom.

8. Perform allele calling and chart data in a Microsoft Excel sheet into four columns: Sample name (i.e., recombinant), marker (i.e., FLP-assay), allele 1, and allele 2. Store the data in the Excel "Text (Tab-delimited) (*.txt)" format.

3.7. FLP Mapper Software

FLP mapper is most useful for complex crossings involving three different genotypes: The mutant genetic background, the wild-type background recombined with the mutant, and a third genetic background into which recombinant chromosomes are crossed. This third genotype may be a mixture of both parental genotypes. For mapping projects involving many markers it is therefore laborious to delineate the genotype of the recombinant chromosome in the presence of the third chromosome *(9)*.

FLP mapper first accesses the three different genotypes stored in an Excel-compatible database. Next, it extracts the genotype of the recombinant chromosome from a spreadsheet listing both alleles by subtracting the genotype of the third chromosome. Recombinant genotypes are then listed and colored such that the most informative recombinants can be easily spotted. This algorithm is tailored for the needs of *Drosophila* and mammalian genetics.

A second algorithm is optimized for the genetics of the self-fertilizing organism *C. elegans*. As both chromosomes are recombinant, subtraction is not necessary and both genotypes are graphically displayed.

1. Download software from http://bio.mcsolutions.ch.
2. To generate a project click "new" in the upper left panel. Existing projects can be opened, closed, and saved by pressing the respective adjacent buttons.
3. After a program has been created, "import-options" appear in the main panel. This option allows to define the content of the raw mapping data in Excel .txt format. The program will cross-reference to four columns of the Excel sheet containing the sample name ("sample"), the name of the assay ("marker"), and the genotypes of the two loci for a given marker and sample ("allele 1" and "allele 2").
4. To import raw data press "files." In the middle upper panel an "import" symbol will appear that allows to choose a file when activated. The file name now appears in the left window section (*see* **Fig. 2A**).
5. The strain genotype and mutant selection options can be chosen by activating the "compiling-options" in the task-bar. By choosing this option the middle upper panel will change to reveal a "new" option symbol. Selection of a new option reveals four pull-down menus from which the database of strain genotypes (*see also* **step 8**), the genotype of the mutagenized strain, the genotype of the wild-type strain used for recombination, and the genotype of the strain into which the recombinant chromosome is crossed can be chosen. Finally, there is an option to indicate whether the recombinants have been selected to exhibit or not to exhibit

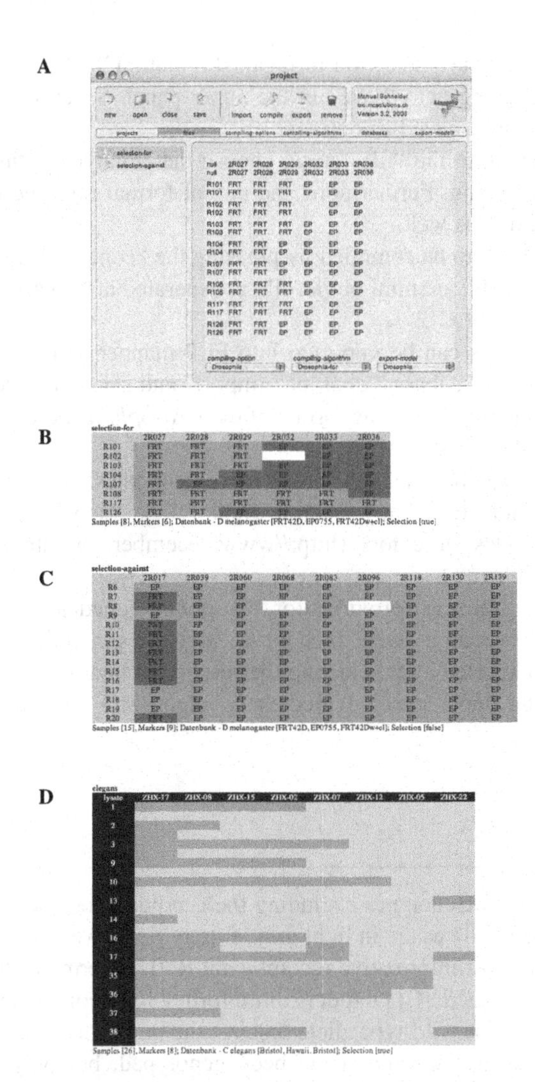

Fig. 2. Graphical display of raw mapping data by FLP mapper. (**A**) Screen-shot of the FLP mapper window exhibiting raw mapping data of the file "selection-for" in the central window. Compiling-options, compiling-algorithm, and export-model have been defined and then chosen from the pulldown-menus. The data can now be compiled and exported as .html file by clicking on the icons in the middle upper panel. (**B**) Output in .html format from the example shown in **A**. Recombinants R101 to R126 have been selected for exhibiting the mutant phenotype. Both the *FRT42D/EP0755* recombinant chromosome and the counter-chromosome *FRT42Dw*$^+$*cl* have been genotyped for FLP-markers *2R027* to *2R036* that are situated from proximal to distal on the right arm of chromosome 2 *(9)*. FLP mapper only displays the genotype of the recombinant

the mutant phenotype ("selection for/against mutation"). These parameters should be saved by clicking onto "save" in the upper panel. Several different parameter permutations can be saved as options.

6. Next, by returning to the "files" task, choose the genotype option defined before on a pull-down menu. Furthermore, the output format (and the model organism) can be determined as well.
7. The data now have to be compiled by pressing the "compile" button in the middle upper panel. Finally, an .html output file is generated after activating the "export" symbol (*see* **Fig. 2B,C**).
8. Genotype databases can be imported into FLP mapper as Excel .txt files. In the menu bar choose "database," click on "import" and select the file to be imported. To create a database refer to the format of the *Drosophila* and *C. elegans* databases supplied with the program.
9. Output colors can be changed. Click "export-models" in the task bar, select respective default file, copy central field, click "new," paste data, change hexadecimal codes of colors (http://www.december.com/html/spec/color.html), click "save."
10. Compiling algorithms are provided for *Drosophila* genetics, which may be suited to vertebrate genetics as well, and for *C. elegans* genetics. Algorithms can be imported and exported after selecting the respective button in the task-bar via the icons in middle upper panel. If necessary, algorithms can be self-programmed in JAVA.

Fig. 2. chromosome. Genotypes excluding the candidate locus are in dark gray and genotypes including the locus are in light gray. Empty fields were not genotyped owing to PCR-failure. The most informative recombinant R107 determines that the candidate region is proximal to *2R028*. **(C)** Output in .html format for recombinants R6 to R20 that have been selected for the wild-type allele (against the mutant allele). Both recombinant and non-informative chromosomes have been genotyped, but only the recombinant chromosomes are listed by FLP mapper. Again, genotypes excluding the candidate regions are dark gray. The informative recombinants place the candidate region distal to *2R017*. The combination of both mapping data localizes the mutation between *2R017* and *2R028*. **(D)** FLP mapper output in .html format from a typical *Caenorhabditis elegans* FLP-mapping experiment. The *C. elegans* compiling algorithm depicts genotyping data of both recombinant chromosomes, Bristol alleles are represented in light gray and Hawaii alleles are in dark gray. The most informative recombinants—lysates 35, 13, 17, and 38—place the mutation between the X-chromosomal markers ZHX-5 and ZHX-22.

4. Notes

1. Water for molecular biology applications should have a resistivity of 18.2 MΩ, a total organic carbon content of less than five parts per billion, and should be pyrogen- and nuclease-free (less than 0.001 EU/mL) (e.g., from a Milli-Q water purification system, Millipore).
2. SYBR green and SYBR gold are extremely light sensitive and thus shall not be frequently thawed and frozen again. Therefore it is advisable to aliquot fully thawed stock solution into 5-μL portions at dim light. One aliquot, diluted 10,000 times in 10 m*M* TAE is sufficient to stain four gels at once. Do not reuse staining solution.
3. EtBr is mutagenic and should be removed from the staining solution with an appropriate removal system (e.g., charcoal filter, Schleicher and Schuell).
4. The gel tanks must be filled with TAE and may never get into contact with TBE. A single TBE filling spoils the submarine gel system for use with Elchrom gels forever.
5. Usually, primers designed by hand according to the formula $T_M = 4 \times (G+C) + 2 \times (A+T) - 5\,°C$, devoid of repetitive or palindromic sequences, and ideally ending with a G or C work well.
6. Heat-inactivation is absolutely crucial, as one of the most common causes of PCR failure are trace-amounts of proteinase K. Some protocols only include a 10-min heating step, which may be insufficient. Incubation, heating, and cooling can be performed in a thermocycler.
7. Dilution of crude extracts improves PCR success-rate and also reduces the risk of spoiling capillaries by proteins.
8. Tungsten carbide beads can be reused after an overnight incubation in 0.1 *M* HCl and subsequent thorough washing in water.
9. Phusion polymerase is expensive and thus is not recommended for general use. However, it is very robust and yields good results and is preferable to Pfu polymerase for applications where A-addition is unwanted.

References

1. Sachidanandam, R., Weissman, D., Schmidt, S. C., et al. (2001) A map of human genome sequence variation containing 1.42 million single nucleotide polymorphisms. *Nature* **409**, 928–933.
2. Weber, J. L., David, D., Heil, J., Fan, Y., Zhao, C., and Marth, G. (2002) Human diallelic insertion/deletion polymorphisms. *Am. J. Hum. Genet.* **71**, 854–862.
3. Shen, Y. J., Jiang, H., Jin, J. P., et al. (2004) Development of genome-wide DNA polymorphism database for map-based cloning of rice genes. *Plant Physiol.* **135**, 1198–1205.

4. Schmid, K. J., Sorensen, T. R., Stracke, R., et al. (2003) Large-scale identification and analysis of genome-wide single-nucleotide polymorphisms for mapping in Arabidopsis thaliana. *Genome Res.* **13**, 1250–1257.
5. Syvanen, A. C. (2005) Toward genome-wide SNP genotyping. *Nat. Genet.* **37**, S5–S10.
6. Nairz, K., Stocker, H., Schindelholz, B., and Hafen, E. (2002) High-resolution SNP mapping by denaturing HPLC. *Proc. Natl. Acad. Sci. USA* **99**, 10,575–10,580.
7. Weissenbach, J. (1993) A second generation linkage map of the human genome based on highly informative microsatellite loci. *Gene* **135**, 275–278.
8. Brownstein, M. J., Carpten, J. D., and Smith, J. R. (1996) Modulation of non-templated nucleotide addition by Taq DNA polymerase: primer modifications that facilitate genotyping. *Biotechniques* **20**, 1004–1010.
9. Zipperlen, P., Nairz, K., Rimann, I., et al. (2005) A universal method for automated gene mapping. *Genome Biol.* **6**, R19.

Index